BIOEVALUATION OF WORLD TRANSPORT NETWORKS

BIOEVALUATION OF WORLD TRANSPORT NETWORKS

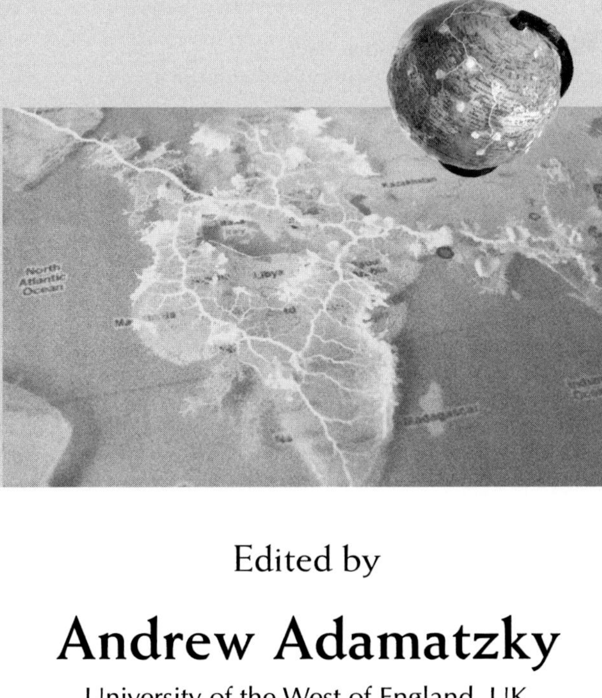

Edited by

Andrew Adamatzky

University of the West of England, UK

World Scientific

NEW JERSEY • LONDON • SINGAPORE • BEIJING • SHANGHAI • HONG KONG • TAIPEI • CHENNAI

Published by

World Scientific Publishing Co. Pte. Ltd.
5 Toh Tuck Link, Singapore 596224
USA office: 27 Warren Street, Suite 401-402, Hackensack, NJ 07601
UK office: 57 Shelton Street, Covent Garden, London WC2H 9HE

British Library Cataloguing-in-Publication Data
A catalogue record for this book is available from the British Library.

BIOEVALUATION OF WORLD TRANSPORT NETWORKS

ISBN-13 978-981-4407-03-8
ISBN-10 981-4407-03-8

Printed by FuIsland Offset Printing (S) Pte Ltd Singapore

Preface

In 2006 I got a package from Soichiro Tsuda. He sent me crooked pieces of paper with dark yellowish stuff sticking to them. "Wet the paper. Sprinkle it with oat flakes. And your life will change forever", he wrote. I did as he told. On the next day I saw that the oat flakes were spanned with a network of protoplasmic tubes. The tubes looked like roads.

"What if we cut a piece of filter paper in the shape of some island and place oat flakes where cities are? Will the protoplasmic network match the existing roads?", I thought. There is only one way to find out. I have chosen the Isle of Wight as my first 'Physarumland'. Results were intriguing yet inconclusive. Some roads were matched by the slime mould, others not. "Minor roads are too messy. Let us try to imitate UK motorways", I thought. First experiments showed that I hit a gold mine. I placed oat flakes in major urban areas of the United Kingdom and inoculated the slime mould in London. The slime mould grew the motorways in a couple of days. The match was almost perfect.

With the help of friends, who shared my delusional belief in superiority of the slime mould, I undertook experiments with Mexico, the Netherlands and other countries. We all got charmed with the wide spectrum of network topologies produced by the slime mould and decided to share our amusement with others. Thus, this book was born.

In the book we present results of laboratory experiments on imitating road network formation in 14 geographical regions and countries, simulate colonisation of the world and uncover basic principles of the slime mould road building rules. The book is self consistent and does not require any specialist knowledge. It is aimed at a wide audience of computer scientists, engineers, biologists, physicists, chemists and anyone interested in remarkable abilities of the slime mould.

We thank Prof. Leon Chua and Ms Lakshmi Narayanan for their constant support and encouragement. We are grateful to World Scientific for investing their resources in our unconventional research. Many thanks to Michael Jones for editing the manuscript.

I am personally indebted to all my co-authors for showing dedication, patience and empathy.

Bristol,
February 2012

Andrew Adamatzky
Editor

The slime mould grew roads on the Isle of Wight (2007).

The slime mould built the M4 motorway from London to Bristol (2008).

Contents

Contributors

Andrew Adamatzky
Unconventional Computing Centre, University of the West of England, Bristol, United Kingdom

Amar Faiz Zainal Abidin
Universiti Teknologi Malaysia, Johor Darul Takzim, Malaysia

Selim G. Akl
School of Computing, Queen's University, Kingston, Ontario, Canada

Ramon Alonso-Sanz
Universidad Politécnica de Madrid, ETSIA (Estadistica, GSC), Madrid, Spain

René Asomoza-Palacio
Departamento de Ingeniería Eléctrica, Centro de Investigación y de Estudios Avanzados del Instituto Politécnico Nacional, México

Bernard De Baets
Department of Mathematical Modelling, Statistics and Bioinformatics, Ghent University, Ghent, Belgium

Wesley Van Dessel
Scientific Institute of Public Health, Brussels, Belgium

Jeff Jones
Unconventional Computing Centre, University of the West of England, Bristol, United Kingdom

Andrew Ilachinski
Center for Naval Analysis, Alexandria, USA

Pedro P. B. de Oliveira
Faculdade de Computação e Informática, Universidade Presbiteriana Mackenzie, São Paulo, Brazil

Sergio V. Chapa-Vergara
Departamento de Computación, Centro de Investigación y de Estudios Avanzados del Instituto Politécnico Nacional, México

Anne V. D. M. Kayem
Department of Computer Science, University of Cape Town, Cape Town, South Africa

Michael Lees
Nanyang Technological University, Singapore

Genaro J. Martínez
Instituto de Ciencias Nucleares and Centro de Ciencias de la Complejidad, Universidad Nacional Autónoma de México, México

Badaruddin Muhammad
Universiti Malaysia Pahang, Pekan, Malaysia

Mikhail Prokopenko
CSIRO Information and Communication Technologies Centre, Sydney, Australia

Theresa Schubert
Bauhaus-Universität Weimar, Weimar, Germany

Peter M. A. Sloot
University of Amsterdam, Amsterdam, The Netherlands

Christopher R. Stephens
Instituto de Ciencias Nucleares and Centro de Ciencias de la Complejidad, Universidad Nacional Autónoma de México, México

Emanuele Strano
EPFL, Lausanne, Switzerland

Xin-She Yang
Mathematics and Scientific Computing, National Physical Laboratory, Teddington, United Kingdom

Yu-Xin Zhao
School of Automation, Harbin Engineering University, Harbin, China

Ibrahim Zuwairie
Universiti Malaysia Pahang, Pekan, Malaysia

Chapter 1

Introduction

Andrew Adamatzky

1.1 Motorways

Modern motorway networks are based on the millennium-long emergence of roads. First there were prehistoric trackways — mesolithic footways established by early men "seeking the most direct and convenient alternatives by process of trial and error" [Belloc (1924); Davies (2006)] or based on the routes selected by animals [Taylor (1979)], timber trackways and droveways (used primarily by cattle). Further development of roads was country specific. Talking about England, we can speculate that Romans were building roads along pre-existing trackways and possibly along ridgeways. In the 1700s turnpikes were established. They were based on pre-existing roads with few local diversions [Bogart (2007)]. The turnpikes were then substituted by single carriageways, dual carriageways and, finally, motorways [Davies (2006)]. Thus, by backtracking the history of motorways we arrive at pathways developed by living creatures. How were the pathways developed? Were they efficient? More or less accurate answers could be found by imitating the road network development with living substrates.

The increase of long-distance travel and subsequent reconfiguration of vehicular and social networks [Larset et al. (2006)] require novel and unconventional approaches towards analysis of dynamical processes in complex transport networks [Barrat (2008)], routing and localisation of vehicular networks [Olariu and Weigle (2009)], optimisation of interactions between different parts of a transport network during scheduling road expansion and maintenance [Taplin et al. (2005)]

and shaping of transport network structure [Beuthe et al. (2004)].

Motorway networks are designed for efficient vehicular transportation of goods and passengers; protoplasmic networks are developed for efficient intra-cellular transportation of nutrients and metabolites. Is there a similarity between these two networks? In the book we attempt to build viable analogies between biological and man-made transport networks and project behavioural traits of biological networks onto existing vehicular transport networks.

1.2 Imitating road development

Building of any road starts with calculation of a shortest, possibly even obstacle-free, path (to avoid mountains and water features) between the cities to be connected by the road. This is a shortest-path problem. Approximation of shortest paths has already been a hot application for unconventional computing scientists. Nature-inspired computing paradigms and experimental implementations were successfully applied to calculation of a minimal-distance path between two given points in a space or a road network. A shortest-path problem is solved in experimental reaction–diffusion chemical systems [Adamatzky et al. (2005)], gas-discharge analogue systems [Reyes et al. (2002)], spatially extended crystallisation systems [Adamatzky (2009)b], formation of fungi mycelian networks [Jarrett et al. (2006)] and using computer and mathematical models of collective insects [Dorigo and Stutzle (2004)] and *Physarum polycephalum* [Tero et al. (2006)].

While choosing a biological object to imitate the growth of road networks, we want it to be experimental laboratory friendly, easy to cultivate and handle and convenient to analyse. Ants would indeed be the first candidate. They do develop trails very similarly to prehistoric people. A great deal of impressive results has been published on ant-colony-inspired computing [Dorigo and Stutzle (2004); Solnon (2010)]. However, ant colonies require substantial laboratory resources, experience and time in handling them. Actually very few, if any, papers were published on experimental laboratory implementation of ant-based optimisation, the prevalent majority of publications being theoretical. There is however an amorphous living creature which is extremely easy to cultivate and handle, and which exhibits remarkably good foraging behaviour and development of intracellular transport networks. This is the plasmodium of *Physarum polycephalum*.

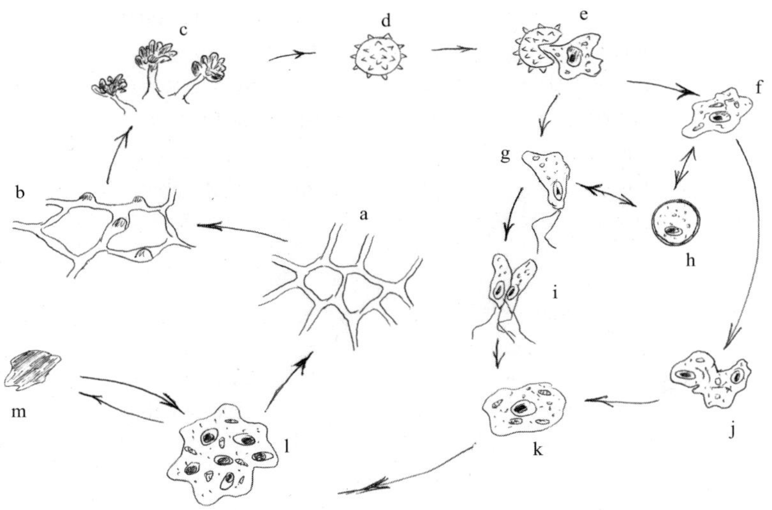

Fig. 1.1 Life-cycle scheme of *Physarum polycephalum*, inspired by [Stephenson and Stempen (1994)]: (a) protoplasmic tree formed by 'vegetative' state — plasmodium, (b) early stage of sporulation, (c) sporangia — fruiting bodies with spores, (d) a single spore, (e) germinated spore — cracked spore with myxamoeba crawling out, (f) myxamoeba, (g) swarm cell, (h) microcyst, (i) fusion of two swarm cells, (j) fusion of two myxamoebas, (k) zygote, (l) plasmodium, (m) sclerotium.

1.3 Slime mould

Physarum polycephalum belongs to the species of order *Physarales*, subclass *Myxogastromycetidae*, class *Myxomycetes*, division *Myxostelida*. It is commonly known as a true, acellular or multiheaded slime mould. The life cycle of *P. polycephalum* is exciting (Fig. 1.1).

Plasmodium — a 'vegetative' phase — is a single cell with a myriad of diploid nuclei (Fig. 1.1a–l). The plasmodium looks like an amorphous yellowish mass (Fig. 1.2) with networks of protoplasmic tubes (Fig. 1.3). The plasmodium behaves and moves as a giant amoeba. It feeds on bacteria, spores and other microbial creatures and microparticles [Stephenson and Stempen (1994)]. When foraging for its food the plasmodium propagates towards sources of food particles, surrounds them, secretes enzymes and digests the food. Typically, the plasmodium forms a congregation of protoplasm covering the food source. When several sources of nutrients are scattered in the plasmodium's range, the plasmodium forms a network of protoplasmic tubes connecting the masses of protoplasm at the food sources. A structure of the protoplasmic networks is apparently *optimal*, in a sense that it covers all sources

Fig. 1.2 Excessively growing mass of plasmodium.

of nutrients and provides a robust and speedy transportation of nutrients and metabolites in the plasmodium's body [Nakagaki (2001); Nakagaki et al. (2001); Nakagaki et al. (2007)].

When plasmodium is deprived of water and/or nutrients and cannot migrate to better places, the plasmodium goes into 'hibernation' mode and forms a hardened mass called sclerotium (Figs. 1.1m and 1.4). Sclerotia survive a range of very harsh conditions, including high temperatures of up to 70–80°C [Blackwell et al. (1984)]. A sclerotium of *P. polycephalum* consists of "crustose deposit containing nucleated spherules of cytoplasm enclosed within a honeycomb-like matrix of organic walls" [Chet and Henis (1975); Andreson (2007)]. When moistened, the sclerotium gradually returns to the state of plasmodium (Fig. 1.1l). Myxamoebas can live as they are for a long time. In the presence of water a myxamoeba is transformed into a swarm cell with two flagellas (Fig. 1.1g). Swarm cells can swim. Myxamoebas and swarm cells can reproduce asexually, by simple division. During changes of environment from good to bad, myxamoebas and swarm cells can form spheroidal microcysts (Fig. 1.1h) with cellulose walls.

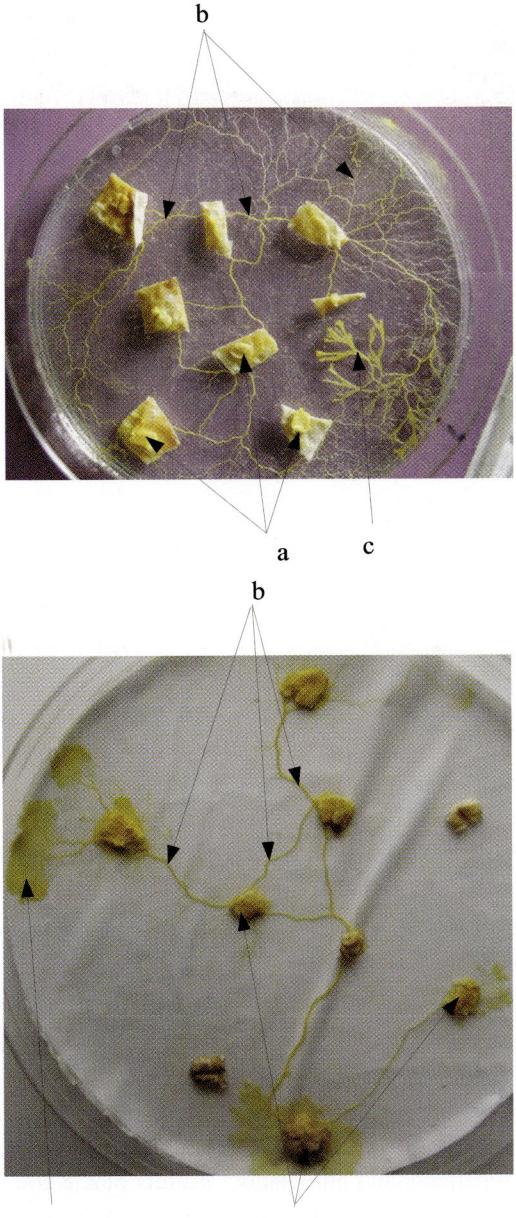

Fig. 1.3 The plasmodium on non-nutrient substrates: agar (top) and filter paper (bottom): (a) oat flakes colonised by plasmodium, (b) protoplasmic tubes, (c) active growing zones.

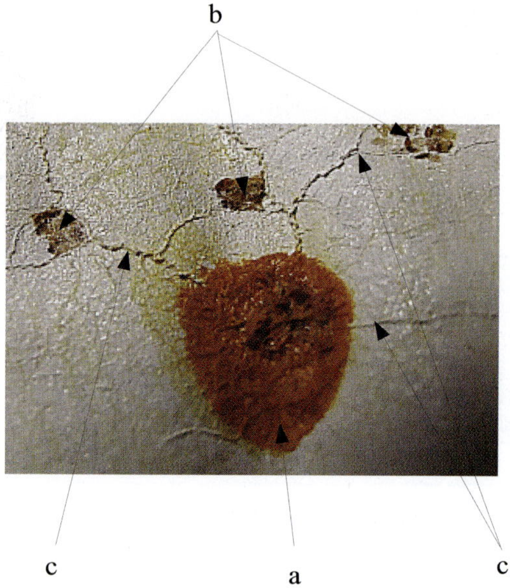

Fig. 1.4 (a) Sclerotium, (b) oat flakes used as sources of attractants to shape the plasmodium network, (c) abandoned and dried-out protoplasmic tubes.

When exposed to bright light and starved, the plasmodium switches to fructification phase. It grows sporangia (Figs. 1.1bc, finger-like globose enclosures of membrane filled with spiny spores. When a spore (Fig. 1.1d) gets into a favourable environment, it cracks and releases a single-cell myxamoeba (Fig. 1.1e). When enough myxamoebas or swarm cells are present in the volume, they begin sexual reproduction (Fig. 1.1i and j) and form a zygote (Fig. 1.1k). The zygote divides mitotically and forms a multinuclear single cell — the plasmodium (Fig. 1.1l).

1.4 Physarum computing

The plasmodium functions as a parallel amorphous computer with parallel inputs and parallel outputs [Adamatzky (2010)b]. Data are represented by spatial configurations of sources of nutrients. A program of computation is coded via configurations of repellents and attractants. Results of the computation are presented by the configuration of the protoplasmic network and the localisation of the plasmodium. When placed on a substrate the plasmodium starts its foraging activity. It follows gradients of chemoattractants and repellents. On non-nutrient agar gel, as used in our experiments, the propagating part of the plasmodium (also

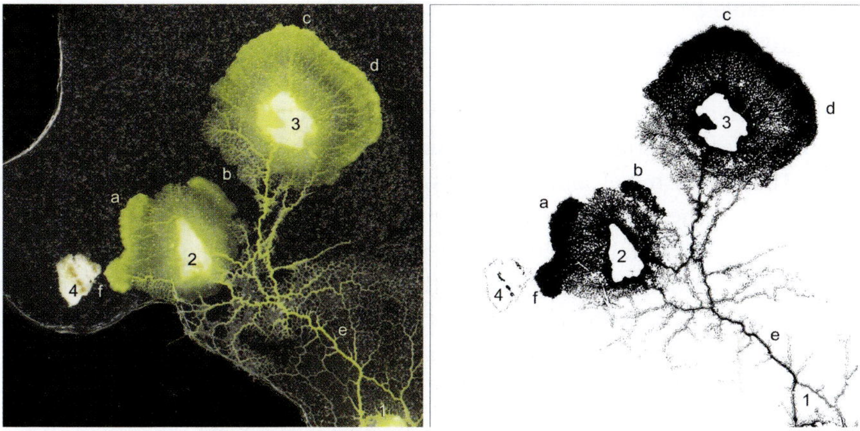

(a) (b)

Fig. 1.5 Local propagation activity and decision making by slime mould: scanned image (a) and its binarised version (b). Oat flakes are marked by 1 to 4, active parts of wave fronts by a–d and f, and protoplasmic tube by e.

called an active zone) is morphologically and behaviourally similar to localised wave fragments in a subexcitable chemical medium [Adamatzky et al. (2009); Adamatzky (2010)b].

An example of Physarum's local activity is shown in Fig. 1.5. Active parts of the plasmodium are seen as domains with high concentration of cytoplasm, visible as dense black zones in Fig. 1.5b. Initially the slime mould colonised the southern flake '1'. The mould then propagated towards the north-west and north, and colonised flakes '2' and '3', respectively. The flakes '2' and '3' become connected with flake '1' by protoplasmic tubes 'e'. When propagating towards flakes '2' and '3' Physarum splits into two active zones. The active zone moving north-west reached flake '2' earlier than the active zone moving north reached flake '3'. Part of the plasmodium in flake '2' then initially started propagating to flakes '3' and '4'. However, when it reached flake '3' a coordinating action occurred and the part propagating from flake '2' to flake '3' (this part is marked by 'b' in Fig. 1.5) cancels its movement. Thus, the following active parts of the plasmodium are actualised: zone 'f' propagating to flake '4', exploratory zone 'a' propagating north-west (this zone will be cancelled later) and exploratory zones 'c' and 'd'.

The plasmodium's behaviour is determined by external stimuli and excitation waves travelling and interacting inside the plasmodium [Nakagaki et al (1999)]. The plasmodium can be considered as a reaction–diffusion [Adamatzky (2007)c] or an excitable [Achenbach and Weisenseel (1981)] medium encapsulated in an elastic growing membrane.

The plasmodium is a network of biochemical oscillators [Matsumoto et al. (1988); Nakagaki et al (1999)]. Waves of excitation or contraction originate from several sources, e.g. induced by external stimuli and perturbations. The waves travel along the plasmodium and interact one with another in collisions. The oscillatory cytoplasm of the plasmodium is a spatially extended nonlinear excitable medium. Growing and feeding plasmodium exhibits characteristic rhythmic contractions with articulated sources. The contraction waves are associated with waves of electrical potential change. The waves observed in plasmodium [Matsumoto et al. (1986); Matsumoto et al. (1988); Yamada et al. (2007)] are similar to the waves found in excitable chemical systems, like a Belousov–Zhabotinsky medium [Adamatzky et al. (2005)].

1.5 What the book is about

To uncover analogies between biological and man-made transport networks and to project behavioural traits of biological networks onto the development of vehicular transport networks, we conducted a series of experimental laboratory studies on evaluation and approximation of motorway networks by *P. polycephalum* in 14 geographical regions: Africa [Adamatzky and Kayem (2012)], Australia [Adamatzky and Prokopenko (2011)], Belgium [Adamatzky et al. (2011)], Brazil [Adamatzky and Oliveira (2011)], Canada [Adamatzky and Akl (2011)], China [Adamatzky at al. (2011)a], Germany [Adamatzky and Schubert (2012)], Iberia [Adamatzky and Alonso-Sanz (2011)], Italy [Strano et al. (2011)], Malaysia [Adamatzky et al. (2012)], Mexico [Adamatzky et al. (2011)b], The Netherlands [Adamatzky et al. (2012)], UK [Adamatzky and Jones (2009)], USA [Adamatzky and Ilachinski (2012)]. We represented each region with an agar plate, imitated major urban areas with oat flakes, inoculated plasmodium of *P. polycephalum* in a capital and analysed structures of protoplasmic networks developed. We found that the network of protoplasmic tubes grown by plasmodium matches, at least partly, the network of man-made transport arteries. The shape of a country and the exact spatial distribution of urban areas, represented by sources of nutrients, may play an important role in determining the exact structure of the plasmodium network.

Chapter 2

Methods: how we conducted experiments and analysed their results

Andrew Adamatzky

The slime mould *Physarum polycephalum* is an ideal biological object for both high-tech laboratories of boffins and kitchen-based citizen scientists. It does not require any specialised microbiological equipment or consumables, or any particular experimental containers or temperature-controlled environment.

2.1 Obtaining *P. polycephalum*

Where to get plasmodium of *P. polycephalum*? You can go to a nearby forest and look under decaying branches and leaves. Have a look in [CalPhotos (2009)] and [Discover Life (2009)] to know what the plasmodium of *P. polycephalum* looks like in natural conditions. Read books [Stephenson and Stempen (1994)] and [Ing (1999)] for helpful hints on collection of samples. There is one book on *Myxomycetes* you can read for free [Morgan (1893)]. 'Wild' plasmodium may not look like as it is pictured in handbooks, thus chances are high, especially if you are living in a mild climate, that you will miss your game. Then you can buy *P. polycephalum* for around ten dollars from Carolina Biological Supply[1]. Finally, you can Google those who do experiment with *P. polycephalum*, select the friendliest looking one and gently e-mail them asking for a sample. Most likely you will receive a dried sclerotium on a piece of kitchen towel or a filter paper. Place the piece with sclerotium somewhere, wet generously for 6–12 h and the plasmodium will wake up.

[1] `http://www.carolina.com`

2.2 Cultivation

You can cultivate plasmodium on any porous substrate, e.g. paper kitchen towels, toilet paper, paper napkins/serviettes, rolls of filter paper. Any container is fine: plastic lunch boxes and empty ice-cream boxes. Moisten the paper every couple of days or when you feel it dries, and replant the slime mould to fresh substrate every week or two. An expert hint from [Ing (1999)]:

> if it is too smelly — it is too wet,
> if it is too hairy[2] — it is too dry.

The plasmodium is usually fed with oat flakes. However, you can try a variety of foods and see what will happen. With some patience and a bit of luck you establish a routine to keep a sustainable growth of Physarum colonies, but often this will go beyond reasonable needs in your production. How you dispose of the excess is your moral choice. The best way is to put excess of paper with slime mould on top of your kitchen unit and let it dry. The plasmodium forms sclerotium. It can stay in this form for years.

2.3 Experiments

For experiments we use 120×120 mm^2 and 220×220 mm^2 square polystyrene Petri dishes. Non-nutrient agar is used in all experiments. On a nutrient agar the plasmodium grows omnidirectionally as a circular wave in an excitable medium while on a non-nutrient agar the plasmodium follows gradients of chemoattractants, emitted by sources of food, more selectively and propagates as travelling localisations [Adamatzky (2010)b]. We used 2% Select agar, by Sigma Aldrich; however, any, even low-grade food agar, will be fine for the experimental purposes. In Chap. 13 we also used plain filter paper as a substrate: results produced were of the same quality as those using agar gel, however it takes more effort to keep the filter paper humid while the agar gel is maintenance free.

When experiments are conducted in the Petri dishes the hot agar gel is poured into the dishes to fill the dishes by 2–3 mm. When the agar cools down we cut the agar plate in the shape of the geographical regions (a country or a continent) and remove the outside parts of the agar. Impassable parts of the country, e.g. seas and lakes, can also be cut our of the plate. In experiments described in Chap. 15, we also used a three-dimensional globe 15 cm in diameter (Stellanova, Germany) as a template. When experimenting with the globe we poured hot gel onto the globe while rotating the globe, so it is covered by agar uniformly. The globe was

[2] Overgrown by bacteria or fungi.

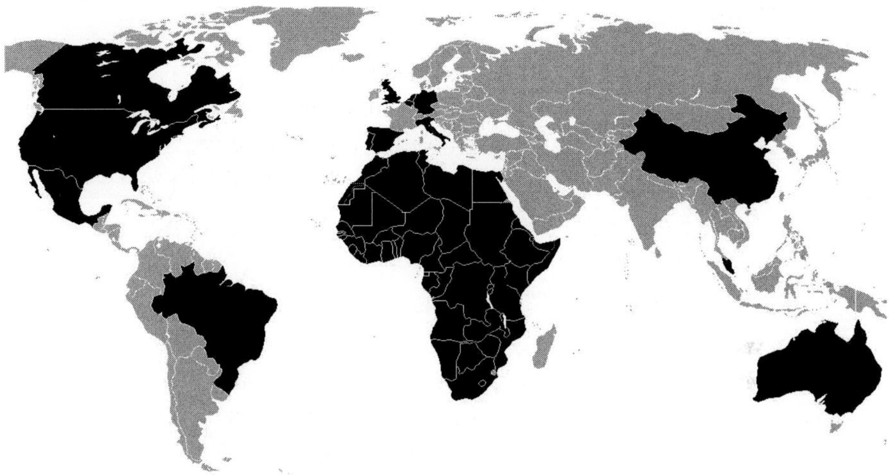

Fig. 2.1 Geographical regions studied in laboratory experiments are coloured in black.

covered by the agar gel layer by layer. When one layer cooled down and settled, the next layer of hot agar was applied. When the agar cooled down, the shapes corresponding to oceans and seas are cut out and removed, and only the dry land is represented by the agar substrate.

The experimental containers with Physarum are stored in a dark place at room temperature. They are only exposed to light during observation and recording of images. Any camera or scanner can be used to take snapshots of protoplasmic networks developed by the slime mould. Illustrations presented in the book are made using Epson Perfection 4490 and C600 scanners, and photographs with a FujiFilm FinePix camera. Scanning is better than photographing: just place the Petri dish on the scanner; the agar gel is transparent, so good-quality images will be produced. Do not have a camera or a scanner? Not a problem. Trace contours on paper.

In the laboratory experiments we studied Australia, Belgium, Brazil, Canada, China, Germany, Italy, Malaysia, Mexico, The Netherlands, UK, USA, one continent (Africa), one peninsula (Iberia) and the whole world (Fig. 2.1)

For each geographical region we identified major urban areas **U**, usually most populated cities selected using open-source national statistics and census resources. To represent areas **U**, we placed oat flakes (each flake weighs 9–13 mg and is 5–7 mm in diameter) in the positions of the agar plate corresponding to the areas. The size of a single oat flake imposes a limit on resolution: urban areas which are closer than 7 mm are merged into a single urban area.

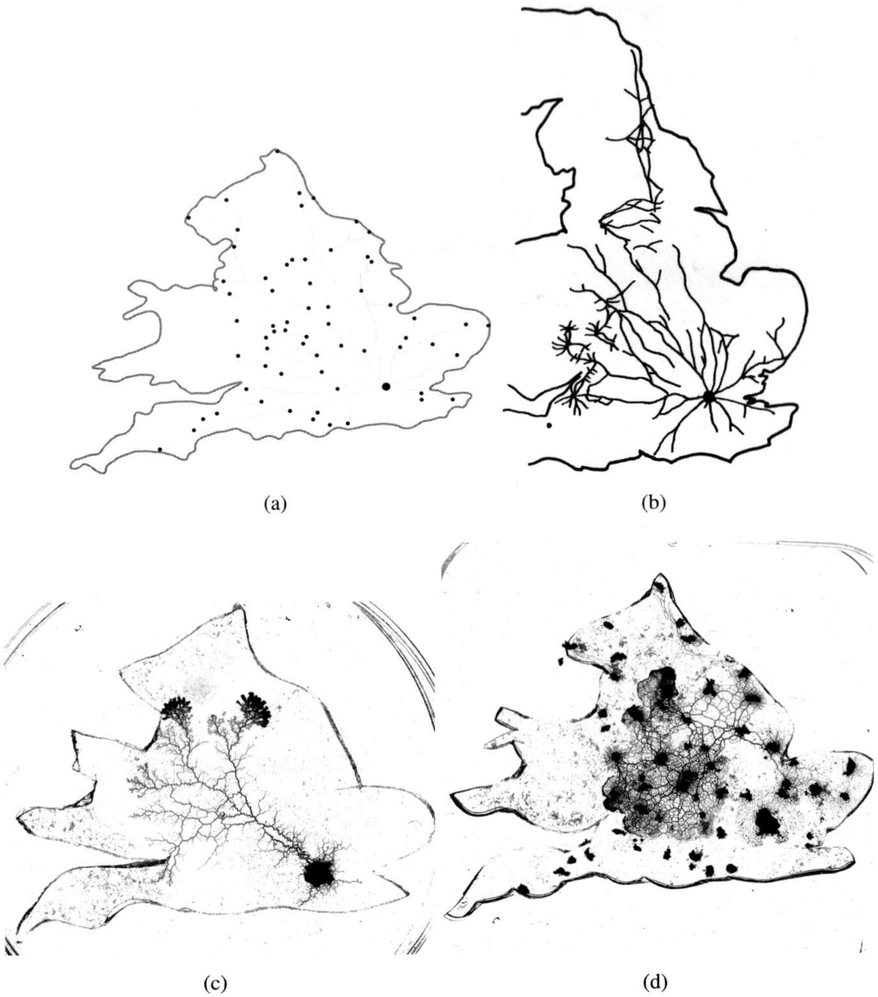

Fig. 2.2 Turnpikes and protoplasmic networks. (a) A scheme of London turnpike network in England, c. 1720. Modified from Bogart (2007). (b) A scheme of the English turnpike network in 1750. Modified from Davies (2006). (c) Plasmodium tree grown on the England-shaped agar plate from a site corresponding to London. (d) Protoplasmic network developed from London and partially spanning small oat flakes, which represent some cities from scheme (a).

At the beginning of each experiment an oat flake colonised by plasmodium (25–30 mg plasmodial weight) was placed in the capital of the country studied (apart from the USA, where the slime mould was inoculated in the New York area, which is covering Washington anyway in terms of the oat-flake resolution). Why in the capital? This is because in the majority of cases capitals are the most populated and industrially developed urban areas, and the road 'diffusion' in ancient times usually originated from the capital city. For example, see a rough scheme of turnpikes in England (Fig. 2.2a), which demonstrates a classical growth pattern, typical for fungi, myxomycetes and bacterial colonies. Moreover, we can even see evidence of secondary growth from other cities, e.g. in Fig. 2.2b we see the main turnpike network growing from London and several subnetworks growing from Bristol, Ross, Leominster, Worcester and Manchester. The turnpike networks express a high degree of resemblance to protoplasmic trees and networks developed by plasmodium on the non-nutrient substrate (Figs. 2.2c and 2.2d). Also, from our previous experimental studies [Adamatzky (2008); Shirakawa et al. (2009)] we know that when plasmodium is inoculated in every point of a given planar set, the protoplasmic network formed approximates the Delaunay triangulation of the set. Neither of the motorway networks considered match the Delaunay triangulation of major urban areas; thus, simultaneous inoculation in all urban areas would not bring any additional benefits.

To study reconfiguration of plasmodium transport networks in response to a disaster with large-scale contamination, we imitated the disaster by placing a crystal of sea salt (SAXA Coarse Sea Salt, crystal weight around 20 mg) in some chosen site of the agar plate. Inorganic salts are chemorepellents for *P. polycephalum* [Ueda et al. (1976); Terayama et al. (1977); Adamatzky (2010)a]; therefore, sodium chloride diffusing in the agar gel causes plasmodium to retreat from a contaminated zone. We studied plasmodium's response approx. 24 h after initiation of contamination. In most cases, we have chosen locations of nuclear power plants as imitated epicentres of large-scale propagating contamination. Open-source locations of nuclear power reactors and plants in all countries can be found in [Nuclear Reactors (2012)]. We did not question the safety of any particular nuclear power plant or nuclear energy in general, we only wanted to choose some more or less realistic scenario of contamination and a hypothetical contamination originated in power plants was the most obvious choice.

2.4 Physarum and motorway graphs

As every living creature does, the plasmodium of *P. polycephalum* rarely repeats its foraging pattern in all details, and almost never builds exactly the same protoplasmic network twice. To generalise our experimental results, we constructed a generalised Physarum graph with weighted edges.

A Physarum graph is a tuple $\mathbf{P} = \langle \mathbf{U}, \mathbf{E}, w \rangle$, where $\mathbf{U}$ is a set of urban areas, $\mathbf{E}$

is a set of edges and $w : \mathbf{E} \to [0,1]$ associates each edge of $\mathbf{E}$ with a frequency (or weight) of the edge occurrence in laboratory experiments. For every two regions a and b from $\mathbf{U}$, there is an edge connecting a and b if a plasmodium's protoplasmic link is recorded in at least one of k experiments, and the edge (a,b) has a weight calculated as a ratio of experiments where the protoplasmic link (a,b) occurred in the total number of experiments k. We do not take into account the exact configuration of the protoplasmic tubes but merely their existence.

For example, if we observed a protoplasmic tube connecting areas a and b in five experiments, the weight of edge (a,b) will be $w(a,b) = \frac{5}{23}$.

In the book we deal mostly with threshold Physarum graphs $\mathbf{P}(\theta) = \langle \mathbf{U}, T(\mathbf{E}), w, \theta \rangle$. A threshold Physarum graph is obtained from a generalised Physarum graph by the transformation $T(\mathbf{E}) = \{e \in \mathbf{E} : w(e) > \theta\}$, where θ is a parameter to be defined in each experiment. That is, all edges with weights less than or equal to θ are removed. Further, we use the term 'Physarum graph' when talking about the threshold Physarum graph.

To compare protoplasmic networks with man-made motorway (highway, expressway, interstate, autobahn) networks, we construct the motorway graph $\mathbf{H}$ as follows. Let $\mathbf{U}$ be a set of urban regions; for any two regions a and b from $\mathbf{U}$, the nodes a and b are connected by an edge (a, b) if there is a motorway starting in a and passing in the vicinity of b and not passing in the vicinity of any other urban region $c \in \mathbf{U}$.

2.5 Proximity graphs

To construct a proximity graph of a set of points, one selects a measure of neighbourhood and closeness $\mathscr{C}$ and then connects those points which are close neighbours in $\mathscr{C}$ [Toussaint (1980); Toussaint (1989); Jaromczyk and Toussaint (1992)] with edges of the graph.

The Euclidean minimal spanning tree (**MST**) [Nesetril et al. (2011)] is a connected acyclic graph which has minimal possible sum of edges' lengths (Fig. 2.3b). To calculate spanning trees, we used the classical Jaromczyk–Supowit method [Jaromczyk and Kowaluk (1980); Supowit (1988)].

In a relative neighbourhood graph (**RNG**) [Toussaint (1980)], Fig. 2.3c, any two points (a,b) are connected by an edge if the intersection of open discs of radius $|ab|$ centred at a and b is empty (Fig. 2.3e): $(ab) \in \mathbf{E}$ if and only if $|ab| \leq \max_{c \in \mathbf{V}-\{a,b\}}\{|ac|, |bc|\}$.

In a Gabriel graph (GG) [Gabriel and Sokal (1969); Matula and Sokal (1984)], Fig. 2.3d, points a and b are connected by an edge if the closed disc having the

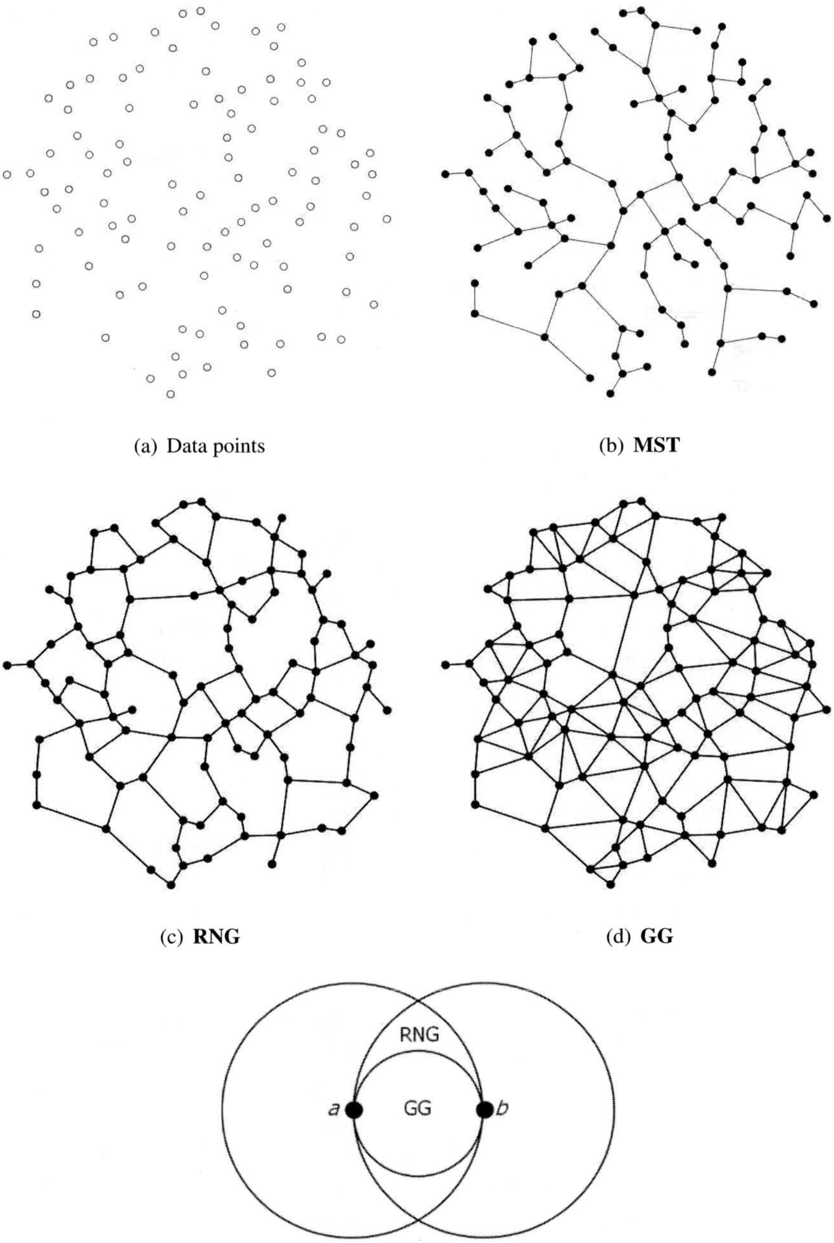

(a) Data points (b) **MST**

(c) **RNG** (d) **GG**

(e) Neighbourhoods of **RNG** and **GG**

Fig. 2.3 Examples of proximity graphs.

segment (ab) as its diameter is empty (Fig. 2.3e): $(ab) \in \mathbf{E}$ if and only if $|ab| \geq \min_{c \in \mathbf{V}-\{a,b\}}\{|\frac{a+b}{2}c|\}$.

A β-skeleton $\mathbf{B}(\beta)$, $\beta \geq 1$, is a planar proximity undirected graph of an Euclidean point set where the nodes are connected by an edge if their lune-based neighborhood contains no other points of the given set [Kirkpatrick and Radke (1985)]. Given a set $\mathbf{U}$ of planar points, for any two points p and q we define β-neighborhood $U_\beta(p,q)$ as an intersection of two discs with radius $\beta|p-q|/2$ centered at points $((1-\frac{\beta}{2})p, \frac{\beta}{2}q)$ and $(\frac{\beta}{2}p,(1-\frac{\beta}{2})q)$, $\beta \geq 1$ [Kirkpatrick and Radke (1985); Jaromczyk and Toussaint (1992)]. Points p and q are connected by an edge in β-skeleton if the pair's β-neighborhood contains no other points from $\mathbf{V}$.

In 1980, Toussaint demonstrated that **MST**$\subseteq$**GG**$\subseteq$**DT** [Toussaint (1980)], where **DT** is a Delaunay triangulation. This hierarchy of the graphs was later enriched with the GG [Matula and Sokal (1984)], and we can add a nearest-neighbourhood graph (**NNG**) at the lower level of the enclosure hierarchy by default:

$$\mathbf{NNG} \subseteq \mathbf{MST} \subseteq \mathbf{RNG} \subseteq \mathbf{GG} \subseteq \mathbf{DT}.$$

Amongst other features, the Toussaint hierarchy represents dynamics of partial closure of a graph; starting from a **NNG**, any next graph in the hierarchy is produced from the previous graph by adding some edges between non-adjacent nodes; moreover, after the transition **MST**$\rightarrow$**RNG**, decondensation starts because more cycles emerge during the partial closure.

Why do we need to compare Physarum graphs with the proximity graphs **MST**, **RNG** and **GG**? Plasmodium of *P. polycephalum* lives in a world of chemical gradient fields. It propagates towards sites with highest concentration of chemoattractants and lowest concentration of chemorepellents. A typical propagating active zone of a plasmodium is shown in Fig. 2.4a; the localised plasmodium pseudopodium propagates north-east. A scheme of interaction between the propagating active zone and chemogradients is illustrated in Fig. 2.4b. We can speculate that intersection of a Physarum active zone and the gradient field is a similar lune neighbourhood of **RNG** (Fig. 2.3e). Moreover, by changing its shape the active zone can intermittently change between a local relative neighbourhood graph and a Gabriel graph formation of edges. Thus, in principle, plasmodium of *P. polycephalum* grows as a spanning tree with simultaneous closure of the tree into a relative neighbourhood graph or a Gabriel graph (see details in [Adamatzky (2010)b]).

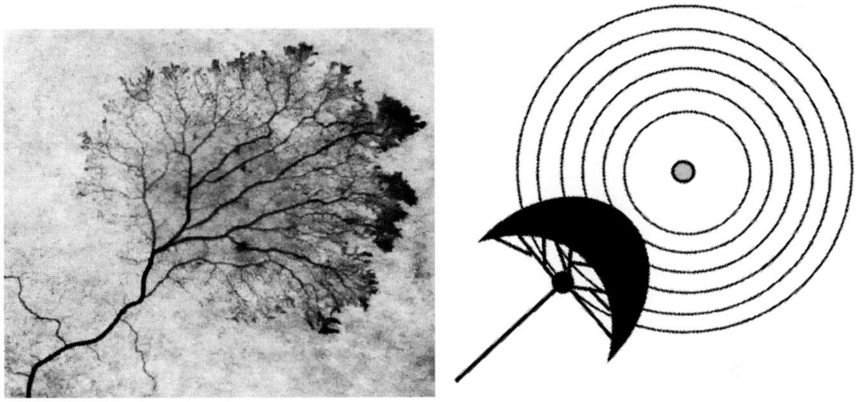

(a) (b)

Fig. 2.4 Example of propagating active zone of plasmodium (a) and its interaction of gradients of chemoattractants (b).

The minimum spanning tree helps to evaluate optimality of the protoplasmic networks: minimal distances of nutrient transportation yet complete covering of the sources of nutrients (sites of **U**). Being an acyclic graph, the spanning tree is sensitive to structural damage. Removal of a single edge might transform the spanning tree into two disconnected trees. The relative neighbourhood graph and the Gabriel graph show a higher degree of fault tolerance (depending on the exact configuration of nodes). They are also considered to be optimal in terms of total edge length and travel distance.

The graphs **RNG** and **GG** are used in geographical variational analysis [Gabriel and Sokal (1969); Matula and Sokal (1984)], simulation of epidemics [Toroczkai and Guclu (2008)] and design of *ad hoc* wireless networks [Li (2004); Song et al. (2004); Santi (2005); Muhammad (2007); Wan and Yi (2007)]. The proximity graphs, especially **RNG**, are invaluable in simulation of man-made road networks; these graphs are validated in studies of Tsukuba central district road networks [Watanabe (2005); Watanabe (2008)]. Gabriel graphs, particularly their relaxed versions [Bose et al. (2009)], are an ideal tool for path finding and online routing on planar graphs [Bose and Morin (2004)].

Chapter 3

Trans-African highways

Andrew Adamatzky and Anne Kayem

Africa has an entirely unique road network infrastructure. The continent has over two million kilometres of roads and only around 20% are paved even though this road network handles over 90% of traffic between the major urban areas [State of Transport Sector Development in Africa (2007)]. As per the report [State of Transport Sector Development in Africa (2007)], the transport infrastructure in Africa bears a significant component of inefficiency evidenced by 15 land-locked countries, whose transportation requirements to reach seaports must be addressed. As indicated in the framework of the state of the art approaches [State of Transport Sector Development in Africa (2007)], there is a need for a re-evaluation of the continental transport system. The connectivity of the African transport system is therefore a priority area in achieving African transport efficiency and sustainable growth [State of Transport Sector Development in Africa (2007)].

A concept of trans-African highways was developed in the early 1970s [The African Development Bank (2003); State of Transport Sector Development in Africa (2007)] to provide direct all-weather vehicular transport links between capitals of Africa and thus increase economic and political integration of the continent. “Article II of the Charter of the Organisation of African Unity (OAU), adopted in 1963, defines economic integration as a prerequisite for political unity. In order to achieve this goal the policies in fields like trade, transport and communications must be properly coordinated and harmonised. Improving the African road network was considered a priority and the trans-African highways were defined as the basic elements of such a network.” [The African Development Bank (2003)].

(a)

(b)

Fig. 3.1 Experimental setup. (a) Outline map of Africa with major urban areas **U** shown by encircled numbers, (b) slime mould *P. polycephalum* colonised urban areas, agar gel with the slime mould is placed on top of a map of Africa for illustrative purposes.

We selected 35 most populated major urban areas **U** listed below (see configuration of the areas in Fig. 3.1a):

(1) Cairo,
(2) Tripoli,
(3) Tunis,
(4) Algiers,
(5) Rabat,
(6) Nouadhibou and Nouakchott,
(7) Tamanrasset,
(8) Dakar, Banjul and Bissau,
(9) Agadez,
(10) Khartoum,
(11) Djibouti,
(12) Bamako,
(13) Ouagadougou,
(14) Kano,
(15) N'Djamena,
(16) Addis Ababa,
(17) Conakry, Freetown and Monrovia,
(18) Abidjan,
(19) Accra, Lome, Cotonou and Lagos,
(20) Yaounde and Douala,
(21) Bangui,
(22) Kisangani,
(23) Kampala and Nairobi,
(24) Mombasa,
(25) Libreville,
(26) Brazzaville and Kinshasa,
(27) Dodoma,
(28) Luanda,
(29) Lobito,
(30) Lubumbashi and Lusaka,
(31) Beira,
(32) Harare,
(33) Gaborone,
(34) Windhoek,
(35) Cape Town.

At the beginning of each experiment an oat flake colonised by plasmodium was placed in the Cairo area. We have chosen Cairo as the inoculation point for two reasons; first, Cairo is arguably one the most populated cities in Africa [Nations Online (2012)] and, second, Cairo is in the proximity of Faiyum, which is one of the oldest inhabited cities in Africa [Hassan (2007)]. Although there is no evidence that the development of all long-distance roads in Africa started from Cairo, we chose to use Cairo as a starting point because trade from Cairo influences a large proportion of the sub-Saharan region that is not easily accessible via a seaport. Also, since the Cairo seaport is the largest that is nearest to Europe and the Middle East, it makes sense to take Cairo into consideration over Lagos, which is another big port, by comparison, but not as close to Europe, North America or the Middle East. We undertook 32 experiments (Fig. 3.1b).

3.1 Propagation from Cairo: three scenarios

As shown in (Fig. 3.2d, being inoculated in Cairo plasmodium usually spans urban areas **U** according to the following three scenarios:

- simultaneous exploration of south-western and south-eastern parts of the African continent by clockwise movements (see scenario AF02, Fig. 3.2a),
- clockwise propagation around the continent (see scenario AX05, Fig. 3.2b),
- clockwise growth south-west followed by clockwise propagation south-east, south and then north (see scenario AC03, Fig. 3.2c).

The scenario AF02 unfolds as follows (indeed, we are discussing only examples of scenarios and every particular experiment may be slightly different). In the first 24 h plasmodium propagates from Cairo to Tripoli, Agadez and Yaounde, see scheme in Fig. 3.2a and experimental snapshots in Figs. 3.3a. In the next 24 h the slime mould occupies Tamanrasset and Kano in the north-west and Yaounde, Kisangani and Brazzaville and Kinshasa in the middle part of Africa (Fig. 3.3b). At this point the spontaneity stops. The plasmodium explores only the north-west of Africa between the 48th and 72nd hours after inoculation. It propagates from Agadez to Tamanrasset and Kano. The plasmodium grows from Tamanrasset to Rabat, Algiers, Tunis, Nouadhibou and Nouakchott and Bamako; and from Kano to Ouagadougou and Accra, Lome, Cotonou and Lagos and further west (Fig. 3.3c). Only when all the urban areas in the north-west are colonised does the plasmodium start occupying urban areas in the south and south-east. In the period between the 72nd and 96th hours of the experiment, the slime mould propagates from Kisangani to Kampala and Nairobi and Dodoma; from Kampala and Nairobi

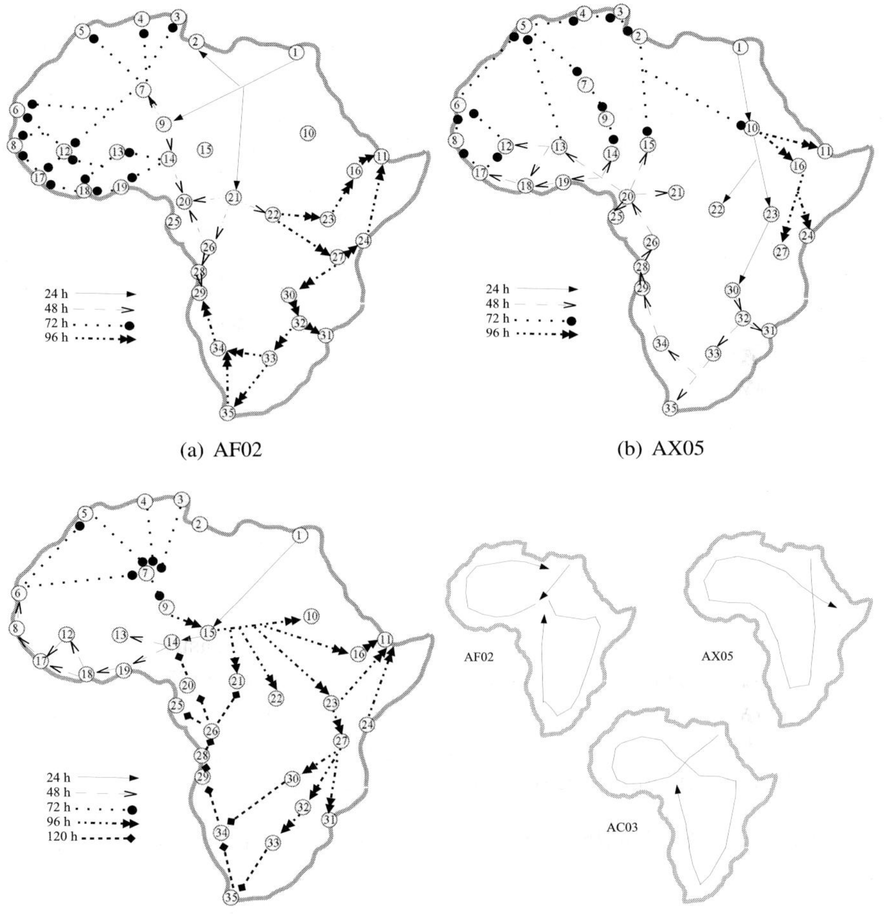

(a) AF02

(b) AX05

(c) AC03

(d)

Fig. 3.2 Scenarios of slime mould colonising urban areas of **U**. Transport links built at various stages are coded as follows (see also schemes in the pictures themselves): 24 h as a solid line with an arrow, 48 h as a fine dashed line with a line arrow, 72 h as a dotted line with a circle end, 96 h as a two dot one dash line with a double arrow, 120 h as a fine dashed line with a square head, (a) AF02 scenario: splitting into two clockwise trajectories, (b) AX05 scenario: clockwise trajectory, (c) AC03 scenario: clockwise figure 8-shaped trajectory, (d) scheme of the three scenarios.

to Addis Ababa and then Djibouti; from Dodoma to Mombasa and Lubumbashi and Lusaka; and from Lubumbashi and Lusaka to Harare and to the rest of the southernmost urban areas of **U** (Fig. 3.3d).

Slime mould propagates clockwise along the African coast in scenario AX05 (Figs. 3.2b, 3.2d and 3.4). In the first 24 h after inoculation the plasmodium con-

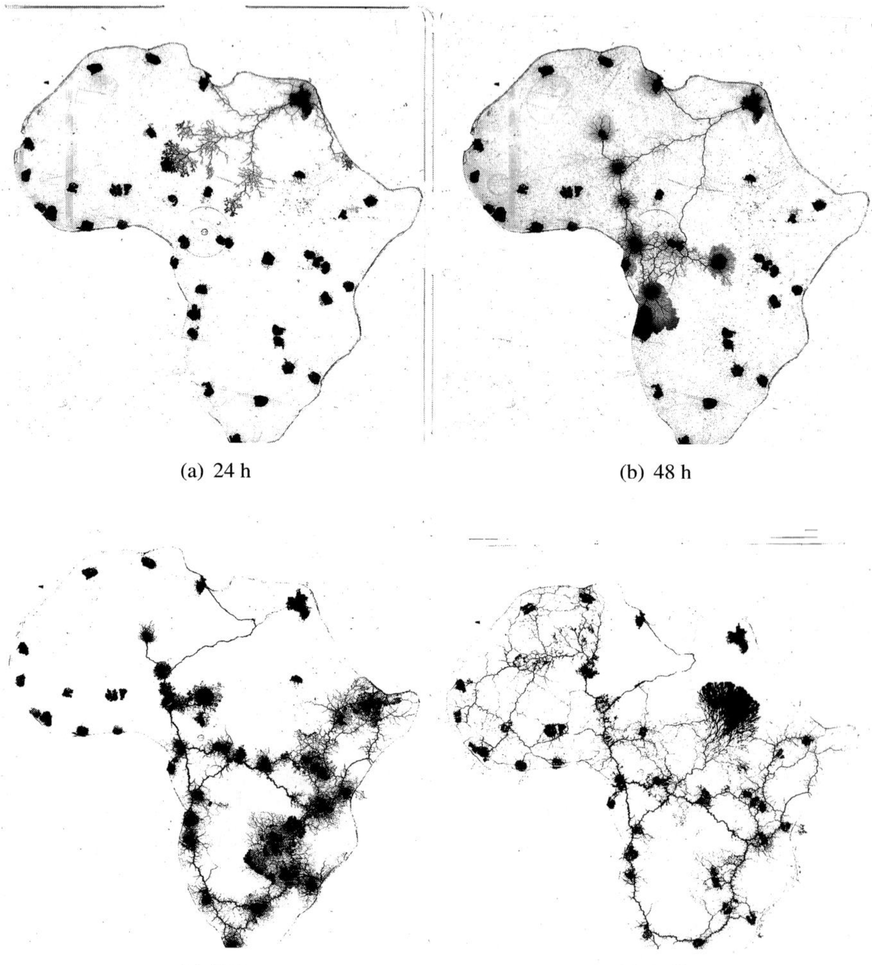

(a) 24 h (b) 48 h

(c) 72 h (d) 96 h

Fig. 3.3 Experimental laboratory example of scenario AF02. Petri dishes with slime mould were scanned every 24 h.

nects Cairo with Khartoum, Khartoum with Kisangani and Kampala and Nairobi, and Kampala and Nairobi with Lubumbashi and Lusaka by protoplasmic tubes (Fig. 3.4a). A chain of transport links between Lubumbashi and Lusaka and Yaounde, also spanning all urban areas in the south of Africa, is developed by the slime mould by the 48th hour of the experiment. Sprouting from Yaounde to Accra, Lome, Cotonou and Lagos, Ouagadougou, Kano, N'Djamena, Bangui and Libreville also takes place during this time interval, 48 h (Fig. 3.4b). The north-

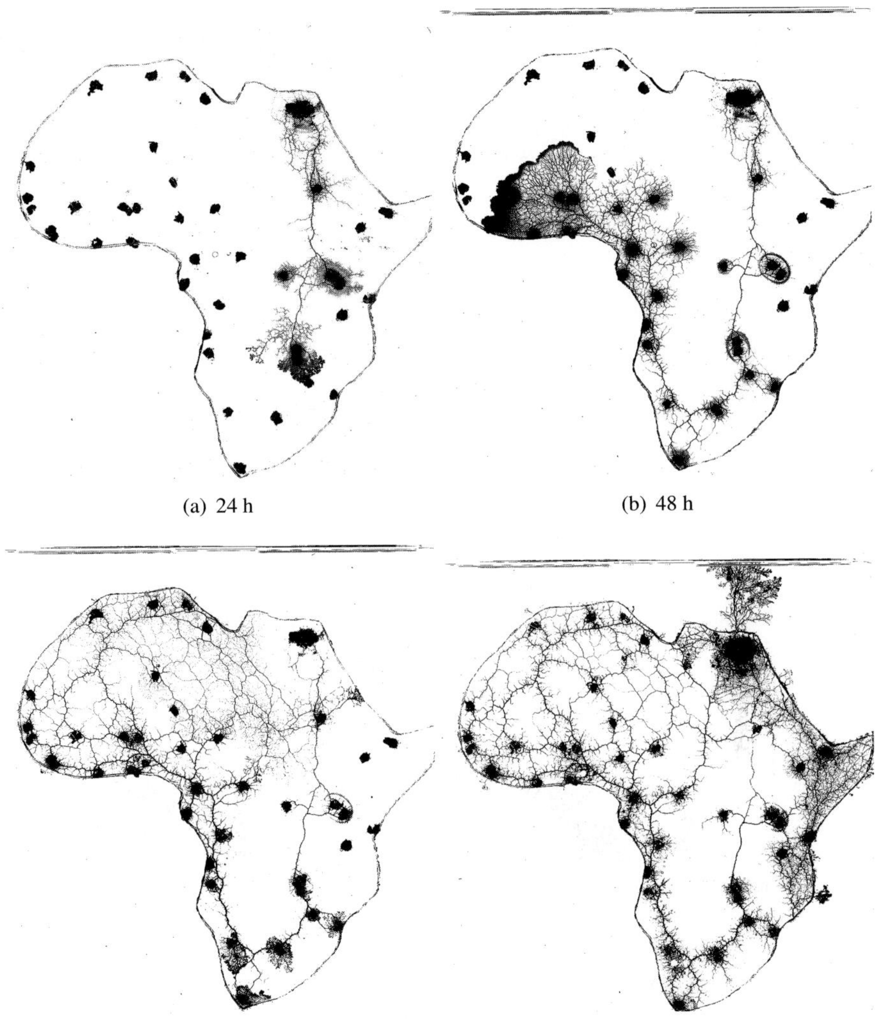

(a) 24 h (b) 48 h (c) 72 h (d) 96 h

Fig. 3.4 Experimental laboratory example of scenario AX05. Petri dishes with slime mould were scanned every 24 h. Images are converted to grey scale and inverted.

west of Africa is colonised by plasmodium between the 48th and 72nd hours of the experiment. Plasmodium propagates towards Rabat and then spans urban areas Rabat, Algiers, Tunis and Tripoli in a chain of protoplasmic tubes that grows from Tripoli to N'Djamena and Khartoum (Fig. 3.4c). Plasmodium intersects its own original trajectory of propagation (Fig. 3.2b and d) by the 96th hour following

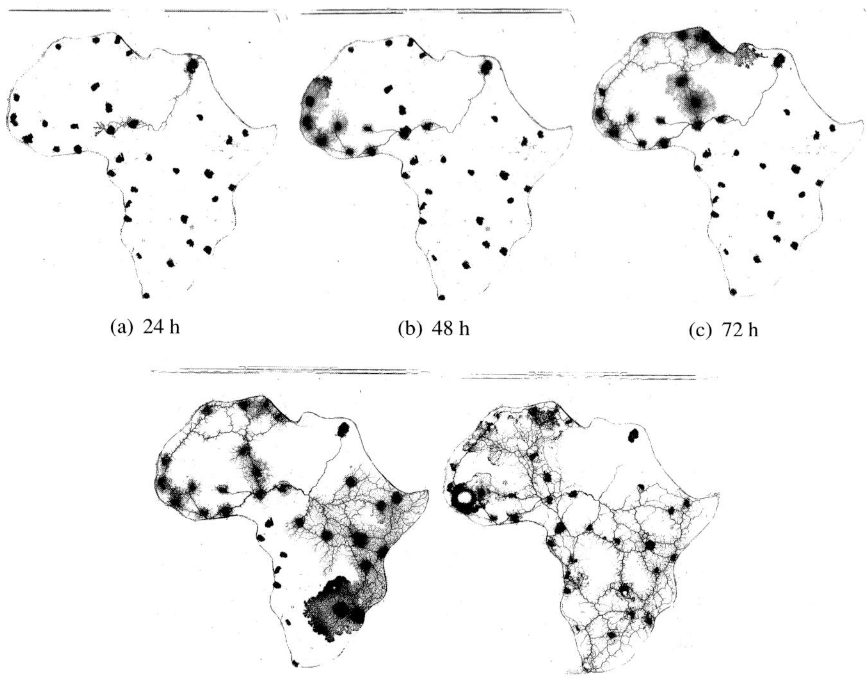

(a) 24 h (b) 48 h (c) 72 h

(d) 96 h (e) 120 h

Fig. 3.5 Experimental laboratory example of scenario AC03. Petri dishes with slime mould were scanned every 24 h. Images are converted to grey scale and inverted.

the inoculation and spreads its protoplasmic tubes from Khartoum to Djibouti and Addis Ababa, as well as from Addis Ababa to Dodoma and Mombasa (Fig. 3.4d).

In the initial phase of scenario AC03 (Figs. 3.2c, 3.2d and 3.5), plasmodium expands in a similar fashion to the trajectories described in AF02 by propagating from Egypt towards Niger and Chad. Just two transport links: Cairo to N'Djamena and N'Djamena to Kano are established in the first 24 h (Fig. 3.5a). Transport chains connecting Kano, Accra, Lome, Cotonou and Lagos, Bamako, Conakry, Freetown and Monrovia, Dakar, Banjul and Bissau and Nouadhibou and Nouakchott are built by the slime mould by the 48th hour of the experiment (Fig. 3.5b). The urban areas Nouadhibou and Nouakchott, Rabat, Algiers, Tunis, Tamanrasset and Agadez are interconnected by transport networks during 48–72 h after plasmodium's inoculation (Fig. 3.5c). By 96 h plasmodium recolonises N'Djamena and propagates to Khartoum, Addis Ababa, Kampala and Nairobi, Kisangani and southward until Lubumbashi and Lusaka, Harare and Beira (Fig. 3.5d). The plasmodium completes its figure 8-shape propagation pattern on the fifth day of the experiment. It propagates from Cape Town to Windhoek and then further north until Yaounde, Bangui and Kano (Fig. 3.5e).

Finding 1. *Being inoculated in Cairo, slime mould never explores Africa and occupies cities by anticlockwise propagation along the African coast.*

We could suggest the following explanation. Slime mould uses gradients of chemoattractants, emitted by oat flakes — urban areas, to navigate in the experimental arena. Plasmodium is inoculated in Cairo at the beginning of the experiments. Three urban areas closest — as the crow flies — are Tripoli, Khartoum and N'Djamena. However, Tripoli is screened from Cairo by the Gulf of Sirte, which might prevent direct diffusion of chemoattractants; therefore, in the prevailing majority of experiments slime mould chooses to propagate to Khartoum and N'Djamena. When propagating to Khartoum, the plasmodium continues clockwise explorations of agar shapes representing Africa. In situations when plasmodium propagates towards N'Djamena and/or Agadez, it continues to spread to Kano, Ouagadougou and Accra, Lome, Cotonou and Lagos and thus explores the south-western part of the African continent clockwise and only then goes north-east.

Examples of Physarum graphs for various values of θ are shown in Fig. 3.6 and special cases, mostly for critical values of θ, are detailed in Fig. 3.7.

A Physarum graph $\mathbf{P}(0)$, where every edge is represented by the slime mould's protoplasmic tube in at least one experiment, is a non-planar well-connected graph. Every urban area has at least two and at most 10 transport links with on average five different transport links serving an urban area (Fig. 3.7a). Some edges can be seen to be unreliable because they were imitated by protoplasmic tubes in just one or two experiments. Let us increase θ to find a structure for a reliable Physarum graph. The Physarum graph becomes planar when $\theta = \frac{6}{32}$ (Fig. 3.7b).

An increase of θ results in the Physarum graph losing edges until every urban area becomes isolated; see dynamics of connectivity in Fig. 3.19. The highest value of θ for which the Physarum graph $\mathbf{P}(\theta)$ remains connected is $\theta = 12$ (Fig. 3.7c). The graph is characterised by a long transport route running along the north coast of Africa: Djibouti–Addis Ababa–Khartoum–Cairo–Tripoli and two large cycles; one large cycle is situated in the north-west and the other in the south as shown in Fig. 3.7d. The urban area N'Djamena becomes isolated when θ increases to 13 (Fig. 3.7e). At the same time the large north-western and southern transport cycles break up due to removal of links Ouagadougou–Kano and Bangui–Kisangani.

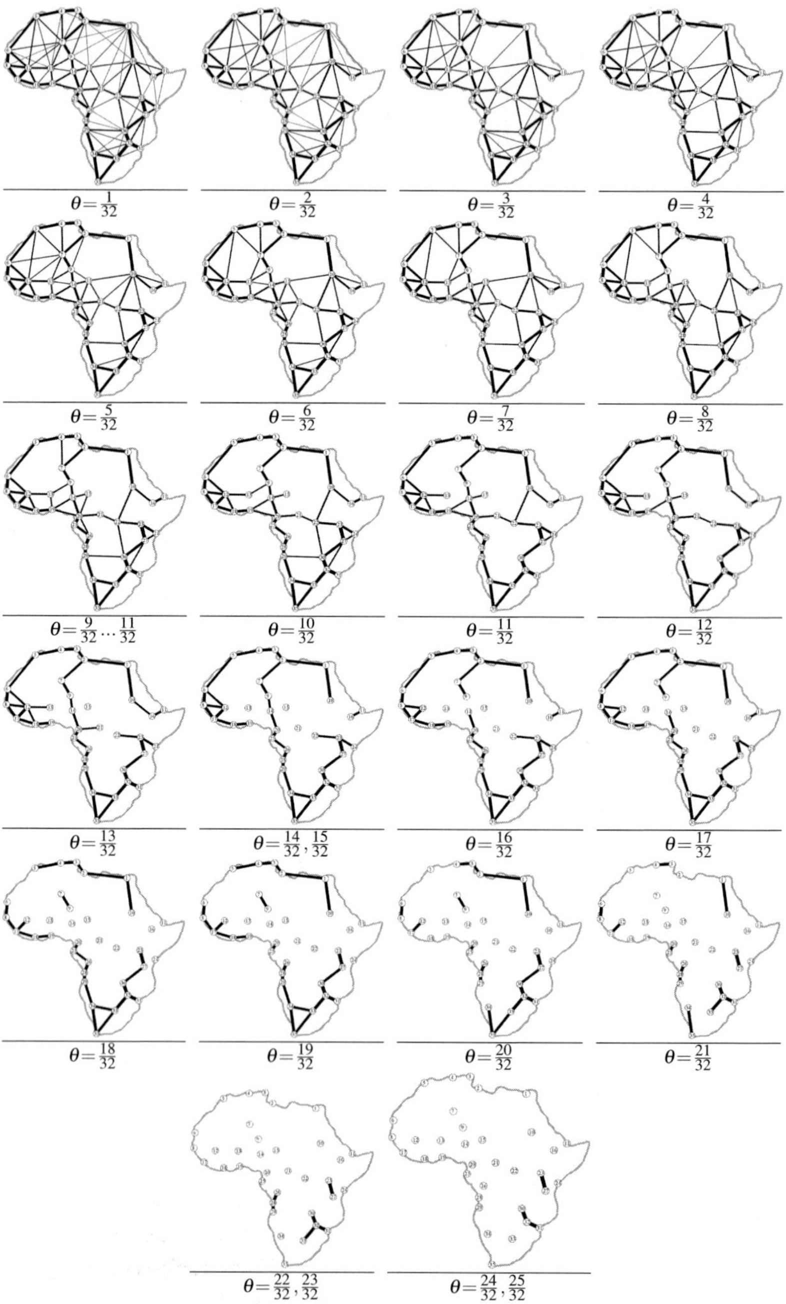

Fig. 3.6 Generalised Physarum graphs $\mathbf{P}(\theta)$ for $\theta = \frac{1}{32}, \ldots, \frac{25}{32}$.

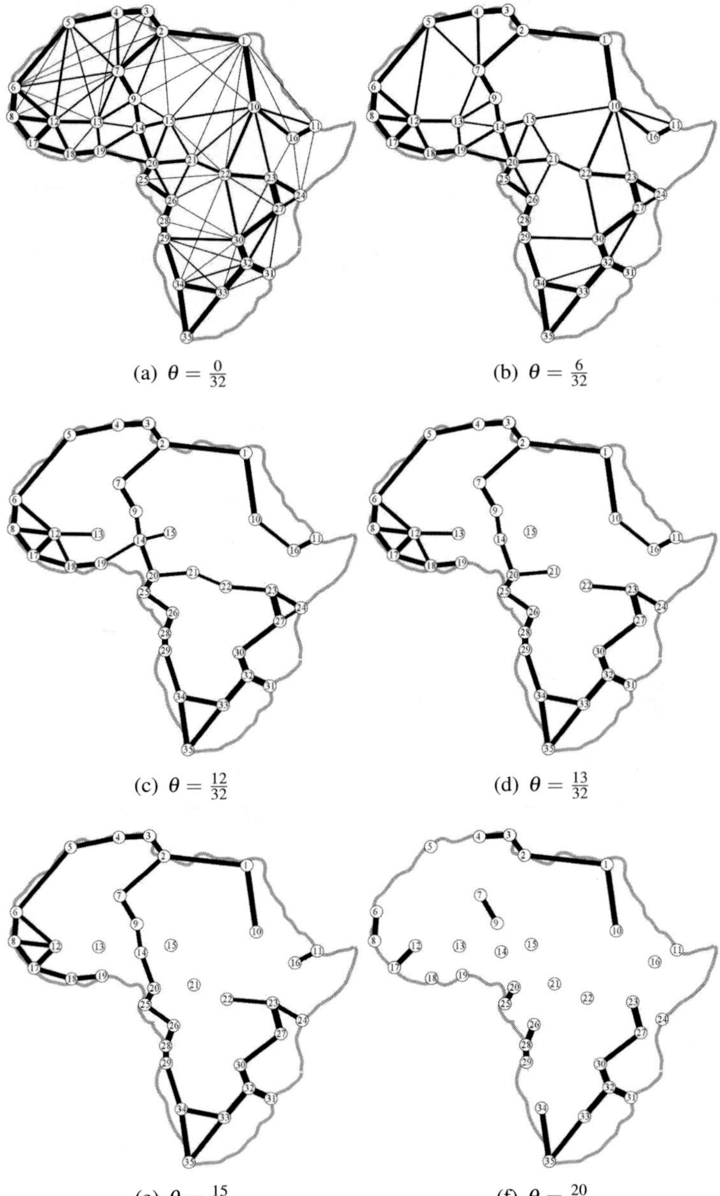

(a) $\theta = \frac{0}{32}$ (b) $\theta = \frac{6}{32}$

(c) $\theta = \frac{12}{32}$ (d) $\theta = \frac{13}{32}$

(e) $\theta = \frac{15}{32}$ (f) $\theta = \frac{20}{32}$

Fig. 3.7 Generalised Physarum graphs $\mathbf{P}(\theta)$ for selected values of θ.

Further increase of θ to $\frac{15}{32}$ leads to the isolation of urban areas Ouagadougou, N'Djamena and Bangui. We also note from Fig. 3.7d and e that an isolated segment Djibouti–Addis Ababa is also formed.

Note. Urban areas Ouagadougou, N'Djamena and Bangui become isolated with an increase of θ not because there are too few transport links but rather due to an abundance of them.

The areas are positioned favourably so that in every experiment, plasmodium of *P. polycephalum* can connect Ouagadougou, N'Djamena and Bangui together in many different ways without jeopardising the alleged optimality of its protoplasmic transport network. Therefore, each route entering Ouagadougou, N'Djamena and Bangui receives a smaller weight and such routes are removed when θ reaches higher values, such as e.g. $\theta = \frac{15}{32}$ (Fig. 3.7e).

The Physarum graph becomes acyclic when $\theta = 20$. Increasing θ to 20, which is equivalent to the fact that all links that are not recorded in at least 62% of the experiments are removed, transforms the Physarum graph into a set of the following components (Fig. 3.7f):

- four one-link chains Nouadhibou and Nouakchott–Dakar, Banjul and Bissau, Bamako–Conakry, Freetown and Monrovia, Tamanrasset–Agadez and Yaounde–Libreville,
- one two-link chain Lobito–Luanda–Brazzaville and Kinshasa,
- one four-link chain Khartoum–Cairo–Tripoli–Tunis–Algiers,
- a tree rooted in Harare with linear branches Harare–Gaborone–Cape Town–Windhoek, Harare–Lubumbashi and Lusaka–Dodoma–Kampala and Nairobi and Harare–Tunis.

Finding 2. *Only the transport links Kampala and Nairobi–Dodoma and Lubumbashi and Lusaka–Harare–Beira are represented by slime mould* P. polycephalum *in over almost 80% of experiments.*

See the Physarum graph $\mathbf{P}(\frac{25}{32})$ in Fig. 3.8. We believe that this is due to the fact that Nairobi is in close proximity to Mombasa, which is the second largest city in Kenya and has a major seaport and a major airport that serve as an entry point for goods into East Africa. Dodoma and Lubumbashi are big cities in close proximity to Nairobi and Kampala respectively and so it makes sense that the slime mould should create a path towards these areas.

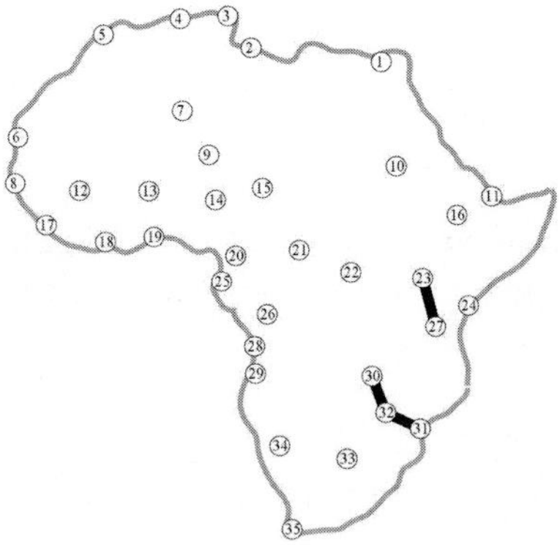

Fig. 3.8 Generalised Physarum graphs $\mathbf{P}(\theta)$ for selected values of θ.

3.2 Protoplasmic networks of trans-African highways

In a perfect world, we would be able to report that all 32 experiments demonstrated perfect matching between protoplasmic networks developed by *P. polycephalum* and trans-African highways. Unfortunately, our real world is far from perfection. Indeed, in some experiments we observed that protoplasmic tubes directly match segments of highways (Fig. 3.9) or even whole highways, such as e.g. the Lagos–Mombasa highway, which provides a transport link between the East African port of Mombasa with the port Douala in Cameroon and port Lagos in Nigeria in West Africa [The African Development Bank (2003)] and the trans-Sahelian highway, which runs from Dakar to N'Djamena, in Fig. 3.10. However, in no single experiment are all highways matched by protoplasmic tubes. This is why we need to compare Physarum graphs with a slightly idealised (with no branching) highway graph.

The graph **H** of the trans-African highway network (Fig. 3.11b) is derived from the original scheme of trans-African highways [Parry (2007)] (Fig. 3.11a)[1]. The highway graph **H** is planar (Fig. 3.11).

Finding 3. *A Physarum graph* $\mathbf{P}(\frac{1}{32})$ *includes the trans-African highway graph but links N'Djamena to Brazzaville and Kinshasa and Addis Ababa to Dodoma.*

This actually reflects the on the ground reality because there is an oil pipeline that runs from the Doha oil fields in southern Chad (country in which N'Djamena

[1] The scheme [Parry (2007)] used in Fig. 3.11a of the present chapter is an updated version of the original scheme 'Map 2.1 Trans-African Highways — Main Links' provided on p. 17 of the report [The African Development Bank (2003)].

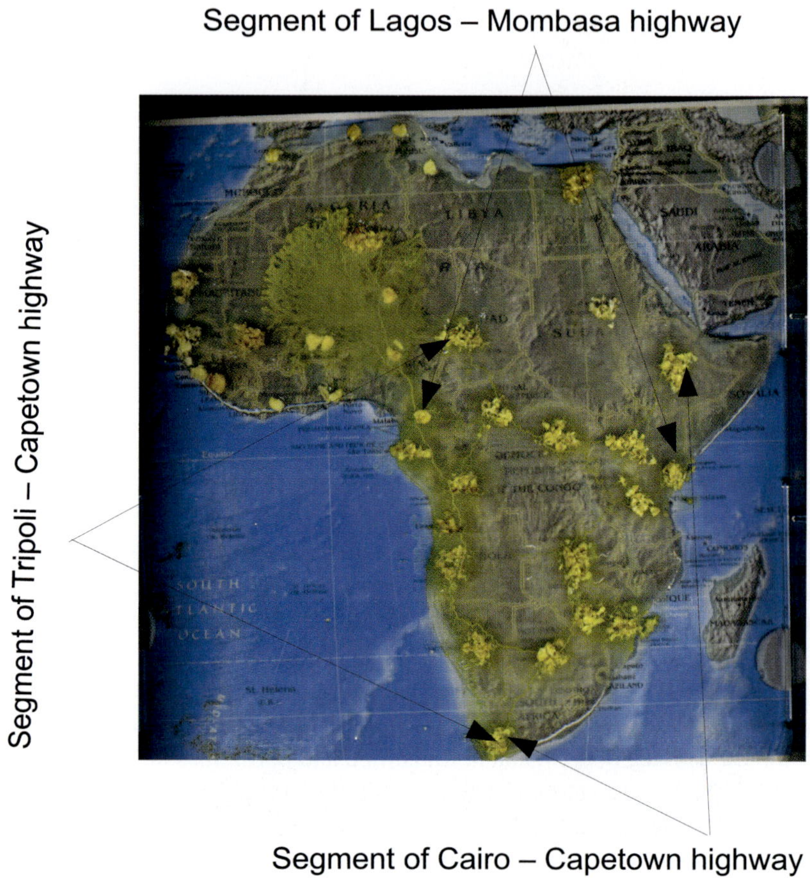

Fig. 3.9 Examples of direct matches of slime mould tubes and segments of trans-African highways.

is situated) to Kribi in Cameroon. Brazzaville in Congo is one of the larger cities with a population of over 4 million that is close to Kribi and also to the forest zone in East Cameroon, so it makes sense that the slime mould picked this path over other potential alternative routes. Referring to Figs. 3.11b and 3.12a, we note that the missing links correspond to the unpaved parts of the Cairo–Gaborone highway and the Tripoli–Windhoek (the longest corridor in the trans-African network) highway. As soon as we increase the threshold θ, the number of highway links that are approximated by the slime mould starts decreasing. Even minor increases of θ from $\frac{1}{32}$ to $\frac{2}{32}$ lead to disappearances, from the slime mould network, of transport

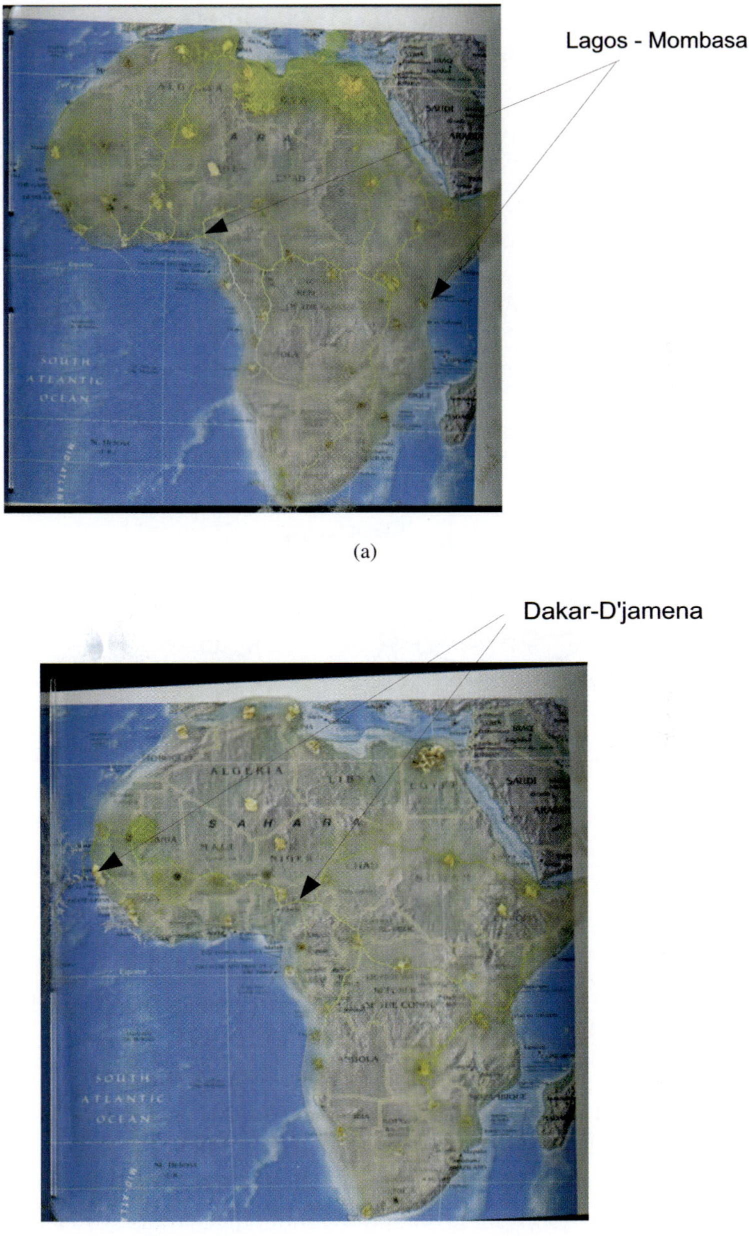

Fig. 3.10 Examples of direct matches of slime mould tubes and some trans-African highways.

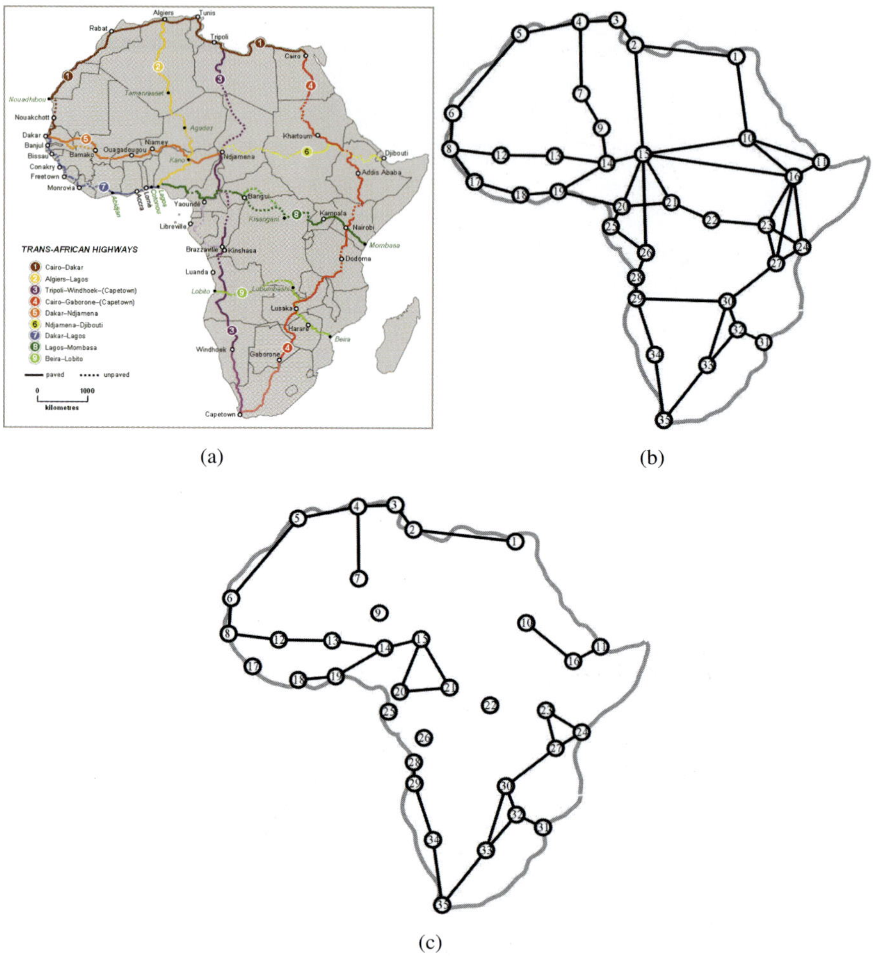

Fig. 3.11 Trans-African highways: (a) original scheme of trans-African highways [Parry (2007)], (b) highway graph **H**, (c) only paved segments of the highway graph are presented.

links like the ones Lubumbashi and Lusaka–Gaborone, Addis Ababa–Kampala and Nairobi and Addis Ababa–Mombasa (Fig. 3.12b).

How efficient is slime mould in approximating trans-African highways? We can define efficiency $\varepsilon(\theta)$ for a given threshold of repeatability θ as a product of completeness $\kappa(\theta)$ and redundancy $\rho(\theta)$: $\varepsilon(\theta) = \kappa(\theta) \cdot \rho(\theta)$. Completeness is the ratio of highway edges represented by the Physarum graph: $\kappa(\theta) = \frac{|\mathbf{H} \cap \mathbf{P}(\theta)|}{|\mathbf{H}|}$. A redundancy of slime mould approximation shows how many Physarum edges

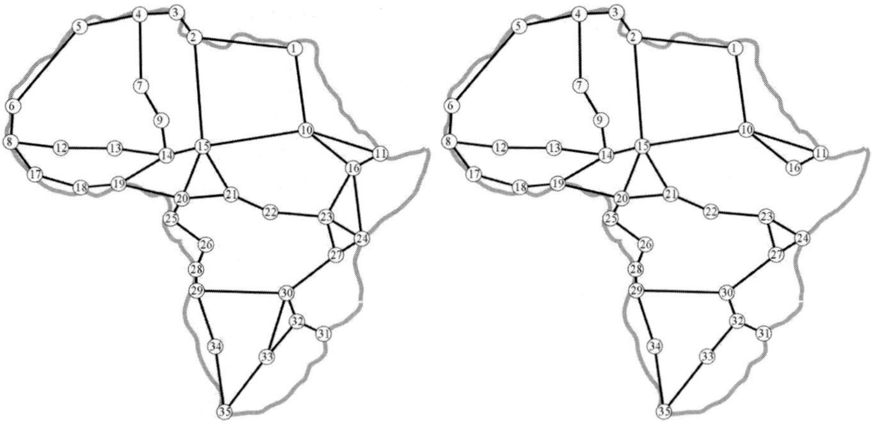

(a) $\mathbf{H} \cap \mathbf{P}(\frac{1}{32})$ (b) $\mathbf{H} \cap \mathbf{P}(\frac{2}{32})$

Fig. 3.12 Intersections of trans-African highway graph and Physarum graphs: (a) $\mathbf{P}(\frac{1}{32})$ and (b) $\mathbf{P}(\frac{2}{32})$.

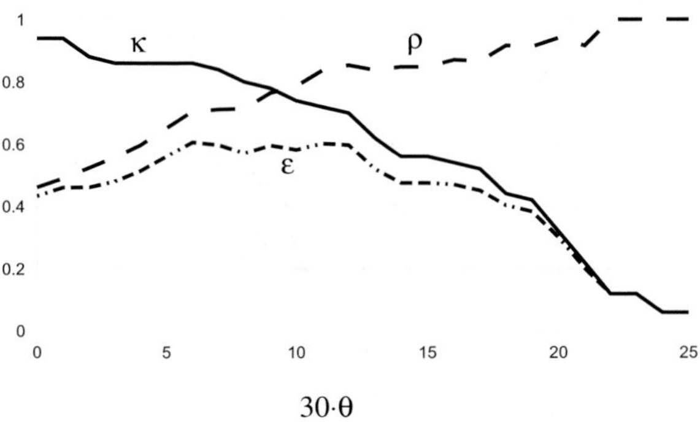

Fig. 3.13 Completeness κ, redundancy ρ and efficiency ε, as functions of threshold θ, of slime mould approximations of trans-African highways.

are not representing any highway edges: $\rho(\theta) = 1 - \frac{|\mathbf{P}(\theta)| - |\mathbf{H} \cap \mathbf{P}(\theta)|}{|\mathbf{P}(\theta)|}$. Completeness, redundancy and efficiency of slime approximation of trans-African highways are illustrated in Fig. 3.13. Completeness $\kappa(\theta))$ decreases and redundancy $\rho(\theta))$ increases with increase of repeatability θ. There are two peaks of efficiency: $\varepsilon(\frac{6}{32}) = 0.606$ and $\varepsilon(\frac{12}{32}) = 0.606$ (Fig. 3.13).

Finding 4. *A family of Physarum graphs* $\mathbf{P}(\theta)$ *parameterised by increasing*

(a) $\mathbf{H} \cap \mathbf{P}(\frac{6}{32})$

(b) $\mathbf{H} \cap \mathbf{P}(\frac{12}{32})$

Fig. 3.14 Physarum graph vs trans-African highway graph. Intersections of trans-African highway graph and Physarum graphs (a) $\mathbf{P}(\frac{6}{32})$ and (b) $\mathbf{P}(\frac{12}{32})$.

θ shows highest efficiency of approximation of trans-African highways by the Physarum graph, which firstly became planar, $\mathbf{P}(\frac{6}{32})$*, and by the Physarum graph which was last connected,* $\mathbf{P}(\frac{12}{32})$.

The graph $\mathbf{P}(\frac{5}{32})$ is non-planar while $\mathbf{P}(\frac{6}{32})$ is planar (Fig. 3.7b). The graph $\mathbf{P}(\frac{12}{32})$ (Fig. 3.7b) is connected while $\mathbf{P}(\frac{13}{32})$ has an isolated vertex N'Djamena (Fig. 3.7c).

Intersections of Physarum graphs $\mathbf{P}(\frac{6}{32})$ and $\mathbf{P}(\frac{12}{32})$ with the highway graph are shown in Fig. 3.14. In principle, the graph $\mathbf{P}(\frac{6}{32})$ has more chances to approximate $\mathbf{H}$ because it has more edges than $\mathbf{P}(\frac{12}{32})$. As we see by comparing Fig. 3.14a and Fig. 3.11b, the following edges of $\mathbf{H}$ are not represented in $\mathbf{P}(\frac{6}{32})$: Tripoli–N'Djamena, N'Djamena–Brazzaville and Kinshasa, N'Djamena–Addis Ababa, Djibouti–Mombasa and Djibouti–Dodoma.

Finding 5. *Section Tripoli to Brazzaville and Kinshasa of the Tripoli–Windhoek–Cape Town and section Djibouti to Dodoma of the Cairo–Gaborone–Cape Town highways are not approximated by a planar Physarum graph in any number of experiments.*

Tripoli is further off from N'Djamena in comparison to Cairo, Tunis and Algiers that are cities with even bigger populations than N'Djamena. This is reflected in the slime mould's network formation that creates paths to Cairo, Tunis and Algiers instead of to N'Djamena.

Finding 6. *All segments of trans-African highways not represented by slime mould have components of non-paved roads.*

This follows directly from Fig. 3.11c and Fig. 3.14a. We believe that this is natural because we based our experiments on current demographics; however, historical routes have continued to be used typically because these are either shorter or easier to navigate [Lydon (2009)].

Finding 7. *Only roads connecting Kampala and Nairobi with Dodoma, and Lubumbashi and Lusaka with Harare and Beira, are approximated by slime mould's protoplasmic tubes in almost all experiments.*

This follows from Figs. 3.11b, 3.11c, 3.8 and 3.14. As mentioned earlier, Nairobi is close to the big seaport of Mombasa and is the nearest link to landlocked Kampala. Likewise, Lubumbashi is in close proximity to Kampala and Lusaka. Lusaka's closeness to Harare and Beira results in a complete network.

3.3 Lubumbashi and Lusaka to Harare and Beira is the strongest link

A relative neighbourhood graph, a Gabriel graph and a couple of spanning trees constructed on sites of **U** are shown in Fig. 3.15.

In Fig. 3.16, we see that if the trans-African highway graph **H** had transport links Tripoli–Tamanrasset, Gaborone–Windhoek, Kano–Yaounde, Ouagadougou–Accra, Lome, Cotonou and Lagos and Bamako–Conakry, Freetown and Monrovia, then **H** would be a supergraph of the minimum spanning tree rooted in Cairo. For the graph **H** to include the relative neighbourhood graph, **H** must have the following highways: Tripoli–Tamanrasset, Conakry, Freetown and Monrovia–Bamako, Bamako–Abidjan, Ouagadougou–Accra, Lome, Cotonou and Lagos, Kano–Yaounde, Accra, Lome, Cotonou and Lagos–Libreville and Windhoek–Gaborone.

Intersections of the proximity graphs and the Physarum graph $\mathbf{P}(\frac{12}{32})$, whose edges represent protoplasmic tubes occurring in over 37% of experiments, are shown in Fig. 3.17.

Finding 8. $\mathbf{MST}$*(Cairo)* $\subset \mathbf{P}(\frac{12}{32}) \cup \{$*(Ouagadougou–Accra, Lome, Cotonou and Lagos, Ouagadougou–Kano, Addis Ababa–Kampala and Nairobi)* $\}$.

That is, just three of the 33 edges of **MST**(Cairo) are not represented by the Physarum graph $\mathbf{P}(\frac{12}{32})$ (Fig. 3.17b). The minimum spanning tree is almost a subgraph of the Physarum graph. For a relative neighbourhood graph to be a subgraph of $\mathbf{P}(\frac{12}{32})$, the Physarum graph must also, in addition to the three edges

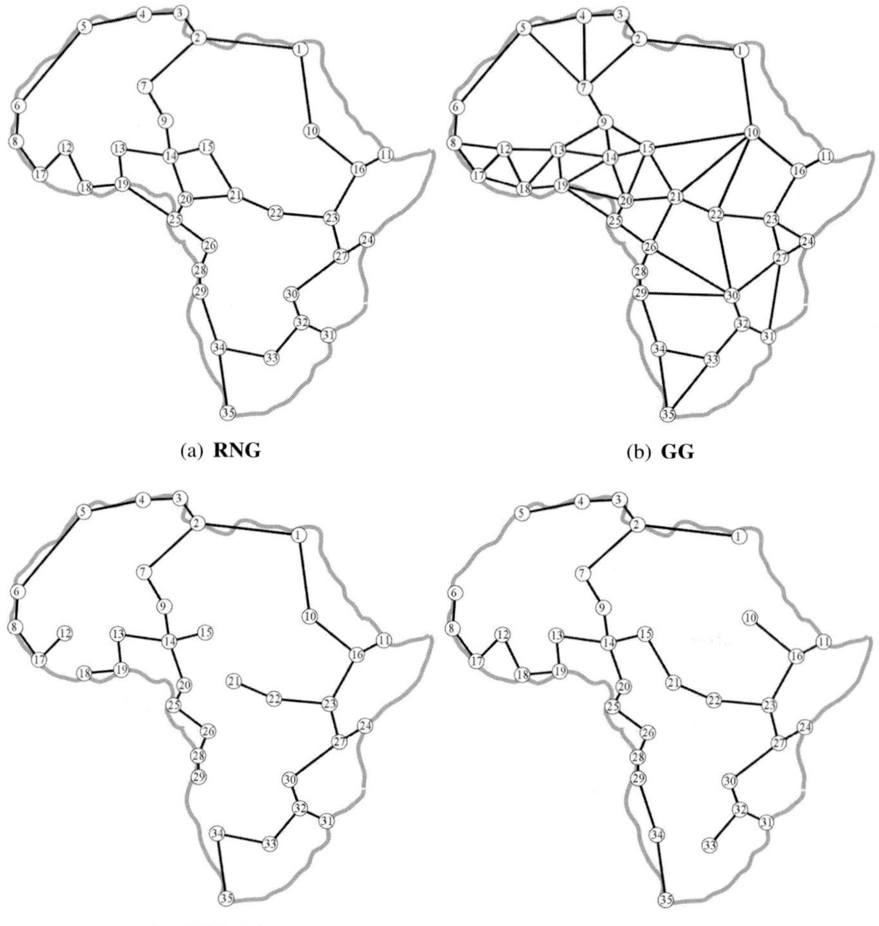

(a) **RNG** (b) **GG**

(c) **MST**(Cairo) (d) **MST**(N'Djamena)

Fig. 3.15 Proximity graphs constructed on sites of **U**: (a) relative neighbourhood graph, (b) Gabriel graph, (c) minimum spanning tree rooted in Cairo.

listed above, represent the edge N'Djamena–Bangui (Fig. 3.17a).

The properties of the family $\mathbf{BS}(\beta)$ on the set **U** are such that the skeleton $\mathbf{BS}(0.9)$ is planar (Fig. 3.20a) and the skeleton $\mathbf{BS}(2.4)$ is 'almost' a tree (Fig. 3.20b). There are no trees which span all sites of the set **U** in the family $\mathbf{BS}(\beta)$. The skeleton $\mathbf{BS}(2.5)$ is a tree and yet it does not span all elements of **U**. For instance, Cape Town becomes an isolated vertex for ≥ 2.4 (Fig. 3.20c), which is quite an interesting observation in that six chains survive an increase

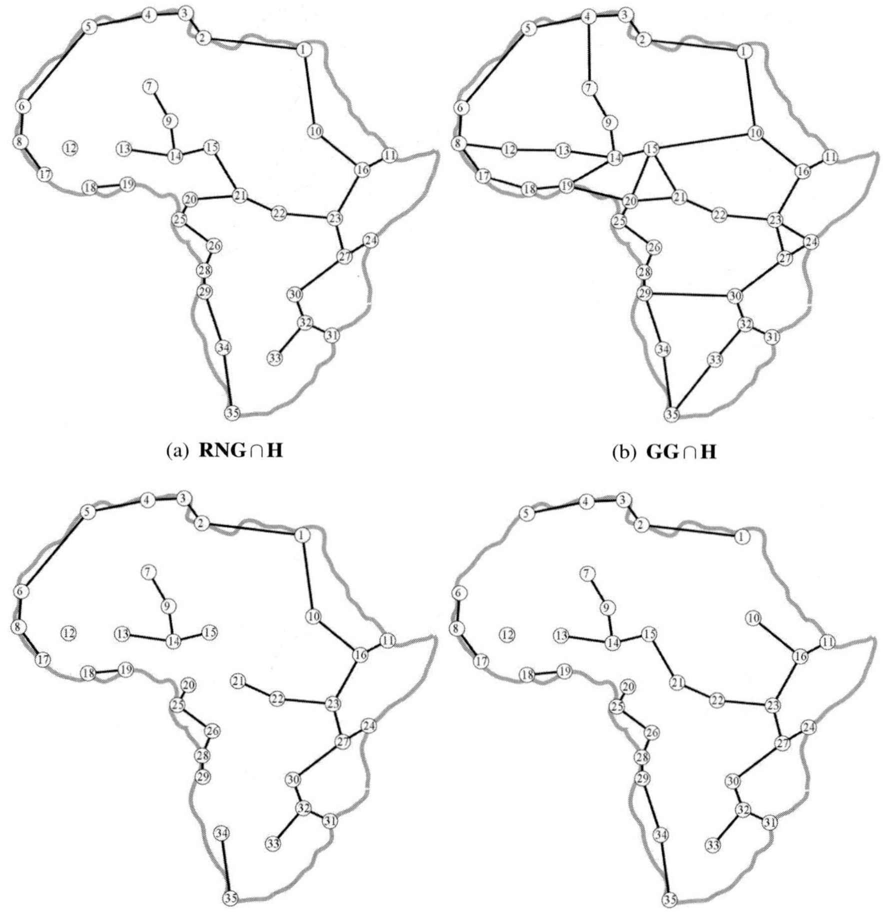

(a) **RNG** ∩ **H** (b) **GG** ∩ **H**

(c) **MST**(Cairo) ∩ **H** (d) **MST**(N'Djamena) ∩ **H**

Fig. 3.16 Intersections of highway graph **H** with (a) relative neighbourhood graph, (b) Gabriel graph, (c) spanning tree rooted in Cairo, (d) spanning tree rooted in N'Djamena.

of β to 20. The following chains are present in $\mathbf{BS}(20)$: Tripoli–Tunis–Algiers, Nouadhibou and Nouakchott–Dakar, Banjul and Bissau, Abidjan–Accra, Lome, Cotonou and Lagos, Bangui–Kisangani, Djibouti–Addis Ababa–Kampala and Nairobi, Yaounde–Libreville, Brazzaville and Kinshasa–Luanda–Lobito, Lubumbashi and Lusaka–Harare–Beira (Fig. 3.20d). The surviving transport chains belong to the highway graph **H**.

Do the same chains persist under an increase of θ in Physarum graphs $\mathbf{P}(\theta)$?

(a) $\mathbf{RNG} \cap \mathbf{P}(\frac{12}{32})$

(b) $\mathbf{GG} \cap \mathbf{P}(\frac{12}{32})$

(c) $\mathbf{MST}$(Cairo)$\cap \mathbf{P}(\frac{12}{32})$

(d) $\mathbf{MST}$(N'Djamena)$\cap \mathbf{P}(\frac{12}{32})$

Fig. 3.17 Intersections of Physarum graph $\mathbf{P}(\frac{12}{32})$ with (a) relative neighbourhood graph, (b) Gabriel graph, (c) spanning tree rooted in Cairo, (d) spanning tree rooted in N'Djamena.

Integrally, β-skeletons and Physarum graphs do lose edges but in a different fashion with an increase of their parameters. Physarum graphs lose their edges linearly on the threshold θ (Fig. 3.19b), while the normalised number of edges in β-skeletons decreases roughly as $\frac{1}{\sqrt{\beta}}$.

The chain Djibouti–Addis Ababa–Kampala and Nairobi is the weakest: it can be found only in the Physarum graph $\mathbf{P}(\frac{1}{32})$ because the link Addis Ababa–Kampala and Nairobi is represented by a protoplasmic tube in only one exper-

Roots	Length
N'Djamena	1.00
Kano, Bangui	1.01
Yaounde, Kisangani, Brazzaville and Kinshasa, Lobito, Gaborone	1.02
Windhoek, Cape Town	1.03
Bamako, Ouagadougou, Luanda	1.04
Cairo, Accra, Lome, Cotonou and Lagos, Libreville	1.05
Tripoli, Tunis, Algiers, Nouadhibou and Nouakchott	1.06
Tamanrasset, Agadez, Conakry, Freetown and Monrovia, Abidjan	
Rabat, Nouadhibou and Nouakchott, Dakar, Banjul and Bissau	1.08
Khartoum, Lubumbashi and Lusaka	1.09
Beira, Harare	1.10
Djibouti, Addis Ababa, Kampala and Nairobi, Mombasa, Dodoma	1.11

Fig. 3.18 Lengths of spanning trees rooted in sites of **U**.

iment. The link Bangui–Kisangani can be found in Physarum graphs for θ up to $\frac{12}{32}$, i.e. the transport link is imitated by slime mould in less than 37% of experiments. Links Abidjan–Accra, Lome, Cotonou and Lagos, Tripoli–Tunis–Algiers and Nouadhibou and Nouakchott–Dakar, Banjul and Bissau can be found in Physarum graphs $\mathbf{P}(\frac{19}{32})$, $\mathbf{P}(\frac{20}{32})$ and $\mathbf{P}(\frac{21}{32})$, respectively.

The strongest transport components are chains Brazzaville and Kinshasa–Luanda–Lobito and Lubumbashi and Lusaka–Harare–Beira that are experimentally imitated by slime mould in over 70% of experimental trials. They are presented by edges of Physarum graphs $\mathbf{P}(\frac{23}{32})$ and $\mathbf{P}(\frac{25}{32})$.

Finding 9. *The transport route Lubumbashi and Lusaka–Harare–Beira is the strongest component of the trans-African networks presented by slime mould and the strongest component of β-skeletons constructed on* **U**.

3.4 Summary

We represented the major urban areas of Africa by oat flakes and inoculated plasmodium of *P. polycephalum* in the agar gel site that corresponds to Cairo. We analysed the dynamics of the plasmodium growth, compared its network of protoplasmic tubes to trans-African highways, and analysed both the slime mould as well as the highway networks as proximity graphs. The results proved very encouraging, especially in the light of *competition among the transit corridors*, that are being considered as one of the key approaches to overcoming the problems of

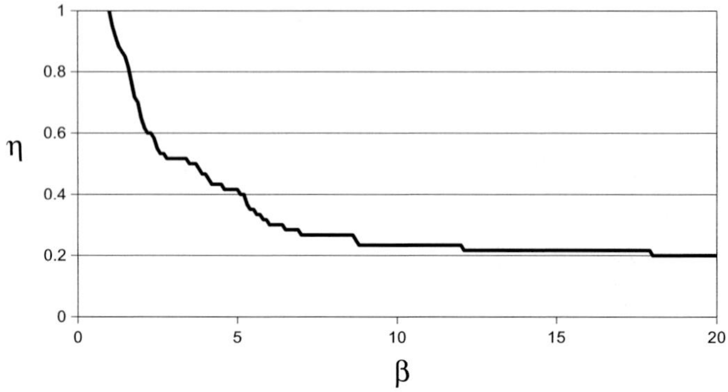

(a)

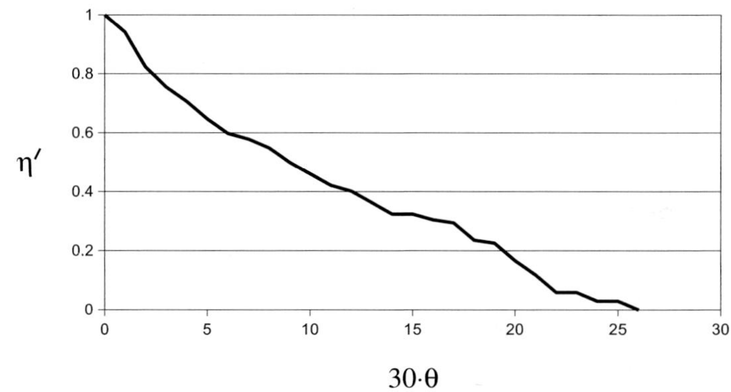

(b)

Fig. 3.19 Parameterisation of connectivity in β-skeletons and Physarum graphs: (a) decrease of β-skeleton's normalised connectivity $\eta(\beta)$ (vertical axis) with increase of β (horizontal axis). The normalised connectivity $\eta(\beta)$ of $\mathbf{BS}(\beta)$ is calculated as number of edges in $\mathbf{BS}(\beta)$ divided by number of edges in $\mathbf{BS}(1)$, (b) decrease of Physarum graph's $\mathbf{P}(\theta)$ normalised connectivity $\eta'(\theta)$ (vertical axis) with increase of θ (horizontal axis). The normalised connectivity $\eta'(\theta)$ of $\mathbf{P}(\theta)$ is calculated as number of edges in $\mathbf{P}(\theta)$ divided by number of edges in $\mathbf{P}(0)$.

non-physical barriers in integrating existing roads into a united network of trans-African highways [The African Development Bank (2003)]. Such competition is 'automatically' addressed in our laboratory experiments with slime mould because its behaviour is based on dynamical interactions and competitions between many active zones of growth and biochemical oscillators in the plasmodium's extended body [Adamatzky (2010)b].

We found that, in principle, slime mould provides an adequate approxima-

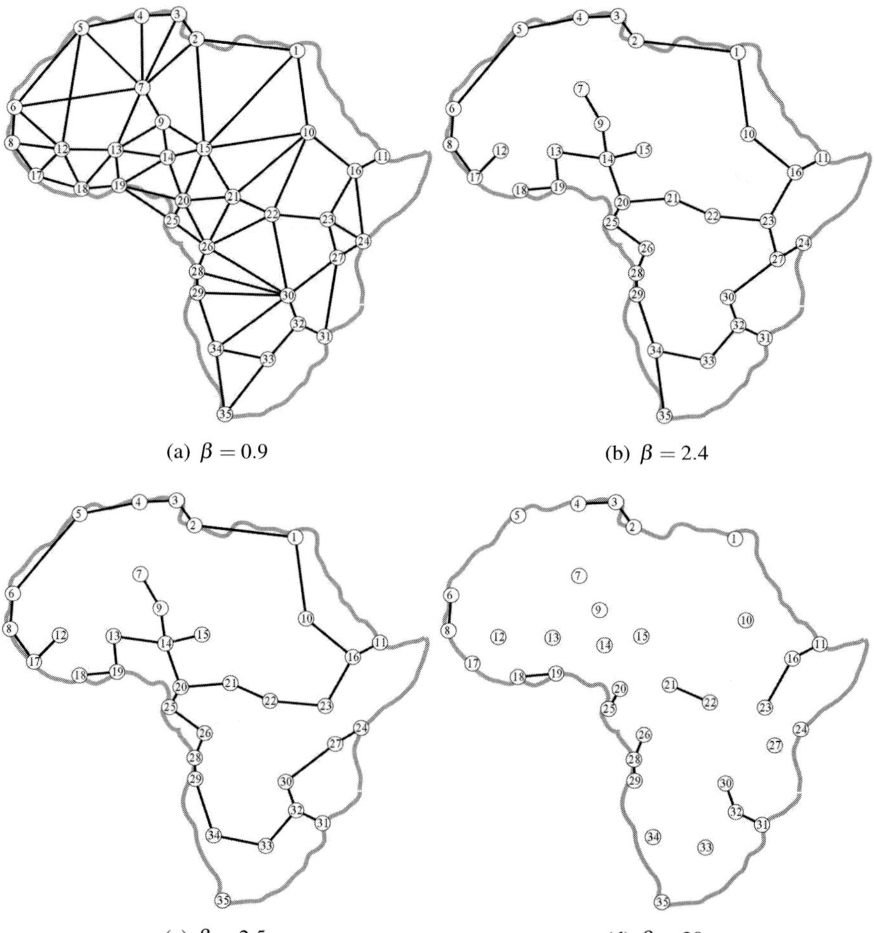

(a) $\beta = 0.9$ (b) $\beta = 2.4$

(c) $\beta = 2.5$ (d) $\beta = 20$

Fig. 3.20 Example of β-skeletons $\mathbf{BS}(\beta)$ for several values of β: (a) $\beta = 0.9$, (b) $\beta = 2.4$, (c) $\beta = 20$.

tion of trans-African highways, and Physarum graphs for low values of θ include almost all the edges of the trans-African highway graph. All segments of trans-African highways that are not represented by the slime mould have components of non-paved roads. Obviously in our experiments slime mould did not get any indication on which roads are paved and which are not paved; thus, the result is surprisingly interesting. A family of Physarum graphs that are parameterised by increasing the threshold of experimental repeatability shows highest efficiency of approximation of trans-African highways by Physarum graphs that first are planar

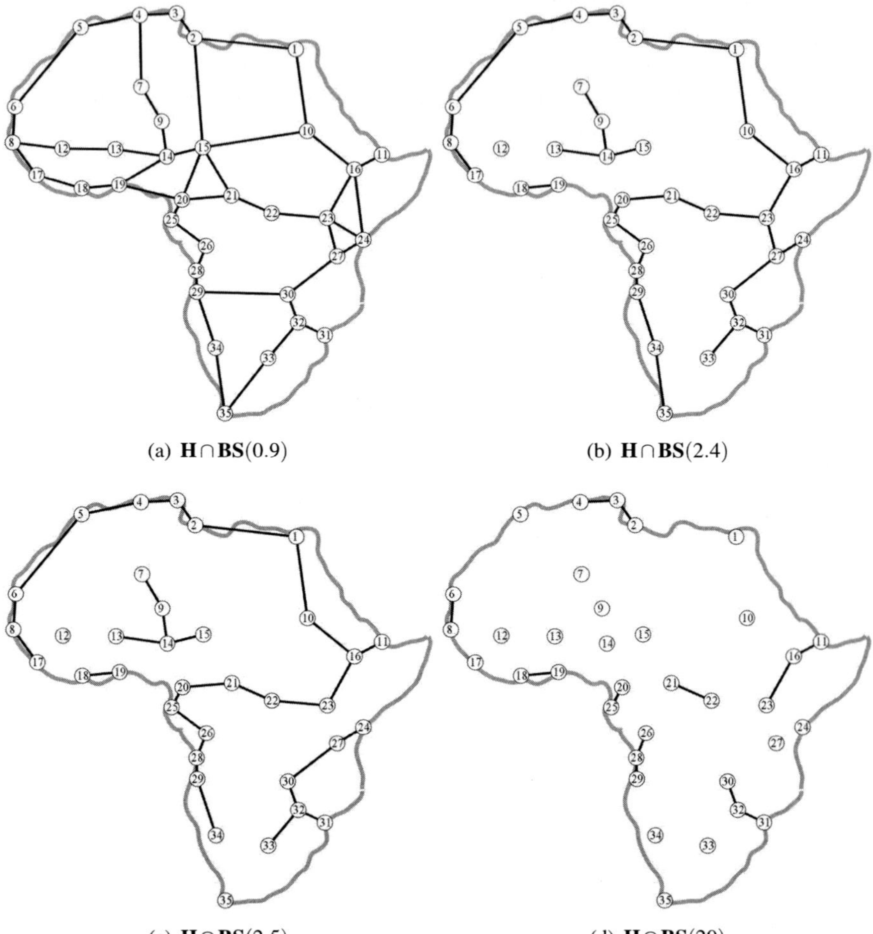

(a) $\mathbf{H} \cap \mathbf{BS}(0.9)$ (b) $\mathbf{H} \cap \mathbf{BS}(2.4)$

(c) $\mathbf{H} \cap \mathbf{BS}(2.5)$ (d) $\mathbf{H} \cap \mathbf{BS}(20)$

Fig. 3.21 Intersections of highway graph $\mathbf{H}$ with β-skeletons $\mathbf{BS}(\beta)$ for several values of β: (a) $\beta = 0.9$, (b) $\beta = 2.4$, (c) $\beta = 20$.

and then by Physarum graphs that are the last connected graphs of the family.

We can classify sections of the trans-African highways into three types: sections which never appear in laboratory experiments, sections which always appear in laboratory experiments and sections which are approximated with some degree of accuracy. In the laboratory experiments we found that the eastern parts of the Beira–Lobito highway, and the route Lubumbashi and Lusaka–Harare– Beira, are the strongest components of trans-African networks in terms of slime mould representations (they appear in most experiments) and β-skeletonisations (they appear for higher values of β than other routes). The section Tripoli to Brazzaville and Kinshasa of the Tripoli–Windhoek–Cape Town highway and the section Djibouti to Dodoma of the Cairo–Gaborone–Cape Town highway are not approximated by the planar Physarum graph in the experiments.

Chapter 4

Tracing historical development of Australian highways

Andrew Adamatzky and Mikhail Prokopenko

There are many unique features of the Australian motorway network. Firstly, one may note a clear dominance of coastal routes that form a network of highways along the entire coast of the Australian continent (i.e. it is a cycle graph or a circuit). This network, called *Australia's Highway 1*, joins all mainland state capitals, being the longest national highway in the world. The total length of Australia's Highway 1 is approximately 14,500 km, exceeding the length of the trans-Siberian Highway ($\approx 11,000$ km) and the trans-Canada Highway ($\approx 8,000$ km). Australia's Highway 1 contains specific highways, such as the Pacific Highway — a major transport route along part of the east coast of Australia, between Sydney and Brisbane. The dominance of coastal routes is due to a range of historic, geographic, economic and geopolitical factors, including the history of early European settlements along the coast (these settlements initially used boats as primary mode of transport); specifics of Australia's main industries and services such as mining and tourism; the harsh continental climate (with arid conditions and significant lack of water) that does not favour sustainable agriculture in regions located further from the coast than 1,000 km on average, and often just a few hundreds of kilometres; etc.

Secondly, the Australian motorway network shows a relative abundance of tributaries of Highway 1: a set of smaller networks interconnecting cities and towns in the vicinity of major cities such as Sydney, Melbourne, Brisbane and so on. Each of these smaller networks reaches a sufficient density around their respective centres, typically attaining the complete subgraph topology. Australia's Highway 1 passes through some of Australia's fastest growing regions, and in turn stimulates regional development making the tributaries necessary.

Thirdly, Highway 1 includes the principal north–south route through the central (mostly desert) interior of mainland Australia, connecting Darwin, Northern Territory, with Port Augusta, South Australia — the Stuart Highway (or simply 'The Track'), as well as the north-eastern route including highways that connect Darwin in the north with Brisbane on the east coast. These two routes provide main 'short-cuts' through the interior of mainland Australia, and also attract tributaries.

In summary, Australia presents a unique case combining the dominant coastal highway circuit, dense tributary routes around main coastal cities and a small number of 'short-cut' routes through the interior of mainland Australia.

We consider the 25 most populated urban areas **U** of Australia (Fig. 4.1a):

(1) Sydney, Newcastle, Central Coast, Wollongong, Maitland, Nowra–Bomaderry and Richmond–Windsor (New South Wales),
(2) Melbourne, Geelong, Ballarat, Melton, Bendigo, Sunbury and Shepparton–Mooroopna (Victoria),
(3) Brisbane, Sunshine Coast, Toowoomba and Gold Coast–Tweed Heads (Queensland/New South Wales),
(4) Perth, Rockingham and Mandurah (Western Australia),
(5) Adelaide (South Australia),
(6) Bundaberg and Hervey Bay (Queensland),
(7) Canberra–Queanbeyan (Australian Capital Territory),
(8) Townsville–Thuringowa (Queensland),
(9) Cairns (Queensland),
(10) Mildura (Victoria),
(11) Mackay (Queensland),
(12) Darwin and Palmerston (Northern Territory),
(13) Bunbury and Albany (Western Australia),
(14) Port Macquarie, Lismore and Coffs Harbour (New South Wales),
(15) Dubbo and Tamworth (New South Wales),
(16) Port Augusta (South Australia)[1],
(17) Warrnambool (Victoria) and Mount Gambier (South Australia),
(18) Gladstone and Rockhampton (Queensland),
(19) Kalgoorlie–Boulder (Western Australia),
(20) Geraldton (Western Australia),
(21) Alice Springs,[1]
(22) Mount Isa and Cloncurry (Queensland)[1],
(23) Broken Hill (New South Wales)[1],

[1]Included only for completeness.

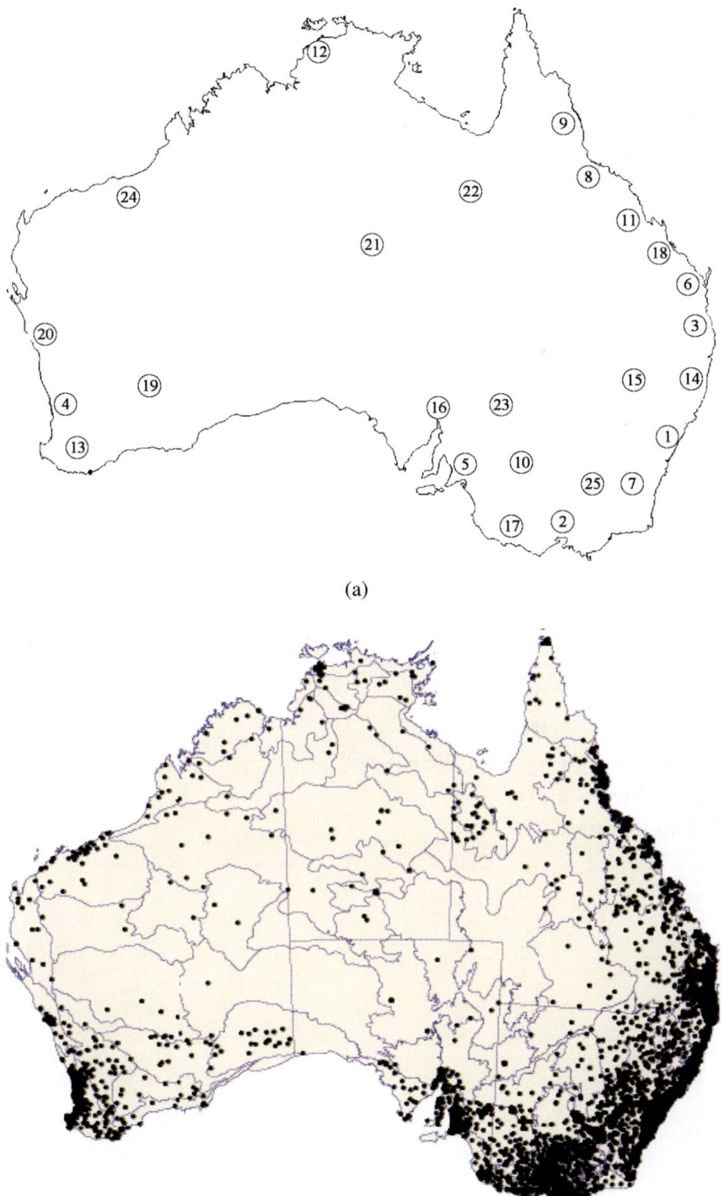

(a)

(b)

Fig. 4.1 Experimental basics: (a) contour map of Australia with 25 urban areas indicated, (b) population density in 1997, from [Williams (2001)].

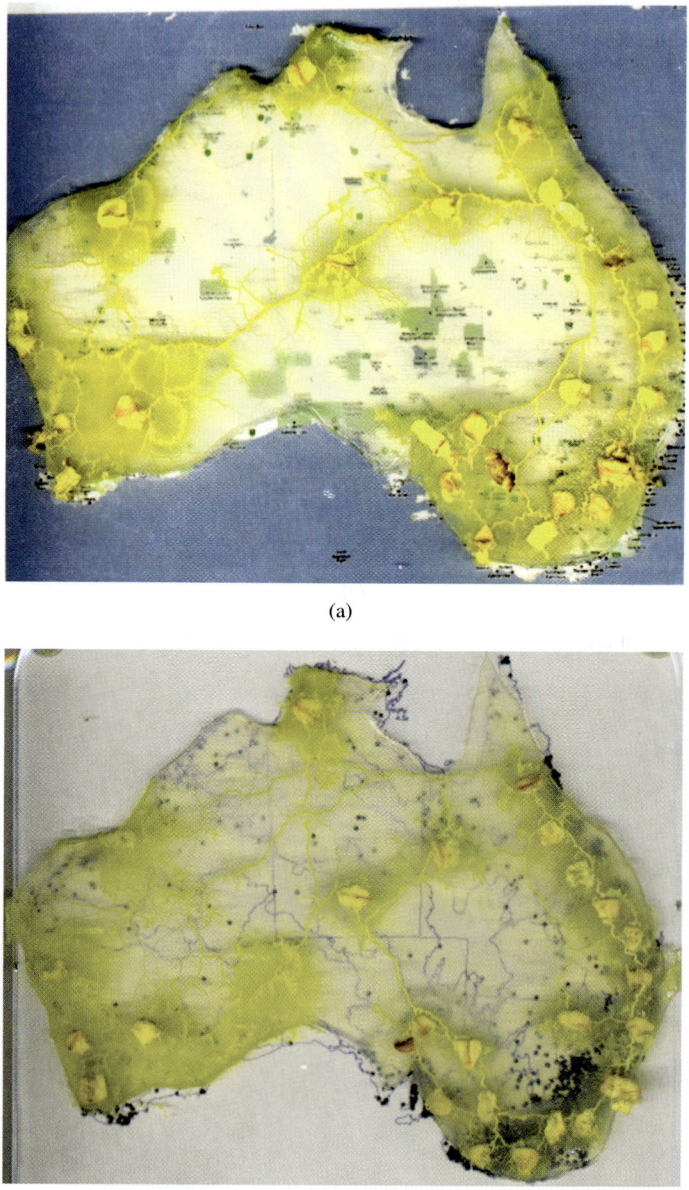

(a)

(b)

Fig. 4.2 Snapshot of a typical setup: urban areas are represented by oat flakes, plasmodium is inoculated in Sydney, the plasmodium spans the oat flakes by protoplasmic transport: (a) on map of Australia, (b) on population density map, from [Williams (2001)].

(24) South Hedland (Western Australia)[1],
(25) Albury–Wodonga (New South Wales/Victoria) and Wagga Wagga (New South Wales).

Our choice of urban areas provides an adequate sampling of population density distribution (Fig. 4.1b). To project regions of **U** onto agar gel we place oat flakes in the positions of the regions of **U** (Fig. 4.2). At the beginning of each experiment a piece of plasmodium, usually already attached to an oat flake, was placed in Sydney (region 1 in Fig. 4.1a). We undertook 31 experiments.

4.1 Slime mould traces gold rush networking

At the beginning of each experiment a piece of plasmodium, usually attached to an oat flake, is placed in the position of the Sydney urban area. The plasmodium develops branching pseudopodia, propagates on the agar gel towards neighbouring oat flakes, occupies them and connects them with protoplasmic networks. The exact way of plasmodium's exploration of its growth substrate and colonisation of urban areas may differ from experiment to experiment. In some experiments, the plasmodium forms several active zones, which propagate simultaneously and colonise urban areas in a concurrent manner, frequently even competing with each other. In the example of concurrent colonisation shown in Fig. 4.3, in 8 h after inoculation in Sydney the plasmodium propagates from Sydney to the Port Macquarie area in the north and to Canberra–Queanbeyan in the south. By the 31st hour since inoculation the plasmodium forms branches from Canberra–Queanbeyan to South Hedland to Mildura and from Canberra–Queanbeyan to Melbourne to the Warrnambool area. The plasmodium also connects the Port Macquarie area with the Dubbo–Tamworth area, and develops a transport chain from the Port Macquarie area to the Brisbane area to the Bundaberg area to the Townsville–Thuringowa area. In the 50th hour of the experiment the plasmodium connects Mildura and Adelaide, Adelaide and Port Augusta, Port Augusta and Broken Hill, and Port Augusta to Alice Springs and Kalgoorlie–Boulder; and also establishes transport routes between Alice Springs and Kalgoorlie–Boulder on one side and the Albury–Wolonga, South Hedland, Geraldton, Perth and Bunbury–Albany areas on another (see scheme of propagation in Fig. 4.5a).

The experiment shown in Fig. 4.4 illustrates a rather sequential colonisation of urban zones by plasmodium. In the first 8 h since inoculation in Sydney the plasmodium builds a connection between the Sydney and Perth areas. In the next 23 h the plasmodium colonises urban areas along the west coast from

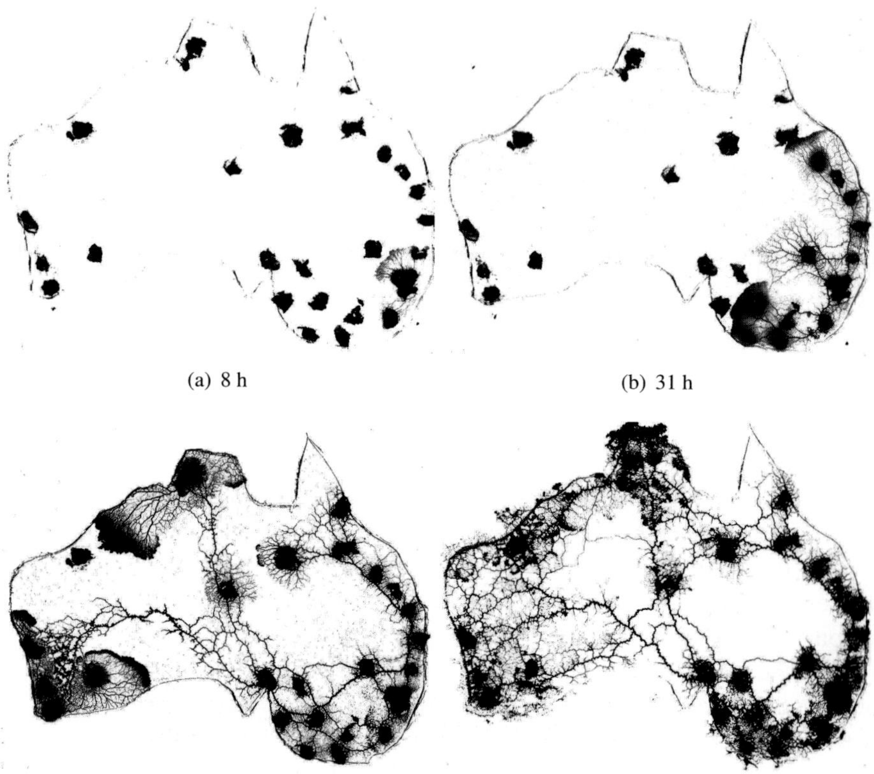

(a) 8 h

(b) 31 h

(c) 50 h

(d) 74 h

Fig. 4.3 Images of plasmodium spanning urban areas **U** with its protoplasmic tubes, scenario of concurrent foraging.

Perth–Rockingham–Mandurah to Townsville–Thuringowa (Figs. 4.4b and 4.5b). Then the plasmodium links Townsville–Thuringowa with Cairns, and develops a chain Mount Isa to Alice Springs to Kalgoorlie–Boulder. It then branches from Kalgoorlie–Boulder to the Bunbury–Albany area and to the Port Macquarie–Lismore–Coffs Harbour area (Fig. 4.4c). At the last stages of colonisation the plasmodium links Kalgoorlie–Boulder to Geraldton and Kalgoorlie–Boulder to South Hedland to the Darwin–Palmerston area (Figs. 4.4d and 4.5b).

Finding 10. *The Dubbo and Tamworth urban area is most sensitive to edge trimming a node of a Physarum graph.*

When θ increases to $\frac{16}{30}$, the Dubbo–Tamworth urban area becomes isolated

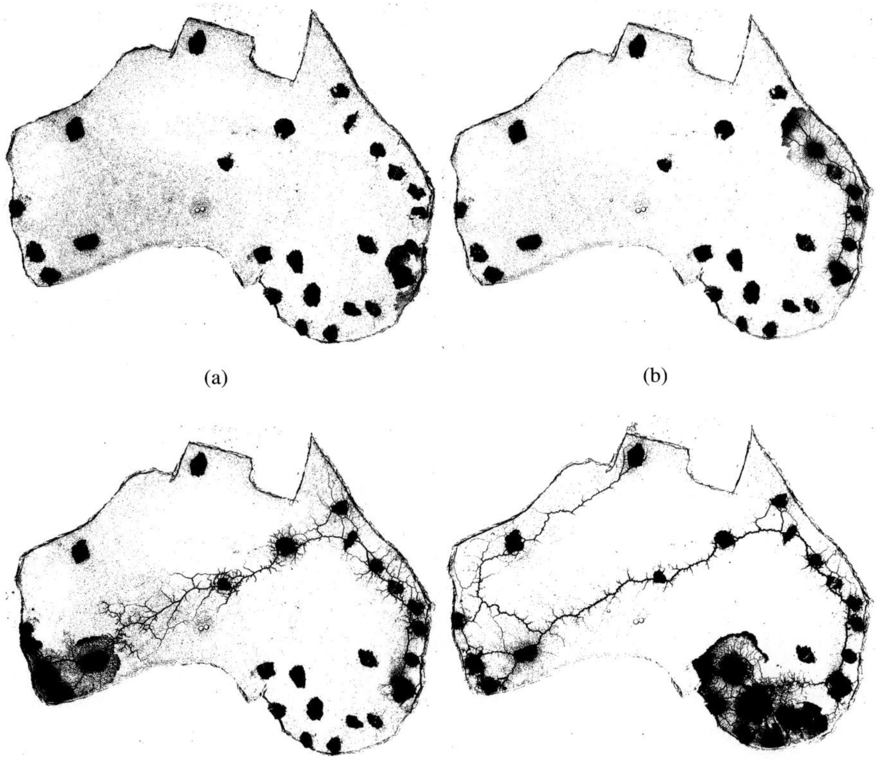

(a) (b) (c) (d)

Fig. 4.4 Images of plasmodium spanning urban areas **U** with its protoplasmic tubes, scenario of serial foraging.

(Fig. 4.6b and c). In a Physarum graph $\mathbf{P}(0)$ this area has highest degree of connectivity, yet a weight of each edge is small compared to weights of edges connecting other nodes (Fig. 4.6a). A possible explanation may be in Physarum strategy of foraging. Inoculated in the Sydney urban area, the plasmodium more likely propagates along the east coast towards Cairns. Oat flakes representing east coast urban areas are so dense that slime mould often just spans this chain of areas without an attempt to branch towards Dubbo–Tamworth.

Finding 10 may be easily correlated with historical developments: when Sydney became the major centre, the transport routes were mostly developed along the east coast. So, it is not a surprise that oat flakes representing the dense east coast urban areas attract the slime mould to span their chain without an attempt

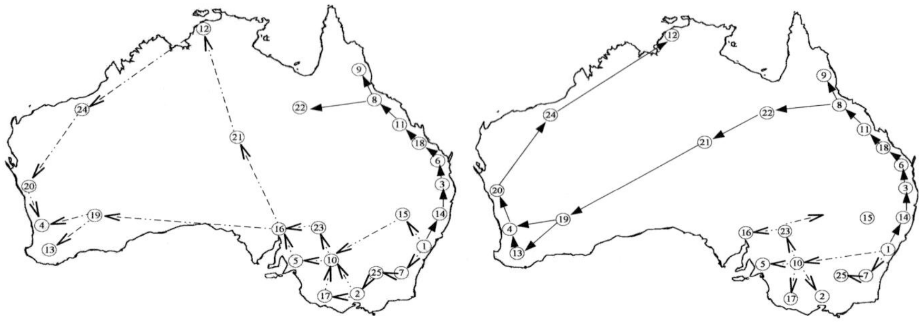

(a) (b)

Fig. 4.5 Schemes of plasmodium propagation extracted from (a) Fig. 4.3 and (b) Fig. 4.4.

to branch towards Dubbo–Tamworth — Dubbo has developed into the crossroads of New South Wales only as a result of the gold rush of the 1860s that brought an increase in north–south trade, becoming a city only in 1966; while Tamworth was reached by the railway only in 1878, becoming a city only in 1946.

Finding 11. *The Physarum graph is split into three components for* $\theta = \frac{18}{31}$.

For $\theta = \frac{18}{31}$, three components are observed (Fig. 4.6e). The first component is a tree T rooted at Alice Springs with three branches:

- Alice Springs–Mount Isa–Townsville–Thuringowa–Cairns,
- Alice Springs–Kalgoorlie–Boulder–Bunbury–Albany,
- Alice Springs–Darwin–Palmerston–South Hedland–Geraldton–Perth area.

The second component consists of a chain Mackay–Gladstone–Rockhampton–Bundaberg–Brisbane–Port Macquarie–Sydney–Canberra-Queanbeyan with two cycles attached: Albury–Wodonga–Melbourne–Warrnambool–Mildura and Mildura–Adelaide–Port Augusta–Broken Hill. The third component is an isolated node Dubbo–Tamworth. Here we observe separation of eastern Australia from the rest of the country.

When θ increases to $\frac{19}{31}$, the chain Darwin–Palmerston–South Hedland–Geraldton–Perth–Rockingham–Mandurah becomes disconnected from the tree in the west (Fig. 4.7a) but the eastern component remains unchanged.

Finding 11, we believe, is also well correlated with historical developments: the second component (the east coast) covers the area developed first of all (driven by farming and mining in New South Wales, Victoria and South Australia), with the third component (Dubbo–Tamworth node) roughly corresponding to the gold

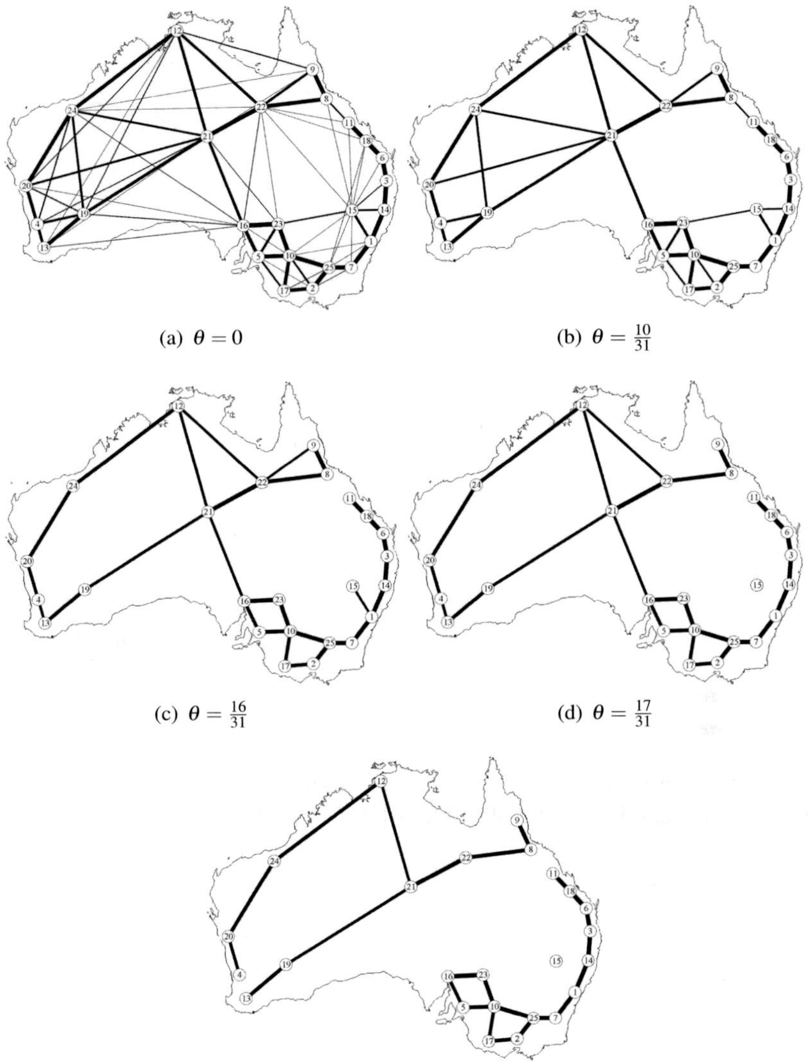

(a) $\theta = 0$ (b) $\theta = \frac{10}{31}$

(c) $\theta = \frac{16}{31}$ (d) $\theta = \frac{17}{31}$

(e) $\theta = \frac{18}{31}$

Fig. 4.6 Configurations of Physarum graph $\mathbf{P}(\theta)$ for various values of θ. Thickness of each edge is proportional to the edge's weight.

rush networking, while the first component (the rest of the country) spans the area that has developed later in time, in response to mining booms during late 19th

(a) $\theta = \frac{19}{31}$

(b) $\theta = \frac{20}{31}$

(c) $\theta = \frac{22}{31}$

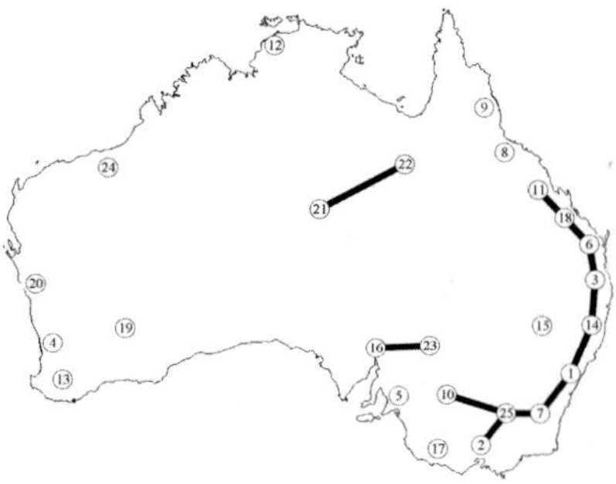

(d) $\theta = \frac{27}{31}$

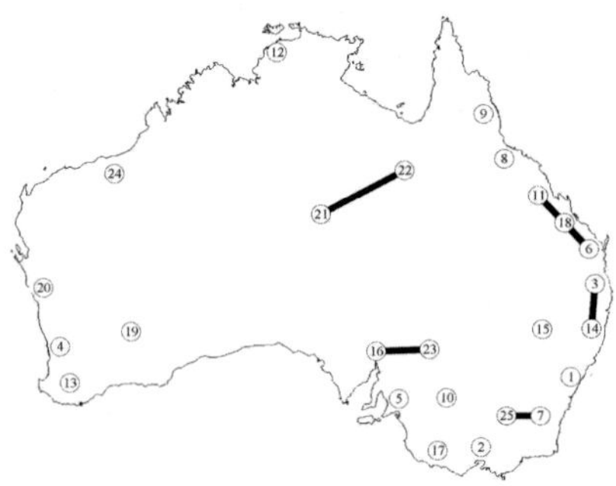

(e) $\theta = \frac{28}{31}$

Fig. 4.7 Configurations of Physarum graph $\mathbf{P}(\theta)$ for various values of θ. Thickness of each edge is proportional to the edge's weight.

and 20th centuries, utilising various mineral resources in Northern Territory and Western Australia.

Finding 12. *The Physarum graph becomes acyclic only for* $\theta = \frac{22}{31}$.

There are several components in the graph $\mathbf{P}(\frac{22}{31})$ (Fig. 4.7c):

- Geraldton–South Hedland–Darwin–Palmerston,
- Bunbury–Albany–Kalgoorlie–Boulder,
- Cairns–Townsville–Thuringowa–Mount Isa–Alice Springs,
- Mackay–Gladstone–Rockhampton–Port Augusta–Brisbane–Port Macquarie –Sydney–Canberra–Queanbeyan–Albury–Wodonga with branches Albury–Wodonga–Melbourne–Warrnambool and a tree with branches Albury–Wodonga–Mildura–Adelaide and Albury–Wodonga–Mildura–Broken Hill–Port Augusta,
- two isolated nodes Perth–Rockingham–Mandurah and Dubbo–Tamworth.

At this stage, we point out that edge trimming may not only reveal historical parallels but also indicate somewhat weaker links in transport networks. Thus, Finding 12 may be interpreted as an indication that the features of the Australian city network make it fairly robust to disruptions. This is due to the dominance of coastal routes complemented by the tributaries and short-cuts — since these three features were reproduced by the slime mould, the resulting network becomes acyclic only when most of the tributaries and all the short-cuts are removed.

Finding 13. *The most stable (often-appearing) components of the Physarum graph are two-node chains Alice Springs–Mount Isa and Port Augusta–Broken Hill and a chain Mackay–Gladstone–Rockhampton–Bundaberg–Brisbane–Port Macquarie–Sydney–Canberra-Queanbeyan–Albury–Wodonga with a fork Mildura–Melbourne attached to South Hedland.*

These components are present in over 27 of 31 experiments (Fig. 4.7d).

Again, Finding 13 indicates both the strongest and historically earlier component(s) — in this instance, Pacific Highway. The two-node chains (Alice Springs–Mount Isa and Port Augusta–Broken Hill) are quite curious and point out that these links may be less dependent on the rest of the network than the first glance would suggest (also, Alice Springs and Broken Hill are main interior-mainland centres).

4.2 Physarum reconstructs the Gabriel graph

Finding 14. **MST** = **RNG**/$\{$*(Mount Isa–Alice Springs + Port Augusta–Broken Hill)*$\}$.

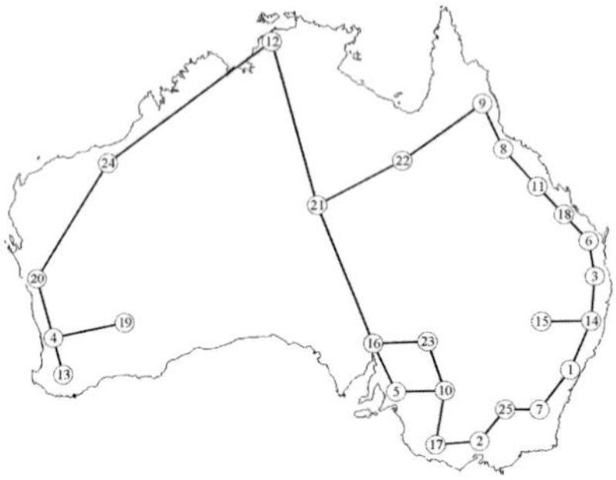

(a) **RNG**

(b) **GG**

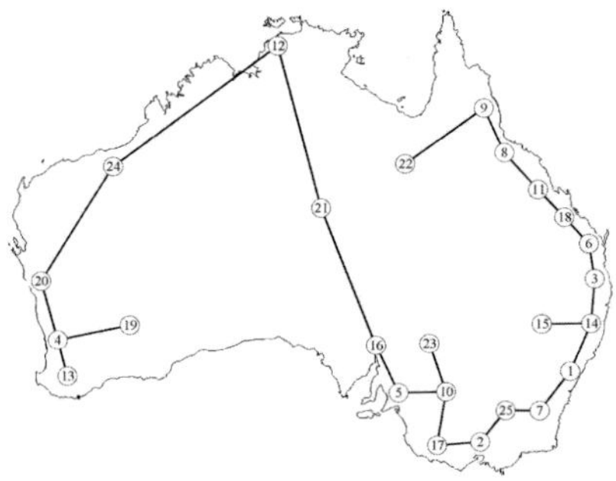

(c) **MST**

Fig. 4.8 Proximity graphs constructed on urban areas **U**: (a) relative neighbourhood graph **RNG**, (b) Gabriel graph **GG**, (c) spanning tree **ST** rooted at Sydney.

Finding 14 indicates that the configuration of urban areas allows for a very efficient spanning. Also, it is worth pointing out that the difference between **MST** and **RNG** is precisely the union of the two-node chains that were identified in Finding 13 as the most stable two-node chains. In other words, adding two stable routes to an efficient **MST** produces precisely the **RNG** where efficiency and stability are combined. These two-node transport links may be seen as bridges between cyclicity and acyclicity of proximity graphs constructed on **U** (Fig. 4.8a and c).

Finding 15. $\mathbf{P}(\frac{10}{31})$ = **GG** + *(Geraldton–Alice Springs).*

See Figs. 4.6b and 4.8b.

Finding 15 presents strong evidence that the graph constructed by the slime mould with limited edge trimming, $\mathbf{P}(\frac{10}{31})$, is almost identical to the Gabriel graph.

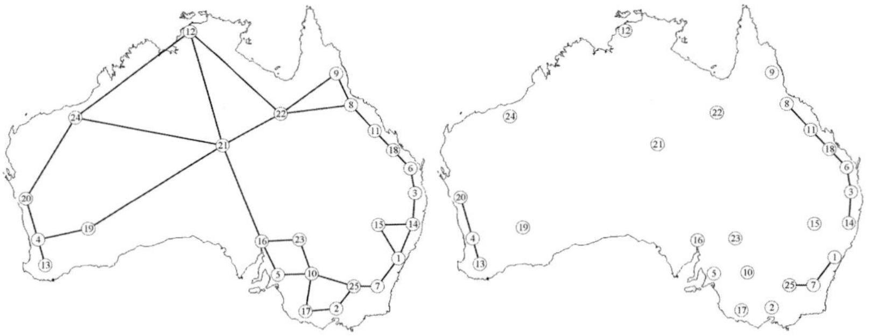

(a) **BS**(1.6) (b) **BS**(20)

Fig. 4.9 β-skeletons constructed on **U** for (a) $\beta = 1.6$ and (b) $\beta = 20$.

To reiterate, the Gabriel graph captures one notion of spatial proximity, connecting the points that are near each other in a certain sense. That is, there are no links in the Gabriel graph that avoid a closer node. The fact that the limited edge trimming almost makes the Gabriel graph indicates that the slime mould is very parsimonious — the resulting graph has no unnecessary connections, with only close neighbours being connected.

Similar conclusions can be reached by analysing the following finding 16.

Finding 16. **BS**(1.6) *is a closest to the Physarum graph's* β*-skeleton.*

To transform **BS**(1.6) to $\mathbf{P}(\frac{16}{31})$, we must carry out the following operations on $\mathbf{P}(\frac{16}{31})$ (Figs. 4.6c and 4.9a):

- remove edges Townsville–Thuringowa–Mackay, Port Macquarie–Dubbo–Tamworth and Perth–Kalgoorlie–Boulder,
- add edge Bunbury–Albany–Kalgoorlie–Boulder.

There is no direct similarity between the θ parameter of the Physarum graph and the β parameter of the β-skeletons apart from the fact that both θ and β define a 'continuous' family of graphs, where the number of edges decreases with increase of the parameter. The Physarum graph becomes a set of isolated nodes when β reaches 1. β-skeletons on **U** are never reduced to a set of isolated nodes because few edges survive for any high value of β. We call a transport link (*a*–*b*) stable if it survives for high values of θ and β.

The links Alice Springs–Mount Isa, Port Augusta–Broken Hill, Canberra–Queanbeyan–Albury–Wodonga, Brisbane–Bunbury–Albany and Bundaberg–Gladstone–Rockhampton–Mackay survive until $\theta = \frac{28}{31}$ (Fig. 4.7). The link Alice

Springs–Mount Isa also survives in the β-skeleton until $\beta = 7.7$.

A comparison shows that the most stable links in β-skeletons are Bunbury–Albany–Perth–Rockingham–Geraldton, Sydney–Canberra–Queanbeyan–Albury-Wodonga and Townsville–Thuringowa–Mackay–Gladstone–Rockhampton–Bundaberg–Brisbane–Port Macquarie. These links are presented in **BS**(20) (Fig. 4.9b).

4.3 Australian highways are a subnetwork of the Physarum network

A scheme of the Australian motorway network is shown in Fig. 4.10a and a graph **H** derived from the network is shown in Fig. 4.10b. The motorway graph **H** does not match the proximity graphs studied (Fig. 4.10c–e): neither of the graphs is a subgraph of **H**. This may give us a first indication that the Australian motorway network is far from geometrical optimality.

Finding 17. *The Australian motorway graph* **H** *is most close to the Gabriel graph* **GG**.

Namely, the intersection of **H** and **GG** is **H** with the following edges removed:

- Perth–Rockingham–South Hedland,
- Melbourne–Canberra–Queanbeyan,
- Albury–Wodonga–Bunbury–Albany,
- Dubbo–Tamworth–Mount Isa,
- Dubbo–Tamworth–Gladstone–Rockhampton,
- Dubbo–Tamworth–Brisbane,
- Gladstone–Rockhampton–Melbourne.

The Dubbo–Tamworth urban area is linked to the highest number of 'non-optimal' edges.

The similarity between the motorway graph **H** and the Gabriel graph **GG** shows that the motorway connections are indeed very parsimonious, without unnecessary connections, and with only close neighbours connected. Hence, Findings 15 and 16 do support the proposition that the graph built by the slime mould, with an appropriate limited edge trimming, is a good approximation for the motorway graph **H** (and the Gabriel graph **GG**).

Finding 18. *The intersection of* **H** *and* **MST** *consists of two connected components and three isolated nodes.*

The connected components are the eastern chain spanning the urban area from

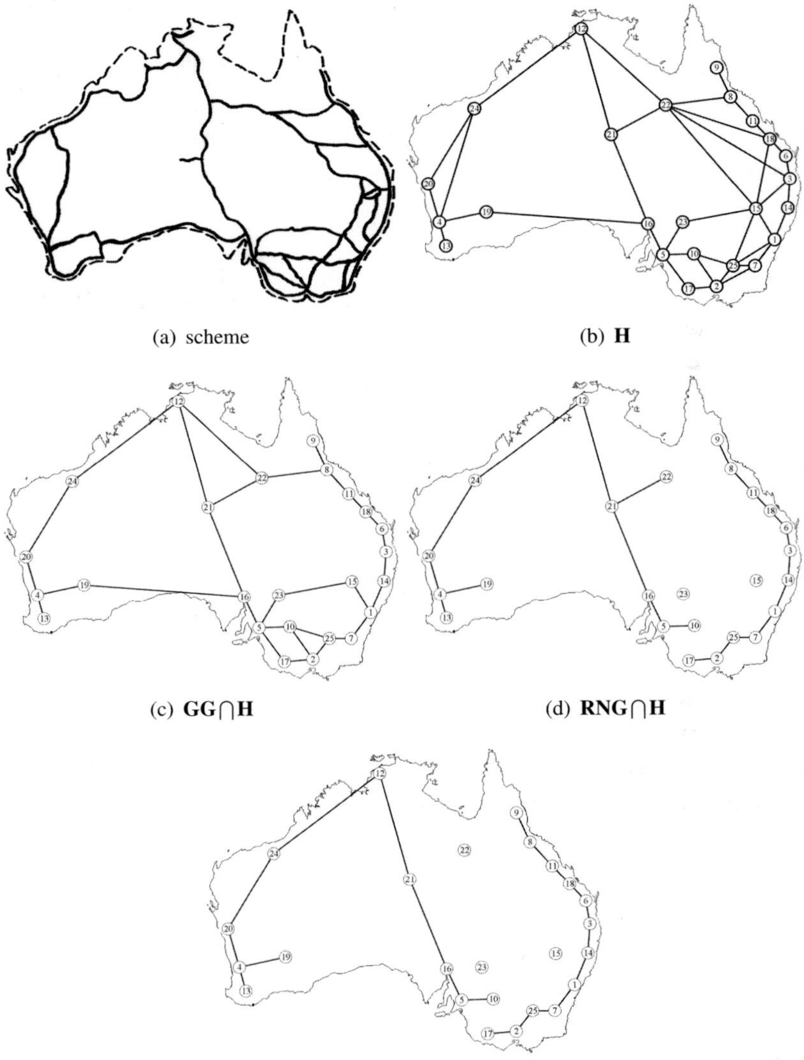

(a) scheme (b) **H**

(c) **GG**$\bigcap$**H** (d) **RNG**$\bigcap$**H**

(e) **MST**$\bigcap$**H**

Fig. 4.10 Structure of Australian motorways: (a) scheme (adapted from [Road Maps of Australian Highways (2011)]) and (b) graph **H** of man-made motorway network in Australia, (c–e) intersections of **H** with Gabriel graph (c), relative neighbourhood graph (d) and spanning tree rooted in Sydney (e).

Warrnambool and Mount Gambier in the south to Cairns in the north and a chain spanning Mildura in the south-east to Darwin–Palmerston in the north-west to

Bunbury–Albany in the south-west. The three isolated urban areas are Dubbo–Tamworth, Mount Isa and Broken Hill (Fig. 4.10e).

Finding 18 points to growing regional centres, Dubbo–Tamworth, Mount Isa and Broken Hill, as the areas around which transport links may have developed in a suboptimal way. That is, locally optimal solutions may have been at odds with the global network optimum Australia-wide.

Finding 19. $\mathbf{P}(0) \bigcap \mathbf{H} = \mathbf{H} /$ *(Broken Hill–Brisbane).*

That is, the Australian motorway graph is almost a subgraph of the Physarum graph $\mathbf{P}(0)$. The intersections of the motorway graph $\mathbf{H}$ with several Physarum graphs are shown in Fig. 4.11. The 'offending' link (Broken Hill–Brisbane) is actually the only edge contributing to non-planarity of $\mathbf{H}$, so the link's removal sounds reasonable. The graph $\mathbf{P}(0)$ includes all protoplasmic tubes that ever occurred in experiments; thus, the fact that it almost includes $\mathbf{H}$ is encouraging but not exciting. Let us compare $\mathbf{H}$ with the more trimmed Physarum graph $\mathbf{P}(\frac{10}{31})$, where only edges recorded in over 30% of experiments are present.

Finding 20. $\mathbf{P}(\frac{10}{31}) \bigcap \mathbf{H} = \mathbf{GG} \bigcap \mathbf{H} /$ *(Kalgoorlie–Boulder–Port Augusta).*

Finding 21. *The only parts of the Australian motorway graph that are represented by Physarum protoplasmic tubes in almost all experiments are a transport link connecting Alice Springs and Mount Isa and Cloncurry, and an east coast transport chain from the Melbourne urban area in the south to the Mackay area in the north.*

This is illustrated in Fig. 4.11h. This finding infers that only the eastern part of the Australian motorway network is substantiated by biological logic of *Physarum polycephalum*.

We believe that, analogously to Finding 11, these findings are mostly related to historical developments of Australia: the east coast had been developed prior to an inception of the overall transport network coverage, while the rest of the country has developed later and in separate phases, in response to various events such as mining booms. One may argue that different levels of edge trimming correspond to stages of the overall network development, with higher levels corresponding to earlier stages, and lower levels corresponding to later stages.

4.4 Famine and large-scale contamination

Finding 22. *In response to exhaustion of natural resources, or crop underproduction, the most active pattern of migration should be observed in the northern and*

(a) $\mathbf{P}(0) \bigcap \mathbf{H}$ (b) $\mathbf{P}(\frac{10}{31}) \bigcap \mathbf{H}$

(c) $\mathbf{P}(\frac{17}{31}) \bigcap \mathbf{H}$ (d) $\mathbf{P}(\frac{18}{31}) \bigcap \mathbf{H}$

(e) $\mathbf{P}(\frac{19}{31}) \bigcap \mathbf{H}$ (f) $\mathbf{P}(\frac{20}{31}) \bigcap \mathbf{H}$

(g) $\mathbf{P}(\frac{22}{31}) \bigcap \mathbf{H}$ (h) $\mathbf{P}(\frac{27}{31}) \bigcap \mathbf{H}$

Fig. 4.11 Intersections of Physarum graphs for various values of θ and motorway graph $\mathbf{H}$.

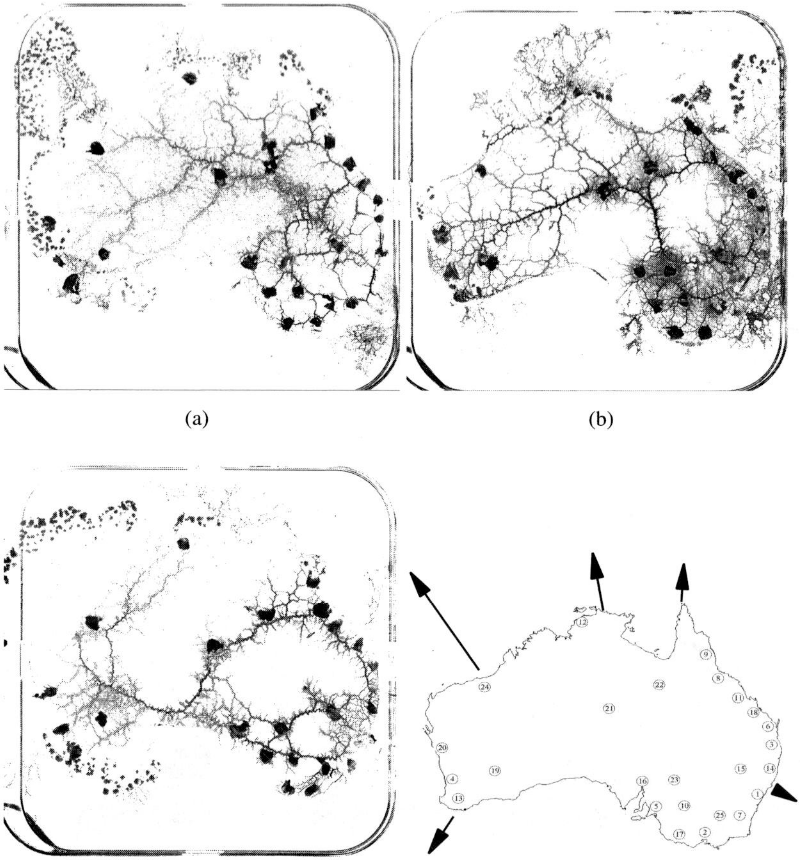

(a) (b) (c) (d)

Fig. 4.12 Plasmodium attempting to explore new territories outside its agar plate: (a–c) images of three experiments recorded 5–6 days after inoculation of plasmodium, (d) a scheme of the plasmodium exodus, lengths of arrows are roughly proportional to the frequency of the directions used by the plasmodium. White/light-coloured protoplasmic network consists of abandoned protoplasmic tubes. Yellow/bright blobs on the Petri dish's bare plastic bottom are parts of escaping plasmodium.

north-western parts of Western Australia.

To simulate a famine, we allow plasmodium to populate an agar plate for over 5–6 days. By that time nutrients, or bacteria, in oat flakes are exhausted, humidity of the substrate decreases and more likely concentration of metabolites excreted by the plasmodium significantly increases. Such living conditions become inappropriate for plasmodium and it starts abandoning its network of protoplasmic

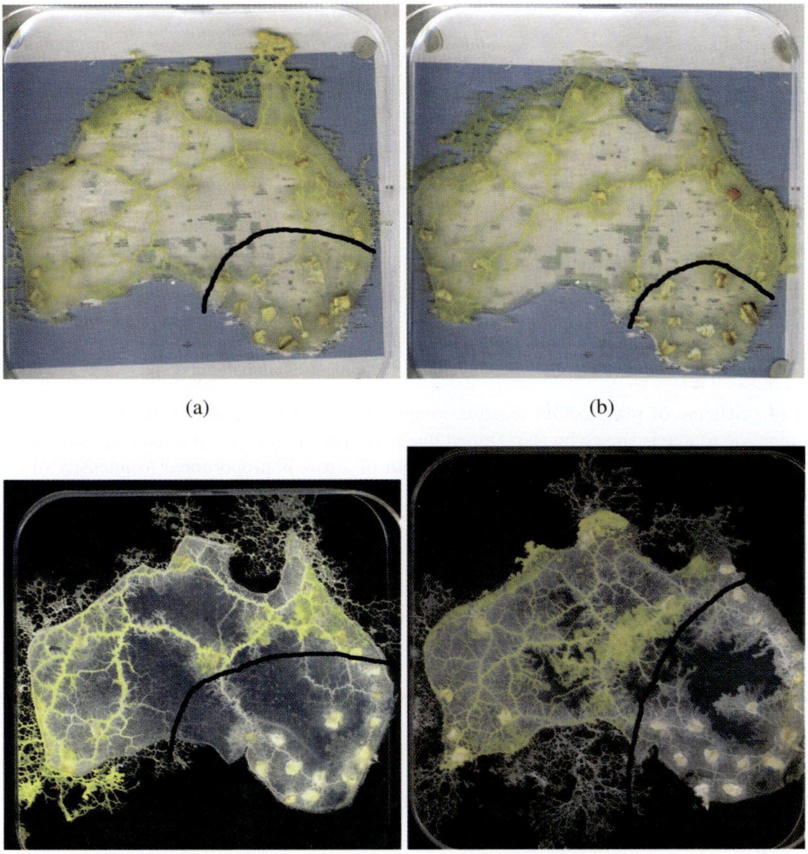

(a) (b) (c) (d)

Fig. 4.13 Images of plasmodium 20–26 h after a crystal of sodium chloride was placed westward of Canberra. Extent of abandoned transport networks is marked by a line.

tubes (Fig. 4.12) and attempts to explore new territories outside its agar plate. Typically, the plasmodium advances in the north-north-western direction from South Hedland in Western Australia. Other yet less frequent directions of emigration include northward migration from the Darwin area in Northern Territory and Rocky Point in Queensland, as well as the south-western direction from Perth, Bunbury and Albany in Western Australia (Fig. 4.12).

To imitate response of plasmodium's protoplasmic network to a spatially extended contamination, we placed a crystal of sodium chloride at the site at the

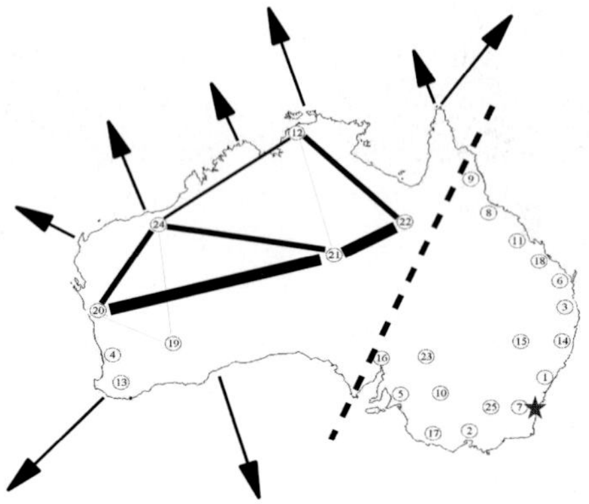

Fig. 4.14 Scheme of plasmodium reaction to spreading contamination. Approximate location of contamination source is shown by star. Dotted line indicates maximum size of contaminated territory. Arrows show directions of attempted escape, length of arrow is proportional to intensity of escape efforts, visible as size of plasmodium sprouting. Transport links usually enhanced in response to contamination are shown by line segments, thickness of segments is proportional to experimental observations.

proposed Jervis Bay nuclear power plant on the south coast of New South Wales (westward of urban area 7 in Fig. 4.1). Sodium chloride diffusing in agar gel makes an unsuitable condition for plasmodium to live. This causes the plasmodium to migrate away (Fig. 4.13), see scheme in Fig. 4.14. Protoplasmic tubes residing in directly affected areas, where concentration of salt exceeds a level tolerable by the plasmodium, become abandoned. They are visible as a white or discoloured network in Fig. 4.13. The area of abandoned transport links can be considered as an area of direct damage. In some of our experiments the area of direct damage was as small as from the epicentre of contamination to Adelaide in South Australia, Broken Hill in New South Wales and Brisbane in Queensland. The largest area of direct damage observed propagated as far as Port August in South Australia and Cairns in Queensland (Fig. 4.13).

Finding 23. *In response to a contamination propagating from the site of Jervis Bay nuclear power plant, plasmodium*

- *substantially increases capacity and traffic through transport links connecting Geraldton to Alice Springs to Mount Isa and Cloncurry,*
- *increases traffic in links Geraldton–South Hedland, Alice Springs–South Hedland, Darwin–Palmerston and Mount Isa–Cloncurry,*
- *shows high level of attempted emigration south-south-west of Bunbury and Albany, south of Kalgoorlie–Boulder, north of Darwin and Palmerston,*

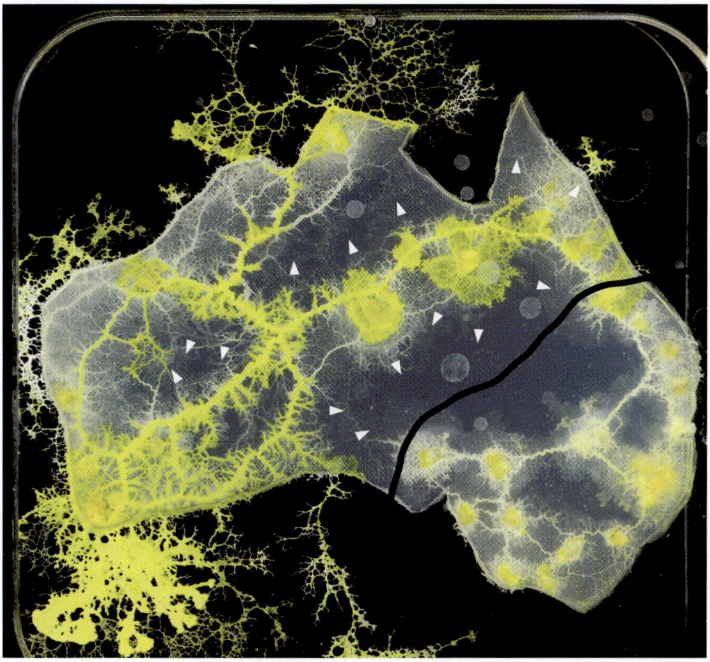

(a)

Fig. 4.15 Example of sprouting in response to contamination. Region with abandoned transport networks is marked by thick black line. Directions of sprouting are shown by white arrows.

north-north-west of South Hedland and north-north-east of Bamaga and Lockhart River.

Transport links enhanced in the result of response to contamination and directions of emigration are shown in Fig. 4.14. They summarise our analyses of 10 experiments on contamination.

It is important to realise that an increase in plasmodium's capacity corresponds to the need to increase transport capacity of the relevant transport links, to mitigate adverse ramifications of such unfortunate events. Some of the transport links may become completely overloaded, e.g. the west-north-eastern routes through the central interior of mainland Australia. This in turn can lead to further fragmentation of the overall network's connectivity, exacerbating the damage.

Indiscriminate sprouting of existing protoplasmic tubes is yet another interesting phenomenon observed in plasmodium networks dealing with a strain of contamination. In a normal condition, a protoplasmic tube usually connects a few sources of nutrients, and branches only in a few places, sometimes similarly to the formation of Steiner points. When a substantial domain of a substrate becomes contaminated, major protoplasmic tubes form branches all over their bodies, as shown in Fig. 4.15. A similar type of sprouting is recorded in plasmodium's response to physical damage, e.g. cutting, of its protoplasmic tubes [Adamatzky (2010)b].

4.5 Summary

We represented 25 major urban areas with oat flakes, positioned the oat flakes on an agar plate shaped in the form of Australia and inoculated plasmodium of *Physarum polycephalum* in Sydney. In a few days after inoculation the plasmodium formed a network of protoplasmic tubes spanning the oat flakes. Structures of protoplasmic networks obtained in 31 experiments were converted to a Physarum graph, where the weight of each edge was set to a frequency of the corresponding protoplasmic tube's occurrence in laboratory experiments. We found that the Physarum graph becomes planar when all edges with weights below $\frac{16}{31}$ are removed. Isolated urban areas, without transport links, appear in over 17 experiments. The Physarum graphs becomes disconnected when we remove all edges with weights less than $\frac{22}{31}$. Also, when comparing Physarum graphs with the most common proximity graphs, we found that in over 70% of experiments slime mould's protoplasmic networks are almost identical to the Gabriel graph.

The Physarum graph does not match the existing motorway network of Australia. Only the untrimmed Physarum graph, $\mathbf{P}(0)$, is almost a supergraph of the Australian motorway graph $\mathbf{H}$. With regard to large-scale disasters, we found that in response to diminished resources increased migration is observed in north and north-western directions. In response to a contamination propagating from the site of Jervis Bay nuclear power plant, plasmodium substantially increases capacity and traffic through transport links connecting Geraldton to Alice Springs to Mount Isa and Cloncurry and exhibits a high level of attempted emigration south-south-west of Bunbury and Albany, south of Kalgoorlie–Boulder, north of Darwin and Palmerston, north-north-west of South Hedland and north-north-east of Bamaga and Lockhart River.

Chapter 5

Belgian transport networks: redundancy and dissolution

Andrew Adamatzky, Bernard De Baets and Wesley Van Dessel

Belgium is a good test bed for the evaluation of slime-mould approximation of motorways because of the following facts.

- Belgium is an artificial country created relatively recently, in 1830.
- It is amongst the most populated areas in Europe.
- There is a density misbalance between two major communities: Flanders is more densely populated than Wallonia.
- The Belgian economy is centred around Brussels, by far the biggest city, with hundreds of thousands of workers commuting to Brussels every day.

In the early days the Belgian highways were constructed to provide a solution against the overcharged national and local roads, caused by the expanding number of cars. In the north of the country, construction was generally based on growing demands from the economic and touristic sectors. The first highway was the one between Brussels and Ostend. Another 'early' highway was the one between Antwerp and Liège (E313) to open up the port of Antwerp's access to the 'hinterland'. At the end of 1972 the most important cities were connected by highways. However, from the point of view of transport economics, only two of them were answering to an economic demand, a need for construction based on increasingly busy roads: the one between Brussels and Antwerp (E19) and the one between Brussels and Liège (E40). The others were intended as an investment trigger. The Autoroute de Wallonie (E42) was aimed at the economic reconversion of the old industrial axis in Wallonia (steel and coal industry). The E17 and E34 motorways provided an additional connection between the port of Antwerp and the French and German inner lands. The purpose of the E314 was to open up the province of

Limburg (coal mining industry), and to provide a short-cut between Antwerp and the German Rhineland (Ruhrgebiet). Highway construction has been the result of political negotiations and the desire or need of the northern and southern partners to balance large investments in both parts of the country [Wegen Routes (2012)].

We consider the 21 most[1] populous urban areas in Belgium **U** (Fig. 5.1a), shown below in descending order of population size:

(1) Brussels area, including Dilbeek and Vilvoorde,
(2) Antwerp area, including Beveren and Brasschaat,
(3) Gent,
(4) Charleroi area, including La Louvière and Chatelet,
(5) Liège area, including Seraing, Verviers and Herstal,
(6) Brugge,
(7) Namur,
(8) Leuven,
(9) Mons,
(10) Aalst,
(11) Mechelen,
(12) Kortrijk area, including Mouscron and Waregem,
(13) Hasselt,
(14) Oostende,
(15) Sint-Niklaas,
(16) Tournai,
(17) Genk area, including Maasmechelen,
(18) Roeselare,
(19) Turnhout,
(20) Arlon,
(21) Sankt-Vith.

At the beginning of each experiment an oat flake colonised by plasmodium was placed in the Brussels area (Fig. 5.2). Our choice of inoculation site does not reflect the historical development of transport routes in Belgium (where inoculation should start in Aalst, Brugge, Kortrijk or Gent); however, it conveys the overwhelming economic power of the capital. We undertook 28 experiments.

[1] Arlon and Sankt-Vith are not amongst the most populated areas but we added them for completeness, to allow the slime mould to propagate towards Luxembourg and Germany.

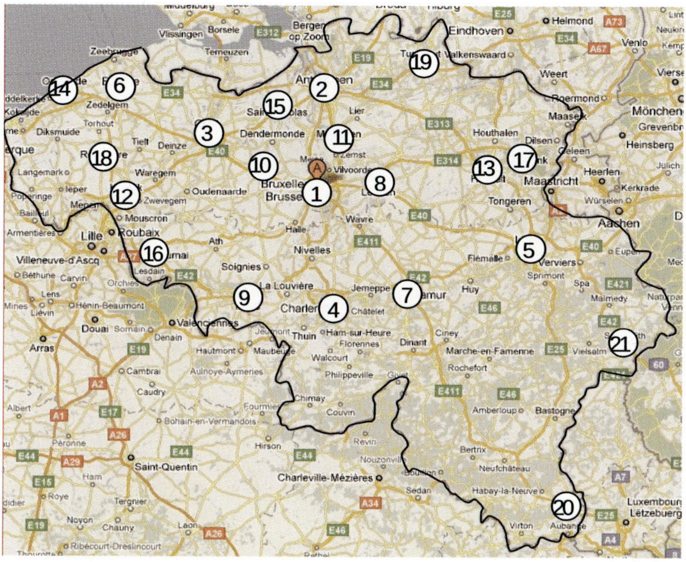

(a)

(b)

Fig. 5.1 Experimental setup: (a) outline map of Belgium [Maps Google (2012)] with major urban areas **U** shown by encircled numbers, (b) urban areas, represented by oat flakes, are colonised by slime mould *P. polycephalum*.

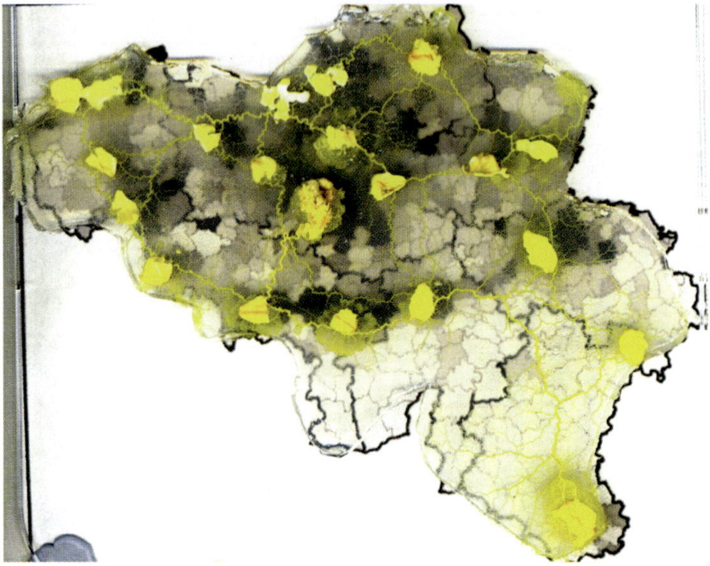

(a)

(b)

Fig. 5.2 Oat flakes, representing urban areas **U**, colonised by slime mould. The growing substrate is on top of the Belgian population density map [Statistics Belgium (2010)].

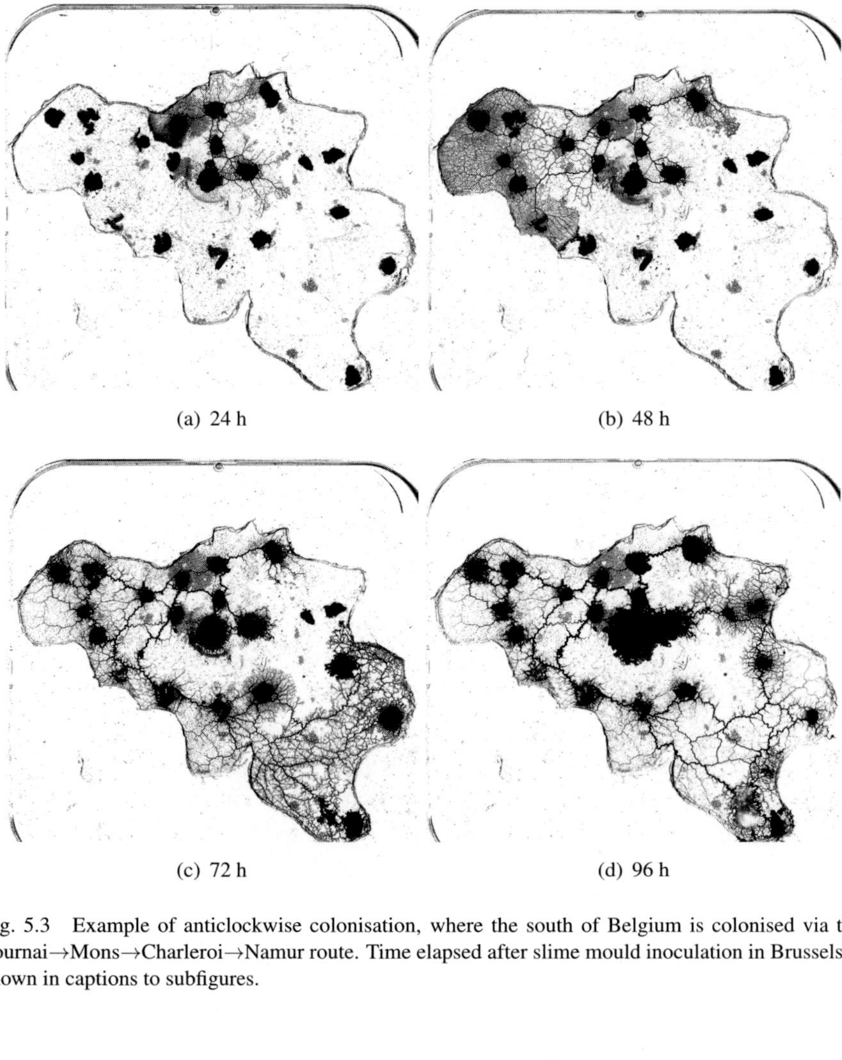

(a) 24 h (b) 48 h

(c) 72 h (d) 96 h

Fig. 5.3 Example of anticlockwise colonisation, where the south of Belgium is colonised via the Tournai→Mons→Charleroi→Namur route. Time elapsed after slime mould inoculation in Brussels is shown in captions to subfigures.

5.1 Bioessential motorways grow from Brussels

Plasmodium is inoculated in the Brussels area. In the first 24 h it propagates towards and occupies Leuven and Mechelen, and then propagates from Mechelen to Antwerp and from Antwerp to Sint-Niklaas (Fig. 5.3a). In the next 24 h the slime mould propagates from Sint-Niklaas to Aalst and Gent, from Gent to Brugge and from Aalst to Kortrijk. Links from Brugge to Oostende and Roeselare are built during the same time interval (Fig. 5.3b). Westward development of plasmodium is somehow stopped. Despite an attempted propagation from Turnhout towards the Hasselt and Genk areas, the slime mould never actually reaches these areas in the first 48 h (Fig. 5.3c).

By the 72nd hour after being inoculated in Brussels almost all urban areas apart from Hasselt and Genk are colonised by slime mould. Namely, in the time interval 48–72 h, plasmodium grows from Aalst to Tournai, from Tournai to Mons and from Mons to Charleroi. The slime mould branches at Charleroi and grows in parallel to Namur and Arlon. It propagates from Namur to Sankt-Vith and from Sankt-Vith to Liège (Fig. 5.3d). By the 96th hour after being inoculated the slime mould propagates from Liège to the Genk and Hasselt areas, and from Hasselt to Leuven. The plasmodium's explorative activities in the Brussels area are 'resumed' when the slime mould propagates from Leuven and reoccupies the Brussels area (Figs. 5.3d and 5.5a).

Colonisation of Belgium by slime mould shown in Fig. 5.4 develops initially according to the scenario described above with the following deviations. Plasmodium propagates from Gent to Brugge and Roeselare and Kortrijk at the same time (Figs. 5.4 and 5.5b). The slime mould grows from Leuven to Namur and then to Charleroi. From Charleroi plasmodium propagates to Mons and from Mons to Tournai (Fig. 5.4a). Arlon and Sankt-Vith are reached by the slime mould via Leuven, Hasselt and Liège (Fig. 5.4b and c). Antwerp is never colonised by the slime mould in this particular experiment (Fig. 5.5b).

Examples of Physarum graphs for various values of θ are shown in Fig. 5.6. A Physarum graph $\mathbf{P}(\frac{1}{28})$ is a non-planar[2] acyclic graph (Fig. 5.7a). In the graph $\mathbf{P}(\frac{1}{28})$, each edge appears in at least one laboratory experiment. We call an edge of a Physarum graph *credible* if this edge is represented by a protoplasmic tube in over 20% of laboratory experiments. Such graph $\mathbf{P}(\frac{6}{28})$ is shown in Fig. 5.7b. It is still non-planar; however, the only intersecting edges are the links Liège area–Arlon and Namur–Sankt-Vith.

Finding 24. *The Physarum graph with all credible edges is a non-planar cyclic graph.*

The graph remains connected while θ grows to $\frac{10}{28}$ (Fig. 5.7c). The Turnhout urban area becomes an isolated vertex when $\theta = \frac{11}{28}$ (Fig. 5.7d). For this value of θ, the Physarum graph becomes planar because the link Liège area–Arlon is represented in more experiments than the link Namur–Sankt-Vith. The graph $\mathbf{P}(\frac{11}{28})$ has the largest (among all Physarum graphs studied here) empty, i.e. not having

[2] A planar graph consists of nodes which are points of the Euclidean plane and edges which are straight segments connecting the points.

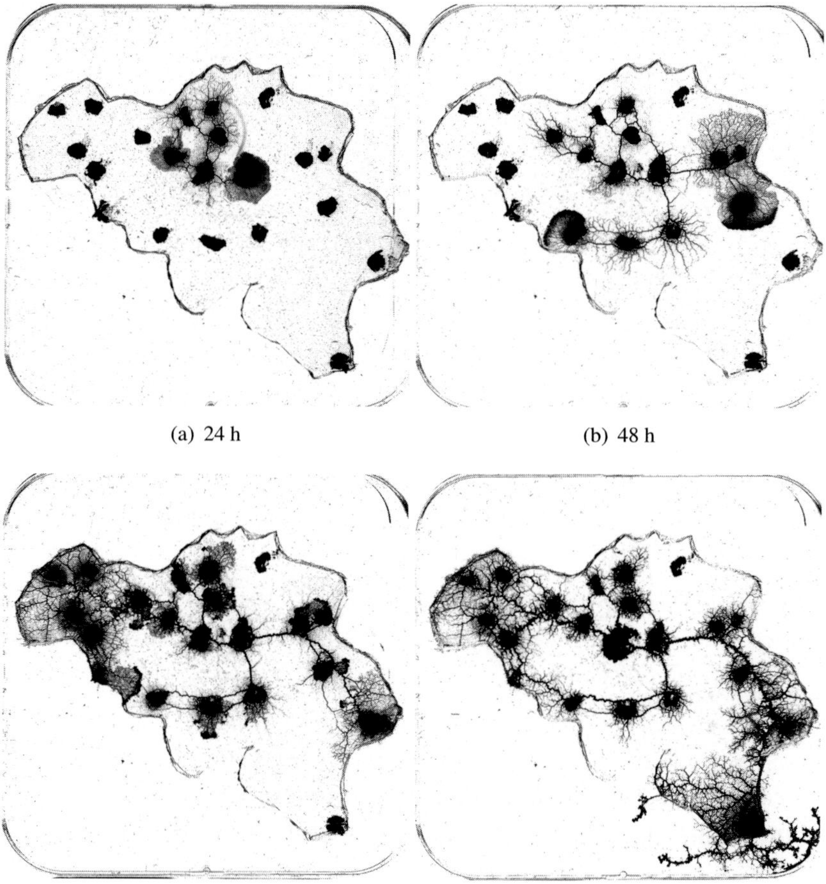

(a) 24 h (b) 48 h (c) 72 h (d) 96 h

Fig. 5.4 Example of clockwise colonisation of south of Belgium with dominating route Leuven→Hasselt→Liège.

any edges inside, circle. Clockwise, starting in the Brussels area, it spans Leuven, Hasselt, Liège, Namur, Charleroi, Mons, Tournai, Kortrijk, Gent, Aalst and finishes in Brussels.

The Physarum graph $\mathbf{P}(\theta)$ splits into three components when $\theta = \frac{16}{28}$. The smallest component is the isolated vertex Turnhout. The medium-size component is a cycle Liège–Hasselt–Genk–Liège attached to a segment Arlon–Sankt-Vith (Fig. 5.7e). The largest component is a proximity graph spanning all remaining urban areas.

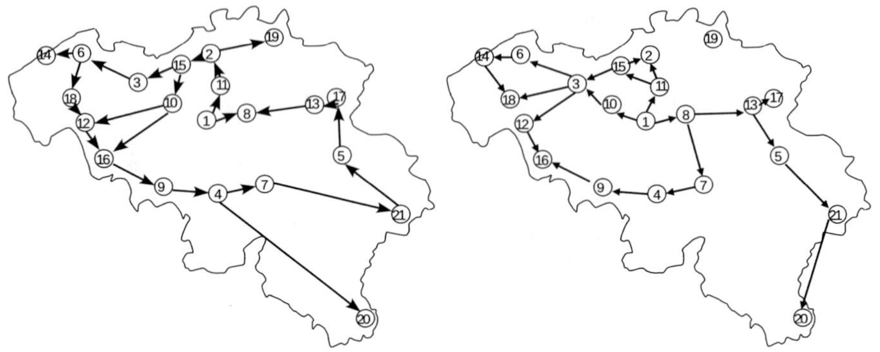

(a) (b)

Fig. 5.5 Scheme of colonisation dynamics illustrated in (a) Fig. 5.3 and (b) Fig. 5.4.

Finding 25. *The following slime mould transport chains appear in almost all experiments (Fig. 5.7f):*

- *the chain Roeselare–Kortrijk–Mons–Charleroi–Namur,*
- *the chain Oostende–Brugge–Gent–Aalst–Brussels–Leuven–Mechelen–Antwerp–Sint-Niklaas,*
- *the chain Liège–Sankt-Vith–Arlon,*
- *the one-link chain Hasselt–Genk.*

5.2 Physarum almost perfectly approximates Belgian motorways

As we can see in the examples of the experimental configurations in Fig. 5.8, plasmodium networks are polymorphic and no two networks are exactly the same. Some motorways are matched by protoplasmic tubes well, others just approximated and some do not have a slime mould representation at all.

For example, Fig. 5.8a demonstrates that slime mould developed protoplasmic tubes corresponding to the motorway A10/E40 between Gent and Brugge, E17 between Gent and Kortrijk, E17 and E403/A17 between Gent and Tournai and E42 connecting Mons to Charleroi to Namur to Liège. Motorways E42 (Tournai–Mons), E46 (Antwerp–Turnhout) and E40 (Leuven–Liège) are represented by slime mould in the experiment illustrated in Fig. 5.8b. At the same time, the transport link E411 from Brussels to Namur to Arlon is not represented in the configurations in Fig. 5.8.

The motorway graph is planar (Fig. 5.9).

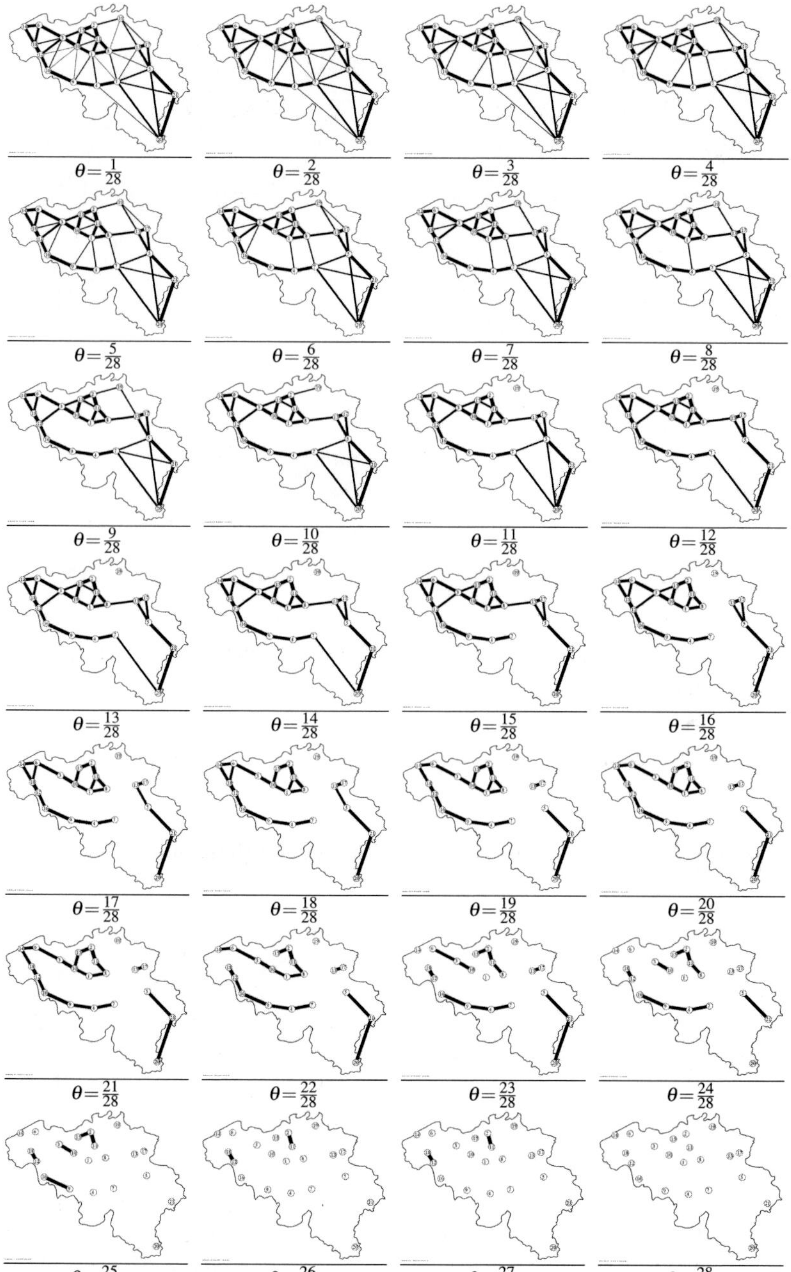

Fig. 5.6 Physarum graphs $\mathbf{P}(\theta)$ for $\theta = \frac{1}{28}, \ldots, 1$.

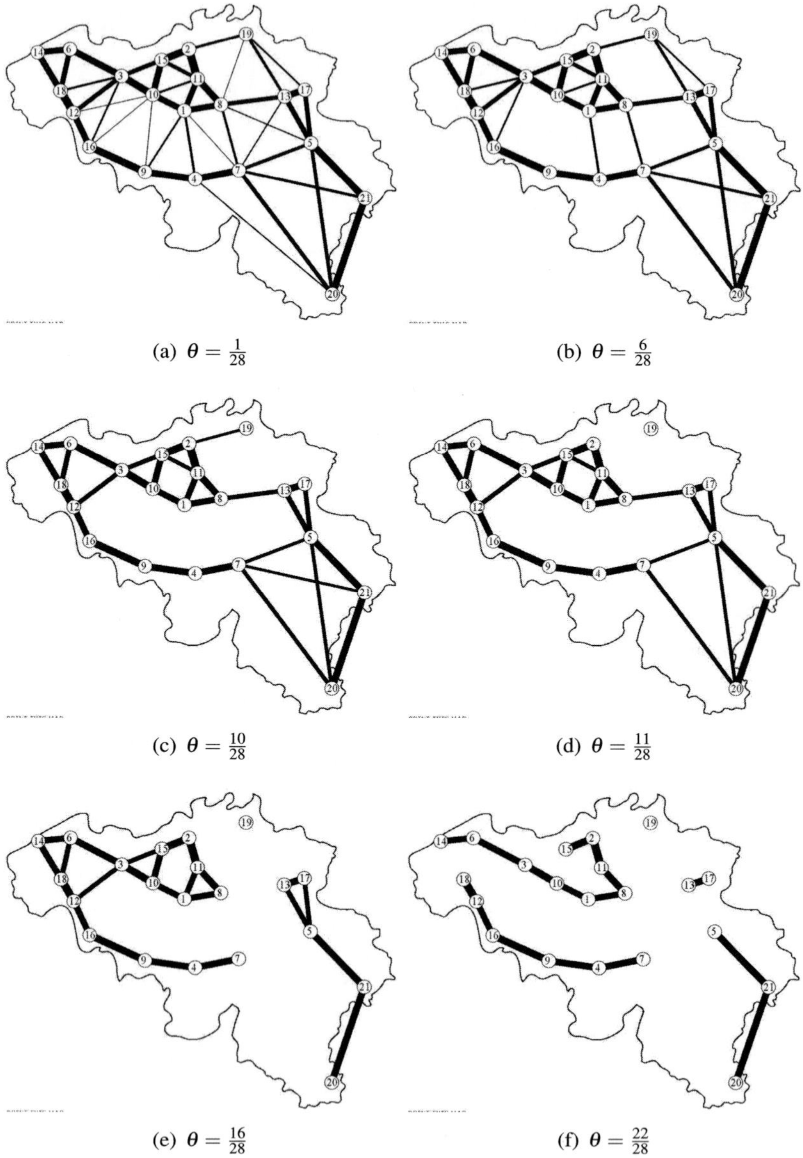

(a) $\theta = \frac{1}{28}$ (b) $\theta = \frac{6}{28}$

(c) $\theta = \frac{10}{28}$ (d) $\theta = \frac{11}{28}$

(e) $\theta = \frac{16}{28}$ (f) $\theta = \frac{22}{28}$

Fig. 5.7 Physarum graphs $\mathbf{P}(\theta)$ for critical values of θ.

Finding 26. Physarum polycephalum *almost completely approximates the Belgian motorway network.*

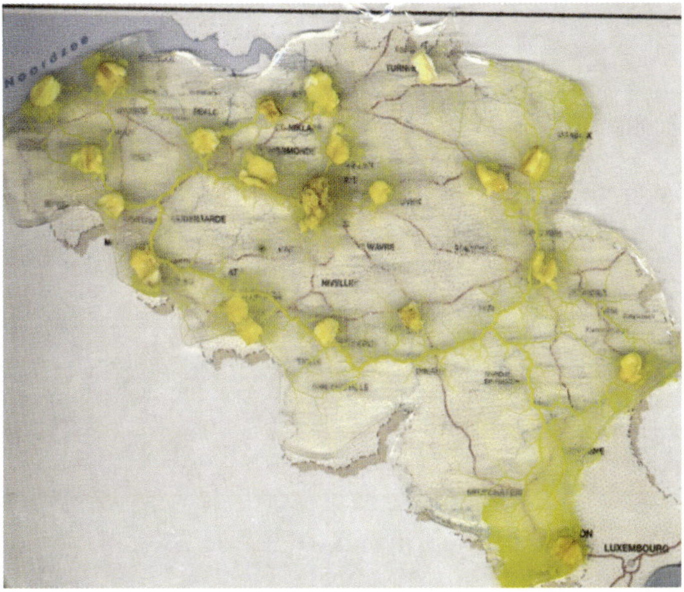

(a)

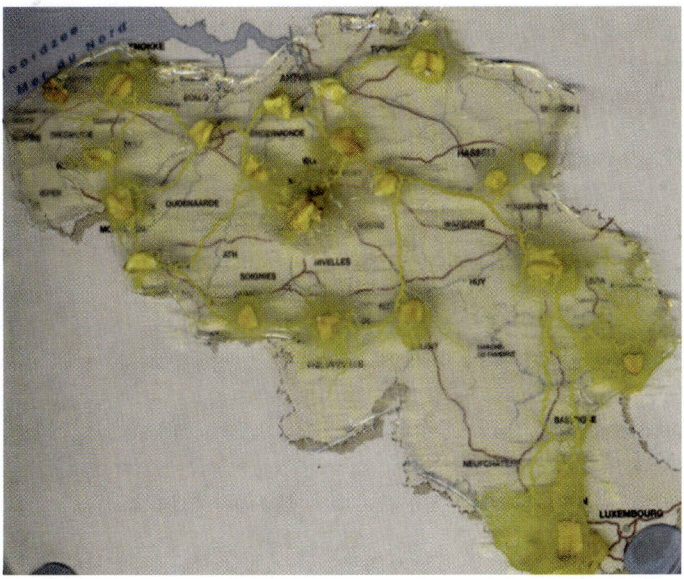

(b)

Fig. 5.8 Examples of *P. polycephalum* protoplasmic networks on the Belgian motorway network.

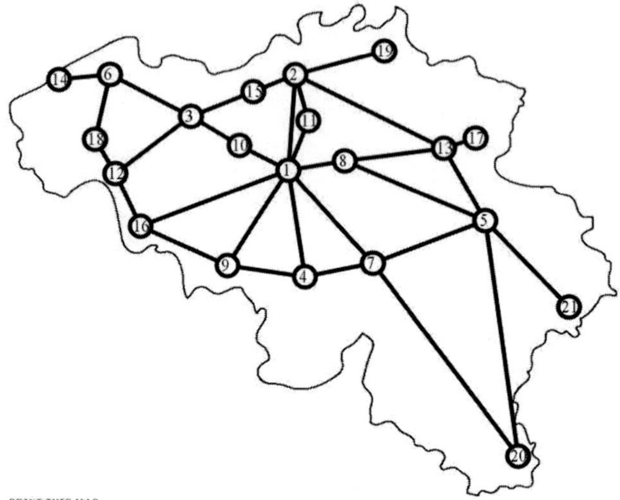

Fig. 5.9 Graph **H** of Belgian motorway network.

Namely, **H** is not a subgraph of $\mathbf{P}(\frac{1}{28})$ but 25 of 28 edges of **H** are edges of $\mathbf{P}(\frac{1}{28})$. The following motorway links are never represented by protoplasmic tubes, not even in a single experiment: Brussels to Tournai, Brussels to Antwerp and Antwerp to Hasselt (Figs. 5.10a and 5.9).

Finding 27. ***On redundancy***. *Motorway links connecting Brussels with Antwerp, Tournai, Mons, Charleroi and Namur, and links connecting Leuven with Liège and Antwerp with Genk and Turnhout, are proved to be redundant components of the Belgian transport system in the slime mould experiments.*

The skeletal motorway network represented by the plasmodium network in laboratory experiments is shown in Fig. 5.10b. This network is represented by protoplasmic tubes in over 35% of experiments, and in almost 40% of experiments without the Antwerp–Turnhout link (Fig. 5.10c and d). By increasing the level of slime mould's 'confidence' to almost 60%, we lose motorway links connecting Leuven with Hasselt, Namur with Liège and Arlon and Liège with Arlon (Fig. 5.10c and d).

Finding 28. *Motorway segments A17 (Antwerp–Sint-Niklaas), A19 (Antwerp–Mechelen), E42 (Liège–Sankt-Vith), A10/E40 (Oostende–Gent–Aalst–Brussels–Leuven), A17/E403 (Roeselare–Kortrijk–Tournai), E42/E19 (Tournai–Mons) and*

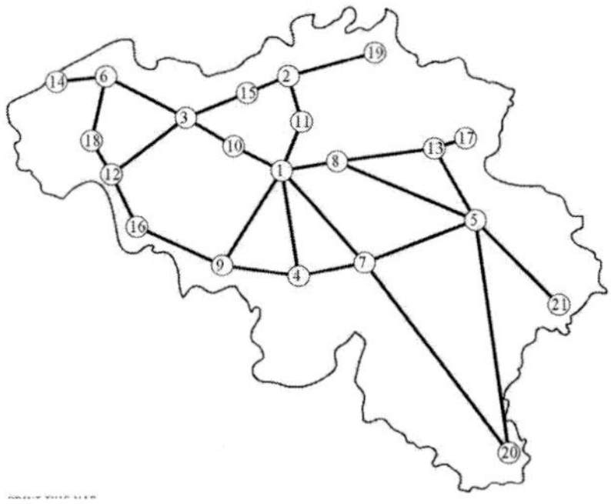

(a) $\mathbf{H} \cap \mathbf{P}(\frac{1}{28})$

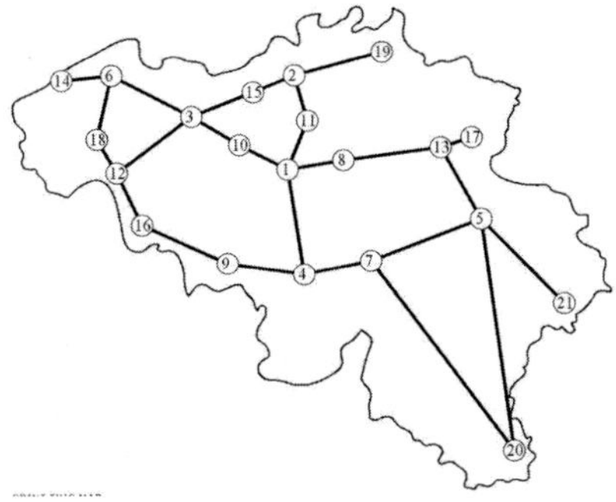

(b) $\mathbf{H} \cap \mathbf{P}(\frac{6}{28}) = \mathbf{H} \cap \mathbf{P}(\frac{10}{28})$

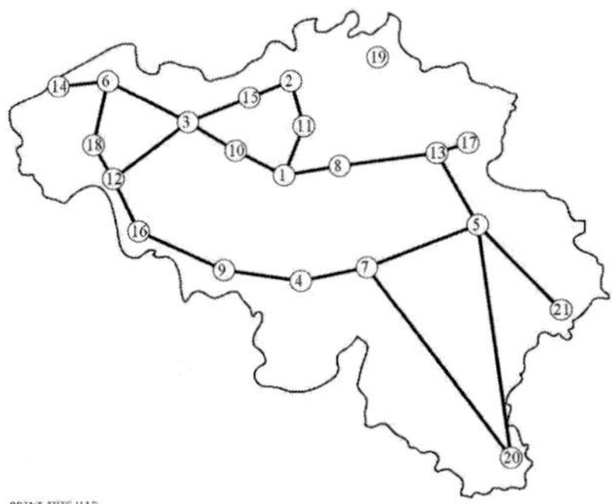

(c) $\mathbf{H} \cap \mathbf{P}(\frac{11}{28})$

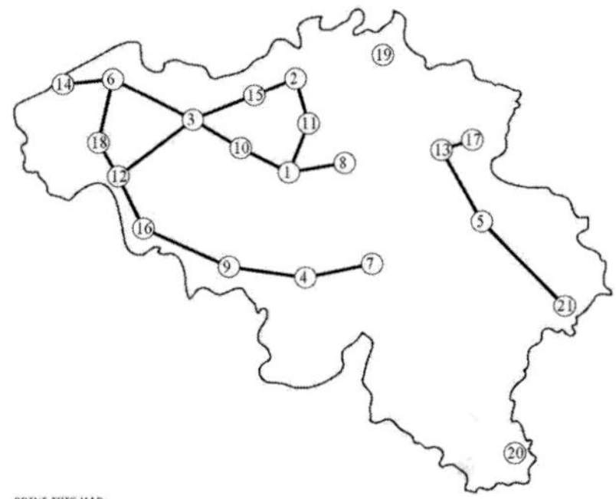

(d) $\mathbf{H} \cap \mathbf{P}(\frac{16}{28})$

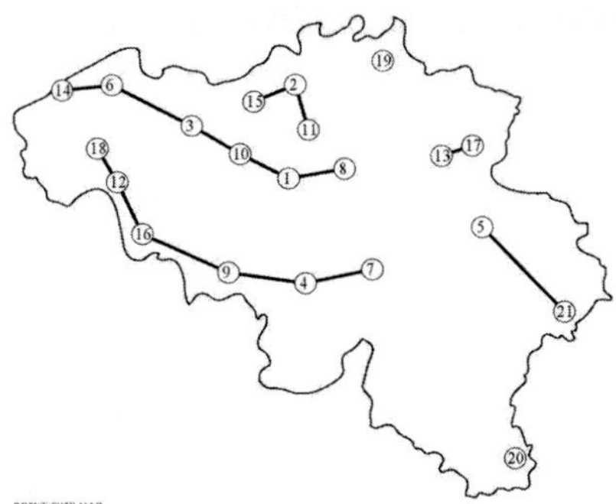

(e) $\mathbf{H} \cap \mathbf{P}(\frac{22}{28})$

Fig. 5.10 Intersections $\mathbf{H} \cap \mathbf{P}(\theta)$ of motorway graph $\mathbf{H}$ with Physarum graph $\mathbf{P}(\theta)$ for critical values of θ.

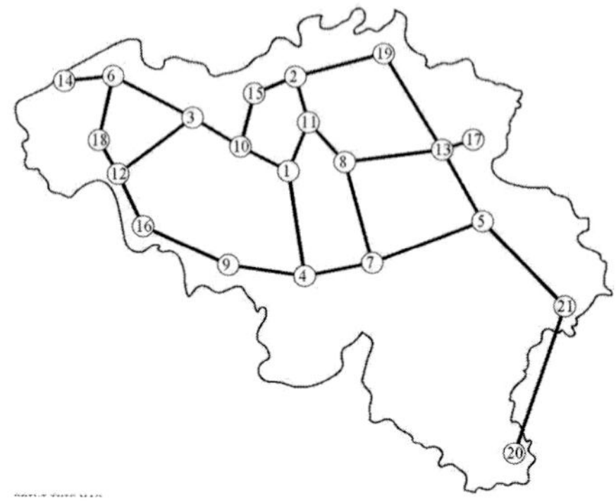

(a) **RNG**

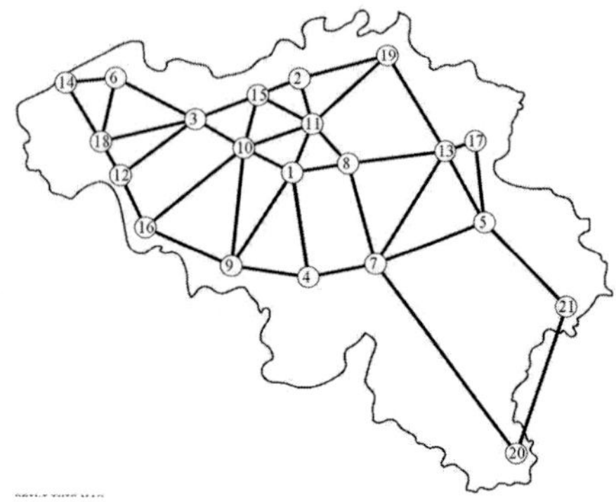

(b) **GG**

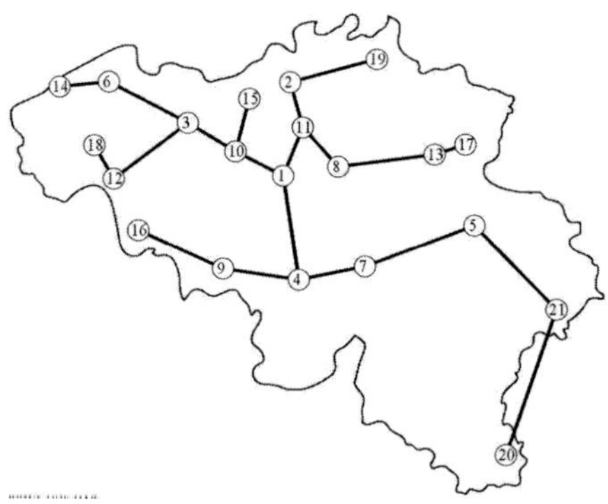

(c) **MST**

Fig. 5.11 Proximity graphs constructed on sites of **U**: (a) relative neighbourhood graph, (b) Gabriel graph, (c) minimum spanning tree rooted in Brussels.

E42 (Mons–Charleroi–Namur) are essential transport links from the slime mould's point of view.

As illustrated in Fig. 5.10f, these transport links are represented by the protoplasmic network in almost 80% of laboratory experiments.

5.3 Minimum spanning tree is not a subgraph of motorway graph

Finding 29. *The Belgian motorway graph is approximated by the Gabriel graph with 80% accuracy and by the relative neighbourhood graph with 70% accuracy.*

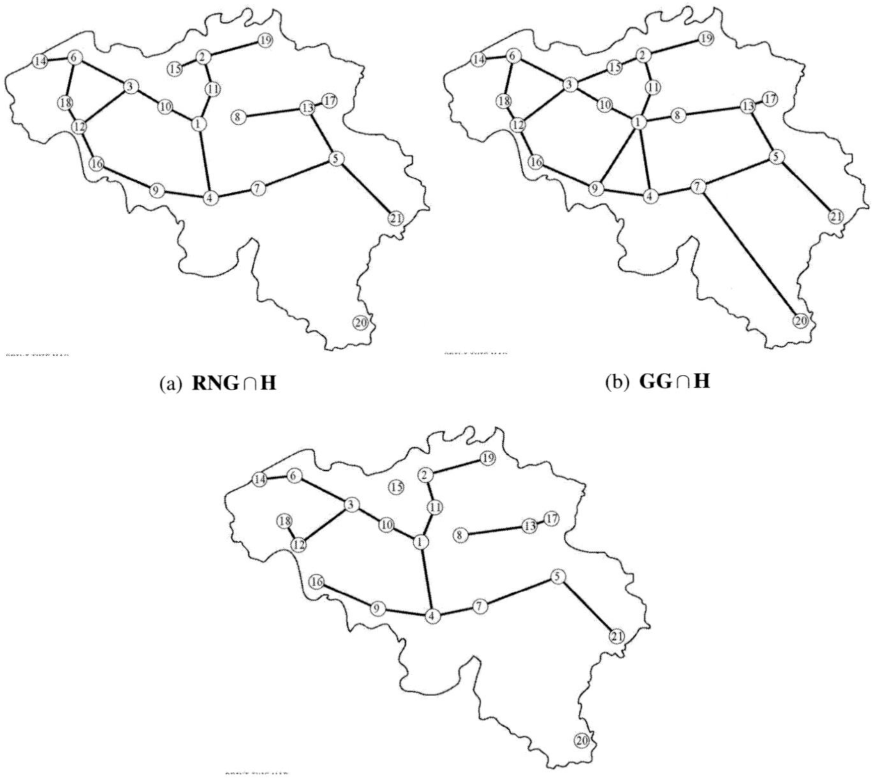

(a) **RNG** ∩ **H** (b) **GG** ∩ **H**

(c) **MST**(**1**) ∩ **H**

Fig. 5.12 Intersections of the Belgian motorway graph with (a) relative neighbourhood graph, (b) Gabriel graph, (c) minimum spanning tree rooted in Brussels.

The following motorway links from **H** are not represented in the Gabriel graph **GG**: Brussels area–Antwerp area, Brussels area–Tournai, Brussels area–Namur, Antwerp area–Hasselt, Leuven–Liège area and Liège area–Arlon (Fig. 5.12a). With regard to the relative neighbourhood graph, a few more transport links from **H** are not a part of **RNG**: Brussels area–Mons, Brussels area–Leuven and Namur–Arlon (Fig. 5.12b). Thus, **GG** represents 23 of 29 edges of the motorway graph **H** and **RNG** 20 of 29 edges of **H**.

Finding 30. *The minimum spanning tree rooted in Brussels is not a subgraph of the Belgian motorway graph.*

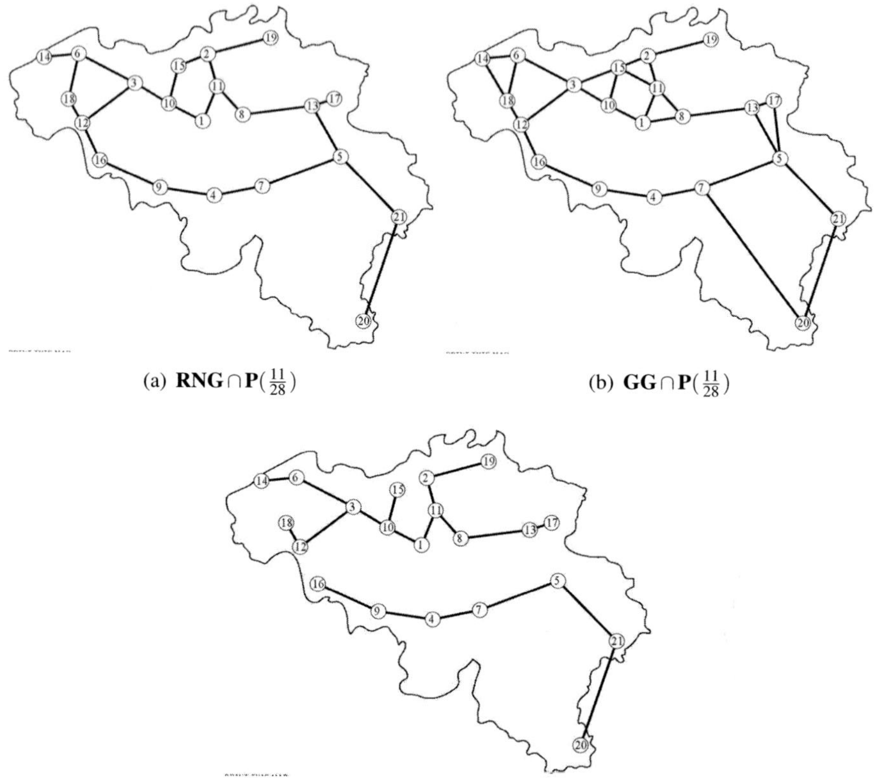

(a) $\mathbf{RNG} \cap \mathbf{P}(\frac{11}{28})$

(b) $\mathbf{GG} \cap \mathbf{P}(\frac{11}{28})$

(c) $\mathbf{MST} \cap \mathbf{P}(\frac{11}{28})$

Fig. 5.13 Intersections of the Physarum graph $\mathbf{P}(\frac{11}{28})$ with (a) relative neighbourhood graph, (b) Gabriel graph, (c) minimum spanning tree rooted in Brussels.

This is because the links Mechelen–Sint-Niklaas and Mechelen–Leuven exist in **MST** but are not present in **H** (Fig. 5.12c). The minimum spanning tree is considered to be an optimal, in a sense of minimality of edge lengths, acyclic planar graph. The fact that two of the spanning tree edges are not represented by man-made motorway links allows us to suggest that the Belgian motorway network is not optimal, at least not optimal in spanning of major urban areas.

Considering that the Physarum graph $\mathbf{P}(\frac{11}{28})$ is credible because its edges are represented by protoplasmic tubes in at least 40% of laboratory experiments, we can compare it with the three basic proximity graphs (Fig. 5.13).

Finding 31. *If slime mould represented A7/A54 Brussels–Charleroi, non-existing*

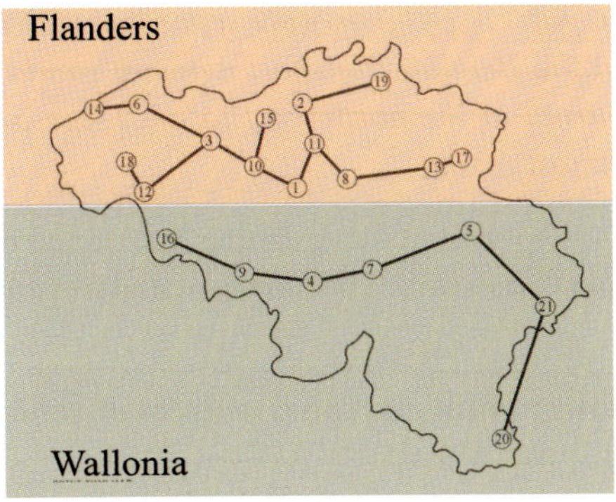

Fig. 5.14 Intersection of Physarum graph $\mathbf{P}(\frac{11}{28})$ with the minimum spanning tree rooted in Brussels.

in reality motorways between Leuven–Namur and Turnhout–Hasselt, its credible transport network would be a supergraph of the relative neighbourhood graph.

This is a direct outcome of comparing Figs. 5.13a and 5.11a. Relative neighbourhood graphs are considered to be optimal cyclic graphs in terms of total edge length and travel distance, and are known to be a good approximation of road networks [Watanabe (2005); Watanabe (2008)]. The Leuven–Namur and Brussels–Charleroi links are represented by slime mould in less than 20% of laboratory experiments (Fig. 5.7b). The intersection of the Physarum graph $\mathbf{P}(\frac{11}{28})$ (Fig. 5.7d) with the minimum spanning tree (Fig. 5.11c) comprises two disconnected components: one lies in Flanders and the other in Wallonia (Fig. 5.13c).

5.4 Dissolution: *snelwegen* or *autoroutes*?

The province Brabant Wallone has no transport routes originating in and passing through it for $\theta = \frac{10}{28}$. When θ increases to $\frac{11}{28}$, the Antwerp province becomes isolated from the other provinces of Belgium. For $\theta = \frac{22}{28}$, we have isolated the Antwerp and Limburg provinces, and clusters of interconnected regions:

(1) West-Vlaanderen, Hainaut and Namur provinces,
(2) West-Vlaanderen, Oost-Vlaanderen, Vlaams-Brabant and Oost-Vlaanderen provinces,
(3) Liège and Luxembourg provinces.

In terms of spanning trees (Fig. 5.11), the transport link between the Brussels area and the Charleroi area is the only means of keeping the Dutch- and French-speaking communities together.

Finding 32. *If the two parts of Belgium were separated with Brussels in Flanders, the Walloon region of the Belgian transport network would be represented by a single chain from Tournai in the north-west to the Liège area in the north-east and down to southernmost Arlon.*

This transport link is not part of the spanning tree rooted in Brussels, and when omitted the two Belgian communities become isolated from each other (Fig. 5.14).

5.5 Doel nuclear power plant and other sources of contamination

To study the reaction of Physarum-grown transport networks to major disasters, we placed crystals of sodium chloride in the approximate positions of the Doel nuclear power plant, near Antwerp (seven experiments) and the Tihange nuclear power station, near Liège (nine experiments). Images of protoplasmic networks reconfigured 24 h after start of contamination are shown in Figs. 5.15 and 5.16.

A typical response to propagating contamination is dissected in Fig. 5.17. In this example, we placed a salt grain on the oat flake representing the Liège area (Fig. 5.17, ‘a’). In 24 h after the imitated accident, the contamination spreads as far as Namur in the west, Sankt-Vith in the east and Hasselt and Genk in the south. Transport links in proximity to the contamination epicentre become destroyed and abandoned. An example of an abandoned transport link is a protoplasmic tube (marked ‘b’ in Fig. 5.17) representing E42/A27 (Liège to Sankt-Vikt motorway). No alternative routes for such destroyed transport links are offered by the slime mould.

Transport links being at a significant distance from the epicentre but yet inside the contamination zone are shifted further away from the contamination. For example, slime mould abandons the protoplasmic tube representing motorway E411, marked ‘c’ in Fig. 5.17, but grows another tube, marked ‘d’ in Fig. 5.17, slightly westwardly. Urban areas not directly affected by contamination show signs of increased explorative and scouting activity, for example the Antwerp and Turnhout areas, marked ‘e’ in Fig. 5.17. Also, transport links not affected by contamination became visibly enhanced (Fig. 5.17, ‘g’). In the situation of continuing contamination, plasmodium ‘considers’ the last opportunity to survive by forming sclerotium (hardened body of the ‘hibernating’ slime mould); the initial stage of the sclerotium formation is labelled ‘f’ in Fig. 5.17.

Based on outcomes of our scoping experiments (Figs. 5.15 and 5.16), we can propose the following scenarios of response to contamination:

- Epicentre of contamination is at the Tihange nuclear power station near

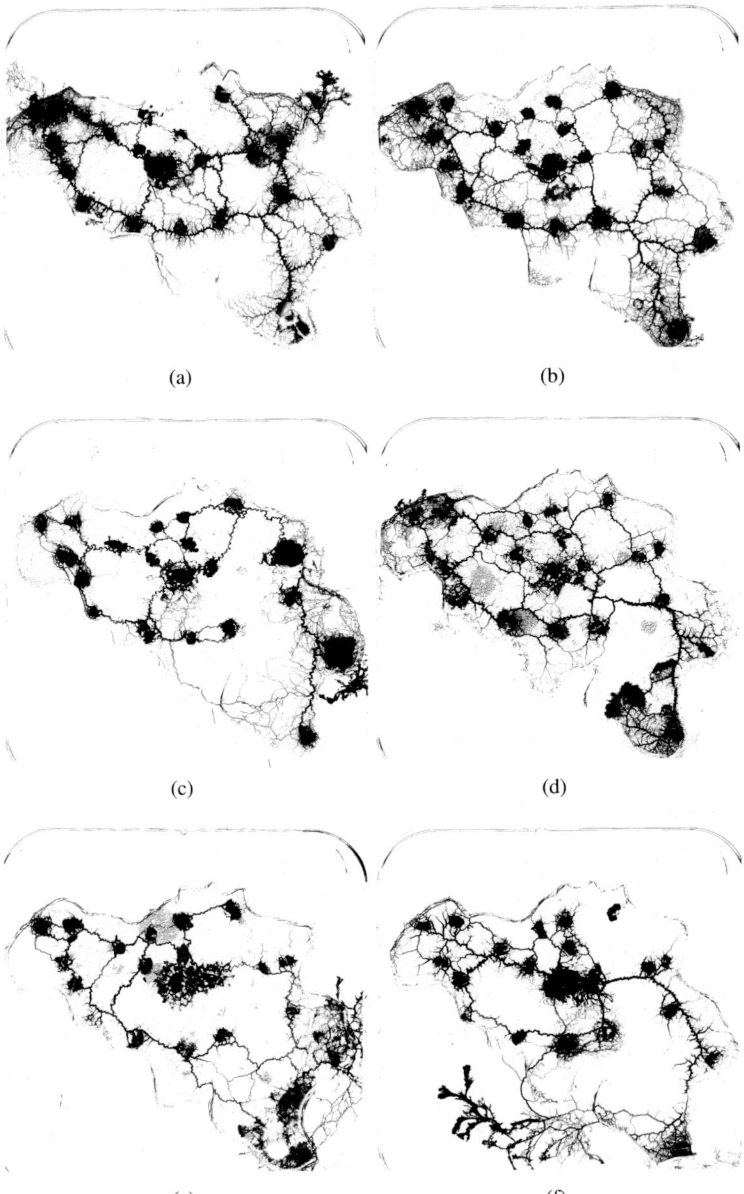

(a) (b) (c) (d) (e) (f)

Fig. 5.15 Response to spreading contamination recorded in laboratory experiments. Epicentre of contamination is in Antwerp area.

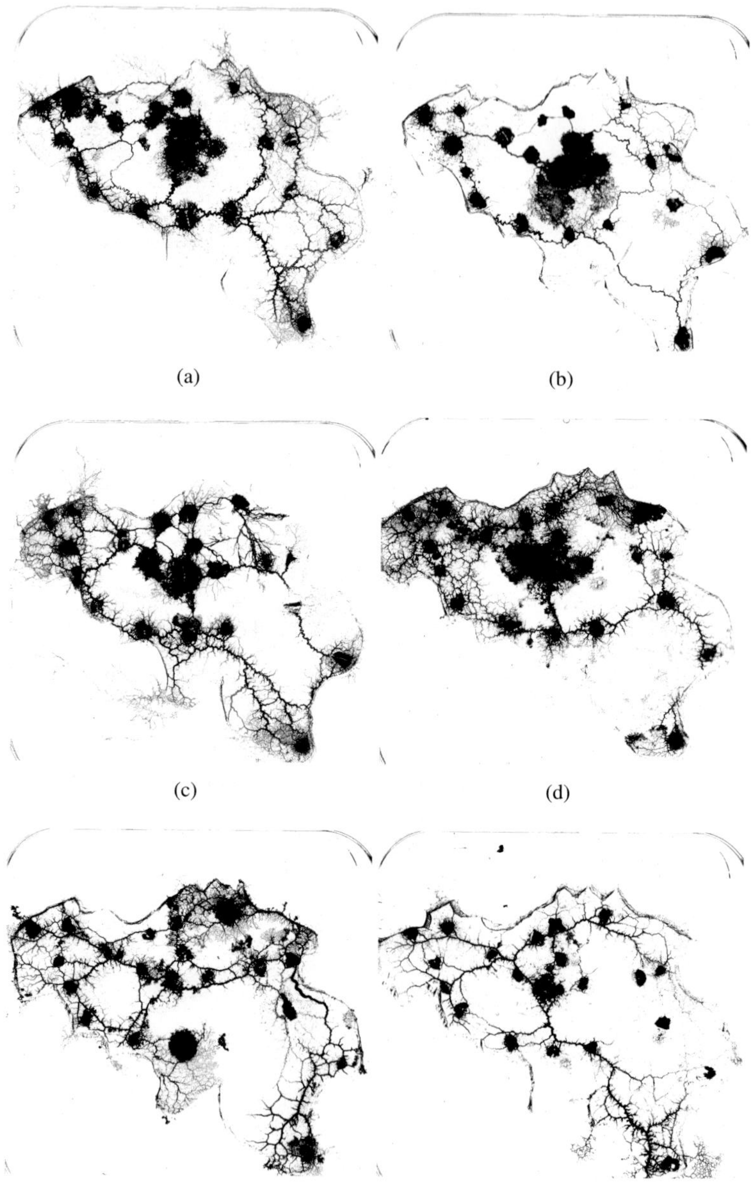

(a) (b) (c) (d) (e) (f)

Fig. 5.16 Response to spreading contamination recorded in laboratory experiments. Epicentre of contamination is in Liége area.

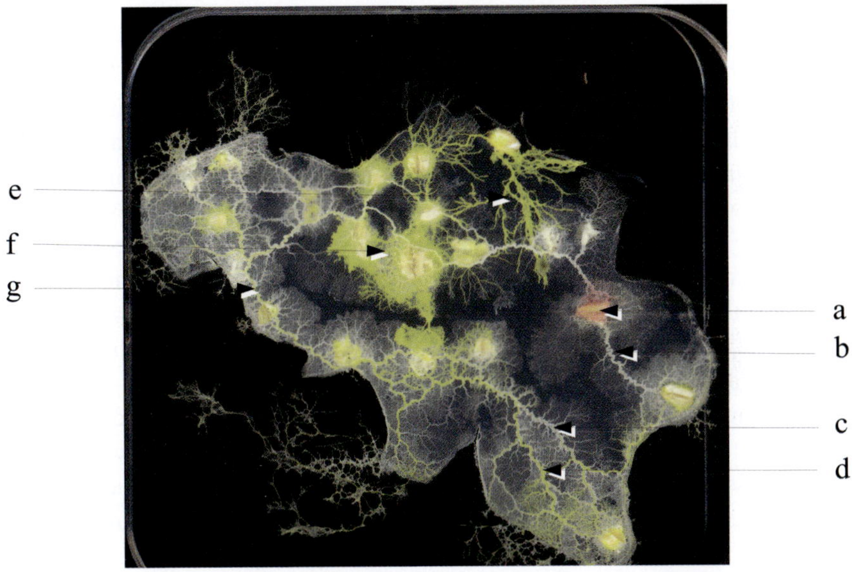

Fig. 5.17 Illustration of the slime mould's response to contamination: (a) epicentre of contamination, (b) abandoned and not repositioned transport link, (c) abandoned and repositioned transport link, (d) new location of the abandoned transport link, (e) explorative activity, (f) emergency preparations, initial stage of sclerotinisation, (g) enhanced transport links. Image is taken 24 h after initiation of contamination in Liège area.

Liège. Segments of the transport network connecting Namur, Liège, Hasselt, Genk, Sankt-Vith and Arlon are destroyed or abandoned. Hyperactivity is observed in domains surrounding Sint-Niklaas, Antwerp and Turnhout. Preparations for emergency hibernation take place at Aalst, Brussels, Leuven and Mechelen. Transport links between Gent, Roeselare, Kortrijk, Tournai, Mons, Charleroi and Namur are hypertrophied (Fig. 5.16).

- Epicentre of contamination is at the Doel nuclear power plant near Antwerp. Domains surrounding Antwerp, Sint-Niklaas, Mechelen and Turnhout become depopulated and transport links are abandoned. Explorative activity is observed in areas of Oostende, Brugge, Roeselare, Arlon and Sankt-Vith. Attempted mass migration is recorded into northern France, the south of The Netherlands, Luxembourg and the west of Germany. Transport links in the chain Brugge–Roeselare–Kortrijk–Tournai–Mons–Charleroi–Namur–Liège–Sankt-Vith/Arlon are significantly hypertrophied (Fig. 5.15). Preparations for emergency hibernation take place in Brussels and Leuven, and in the Hasselt and Genk areas.

5.6 Summary

To evaluate how good Belgian motorways are from an amorphous living creature point of view, we conducted the following laboratory experiments with slime mould *P. polycephalum*. We represented major urban areas with oat flakes and inoculated slime mould in the oat flake corresponding to Brussels. We waited until the slime mould colonised all oat flakes and then analysed the slime mould's protoplasmic network structure and compared it with the man-made motorway network and basic planar proximity graphs. We found that *P. polycephalum* almost but not completely approximates the Belgian motorway network. Transport links Roeselare–Kortrijk–Mons–Charleroi–Namur, Oostende–Brugge–Gent–Aalst–Brussels–Leuven–Mechelen–Antwerp–Sint-Niklaas, Liège–Sankt-Vith–Arlon and Hasselt–Genk appear as protoplasmic tubes in almost all laboratory experiments. Motorway links connecting Brussels with Antwerp, Tournai, Mons, Charleroi and Namur, and links connecting Leuven with Liège and Antwerp with Genk and Turnhout, are 'considered' by slime mould to be redundant and thus almost never appear in our experiments with the slime mould. If slime mould represented A7/A54 Brussels–Charleroi, non-existing in reality motorways between Leuven–Namur and Turnhout–Hasselt, its credible transport network would be a supergraph of the relative neighbourhood graph.

Motorway segments A17 (Antwerp–Sint-Niklaas), A19 (Antwerp–Mechelen), E42 (Liège–Sankt-Vith), A10/E40 (Oostende–Gent–Aalst–Brussels–Leuven), A17/E403 (Roeselare–Kortrijk–Tournai), E42/E19 (Tournai–Mons) and E42 (Mons–Charleroi–Namur) are essential transport links from the slime mould's point of view. If the two parts of Belgium were separated with Brussels in Flanders, the Walloon region of the Belgian transport network would be represented by a single chain from Tournai in the north-west to the Liège area in the north-east and down to southernmost Arlon.

While imitating major disasters leading to contamination propagating from Liège and Antwerp areas, we found several scenarios of major restructuring of transport networks and possible routes of mass migration.

We believe that the results of our scoping experiments and analysis can be used not only in bioinspired unconventional computing but also in more classical fields of urban and transport planning. Possible applications include but are not limited to restructuring of rural landscapes [Froment and Wildmann (1987)], novel approaches towards mapping vehicular accessibility [Vandenbulcke and Thomas (2009)], simulation of traffic dynamics in Belgian motorways [Boel and Mihaylova (2006)], development of trans-European transport networks [De Lathauwer (1995)], bottom-up landscape planning [Sevenant and Antrop (2010)] and modelling relations between urban sprawl and growing transport networks [Poelmans and Van Rompaey (2009)].

Chapter 6

Brazilian highways from slime mould's point of view

Andrew Adamatzky and Pedro P. B. de Oliveira

Brazil is characterised by a continental dimension, high economic and population regional imbalance, yet fast-growing country-wide economy; also, its model of economic development has led to a progressive dismantling of the railway system, in favour of a total dependence upon its highways, for both people and industrial production, which ended up becoming unevenly distributed in quantity and quality. This makes Brazil a unique test platform for evaluation of experimental biological approaches towards simulation and prognostication of transport network evolution.

A multitude of issues have to be faced regarding the implementation of the highway network in any country. This is particularly true for a country like Brazil, that did not have a linear progression of development throughout its history, and due to its continental dimension. Hence, in addition to all the technical decisions related to, say, geographical and geophysical considerations, a number of other factors come into place, such as its history of occupation and internal migrations, political pressures of all sorts, imbalanced economic development and population distribution, etc.

So, choosing Brazil as yet another test bed for Physarum-based road planning clearly represents an experimental instance that has not had a similar or even comparable counterpart in our previous efforts. We hope that our study might lead to tools that would help coping with the enormous demand put on the vehicular transport network of the country, which has undergone continuous restructuring throughout the recent decades and that will require even stronger efforts in the years to come.

We consider the 21 most populous urban areas in Brazil **U** (Fig. 6.1a), shown

(a)

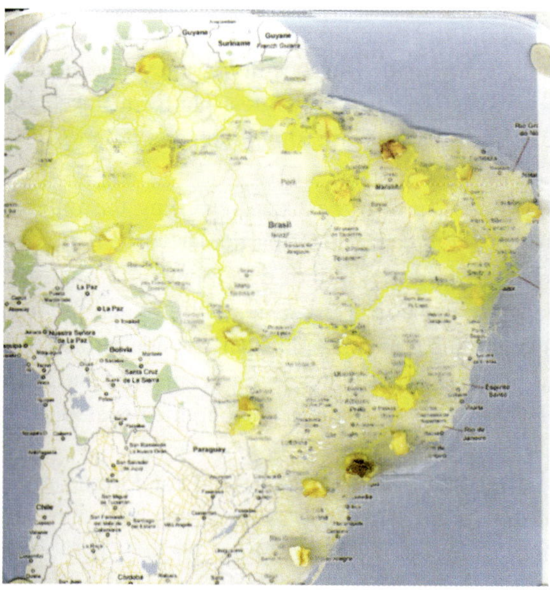

(b)

Fig. 6.1 Experimental setup: (a) outline map of Brazil with major urban areas **U** shown by encircled numbers, (b) urban areas, represented by oat flakes, are colonised by slime mould *Physarum polycephalum*.

below in descending order of population size:

(1) São Paulo,
(2) Rio de Janeiro,
(3) Belo Horizonte,
(4) Porto Alegre,
(5) Salvador,
(6) Recife,
(7) Fortaleza,
(8) Brasília,
(9) Belém,
(10) Manaus,
(11) São Luís,
(12) Teresina,
(13) Cuiabá,
(14) Petrolina-Juazeiro,
(15) Curitiba,
(16) Campo Grande,
(17) Macapá,
(18) Imperatriz,
(19) Boa Vista,
(20) Rio Branco,
(21) Porto Velho.

For the sake of standard geographical analyses, Brazil is divided in terms of five macroregions (Regions, for short [Regions of Brazil (2011)]), which are also used for the analyses herein; the relation of these Regions to the urban areas shown in Fig. 6.1a is as follows: North (9, 10, 17, 19, 20, 21), North-east (5, 6, 7, 11, 12, 14, 18), Central-west (8, 13, 16), South-east (1, 2, 3) and South (4, 15).

To represent areas of **U**, we place oat flakes in the positions corresponding to the areas. At the beginning of each experiment an oat flake colonised by plasmodium was placed in the São Paulo area. We undertook 53 experiments.

To study reaction of Physarum-imitated transport networks on major disasters, we placed crystals of sodium chloride in the approximate positions of Angra nuclear power plant (in the south-east), Guararapes–Gilberto Freyre International Airport (in the north-eastern city of Recife) and Brasília International Airport (in the centre of the country); naturally, this should be regarded as an illustration, since the Brazilian map in the dish does not have resolution for precisely positioning these landmarks. Six scoping experiments have been undertaken for each epicentre of imitated disaster. The salt diffuses in the substrate outwards from its

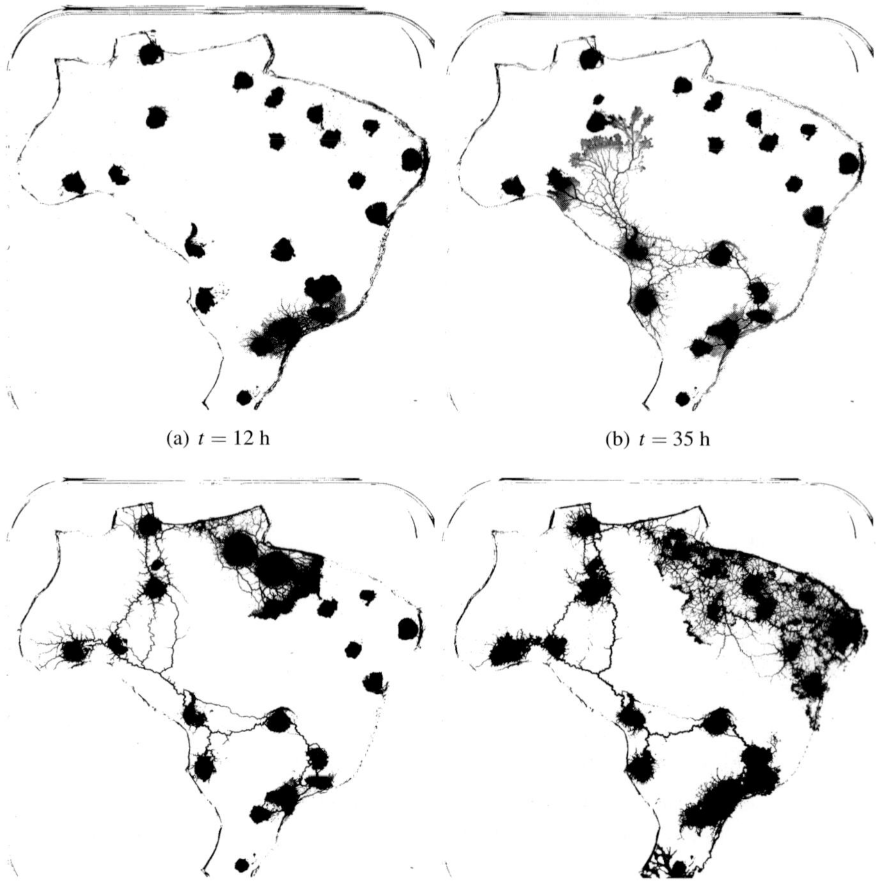

(a) $t = 12$ h (b) $t = 35$ h

(c) $t = 55$ h (d) $t = 79$ h

Fig. 6.2 Illustrative example of plasmodium development on configuration of cities **U** represented by oat flakes, with elapsed time from inoculation shown in the subfigure captions.

original application site (epicentre of disaster). It can therefore imitate radioactive and/or chemical pollution (Angra nuclear power plant) and terrorist or enemy attacks on airports (Guararapes–Gilberto Freyre and Brasília International Airports) and subsequent perturbation spreading along transport networks.

A typical scenario of Physarum behaviour on a Brazil-shaped agar plate with oat flakes representing cities is shown in Fig. 6.2. We placed an oat flake colonised by plasmodium in São Paulo. In 12 h plasmodium occupies Rio de Janeiro, Belo Horizonte and Curitiba (Fig. 6.2a). In a further 23 h the plasmodium spreads from

Belo Horizonte towards Brasília, and then from Brasília to Cuiabá and Campo Grande. The active zone from Cuiabá travels towards Porto Velho and Manaus. Porto Velho becomes colonised by Physarum by the 35th hour from slime mould's inoculation in São Paulo (Fig. 6.2b).

In the next 20 h the following action of events unfolds (Fig. 6.2c). Plasmodium branches from Porto Velho to Rio Branco and Manaus. The plasmodium that colonised Rio Branco stops its exploratory activity while the slime mould based in Manaus spreads to Boa Vista and then from Boa Vista to Macapá and then to Belém. The final branching at this stage occurs after Belém, when the plasmodium occupies São Luís and Imperatriz. In one more day, all urban areas **U** are colonised by plasmodium (Fig. 6.2d). Note that Porto Alegre, despite its proximity to the initial inoculation site — São Paulo — is colonised only at final stages of spanning the experimental space by slime mould. This demonstrates that in some cases the plasmodium makes its decisions on propagating based not only on the distance between current and next sources of nutrients.

Plasmodium's behaviour is determined by many factors and slightly randomised by interactions between a myriad of biochemical oscillators in its cytoplasm. No two experiments produce exactly the same results. To generalise our experimental results, we constructed a probabilistic Physarum graph.

A selection from a series of graphs $\mathbf{P}(\theta)$, $\theta = 0, \ldots, 42$, is shown in Fig. 6.2. A non-pruned graph $\mathbf{P}(0)$ is non-planar (Fig. 6.2a). The Physarum graph becomes planar when the cutoff threshold θ reaches $\frac{7}{53}$, i.e. when only edges appearing in over 13% of experimental trials are present in the graph (Fig. 6.2b). Assuming an acceptable 10% error, we can propose that such Physarum graph is planar.

For $\theta = \frac{22}{53}$, the Physarum graph becomes disconnected. The Brasília urban area loses all its neighbours except for Belo Horizonte (Fig. 6.2c). One component of the graph $\mathbf{P}(\frac{22}{53})$ is a chain c_1 = Porto Alegre–Curitiba–São Paulo–Rio de Janeiro–Belo Horizonte–Brasília. The chain c_1 remains stable up to $\theta = \frac{33}{53}$ in over 62% of experiments (Fig. 6.2d–h).

The second component c_2 of $\mathbf{P}(\frac{22}{53})$ is composed of all other urban areas **U** minus vertices of c_1 (Fig. 6.2c). The component c_2 has three cycles, spanning Belém, São Luís, Teresina and Imperatriz. Cyclic parts of c_2 break down when θ reaches $\frac{26}{53}$ (Fig. 6.2e). At $\theta = \frac{26}{53}$, the Physarum graph consists of the chain c_1, the tree c_3 (derived from component c_2) and an isolated vertex: Boa Vista (Fig. 6.2e). Further increase of θ leads to complete dissociation of the tree c_3 and formation of isolated vertices (Fig. 6.2f–j). A single chain (São Paulo–Rio de Janeiro–Belo Horizonte) is present in almost 80% of the experiments, i.e. up to $\theta = \frac{42}{53}$ (Fig. 6.2j).

(e) $\theta = 0$

(f) $\theta = \frac{7}{53}$

(g) $\theta = \frac{22}{53}$

(h) $\theta = \frac{23}{53}$

(i) $\theta = \frac{26}{53}$

Fig. 6.2 (Continued) Configurations of Physarum graph $\mathbf{P}(\theta)$ for critical values of θ. Thickness of an edge is proportional to the edge's weight.

(j) $\theta = \frac{27}{53}$

(k) $\theta = \frac{28}{53}$

(l) $\theta = \frac{33}{53}$

(m) $\theta = \frac{41}{53}$

(n) $\theta = \frac{42}{53}$

Fig. 6.2 (Continued)

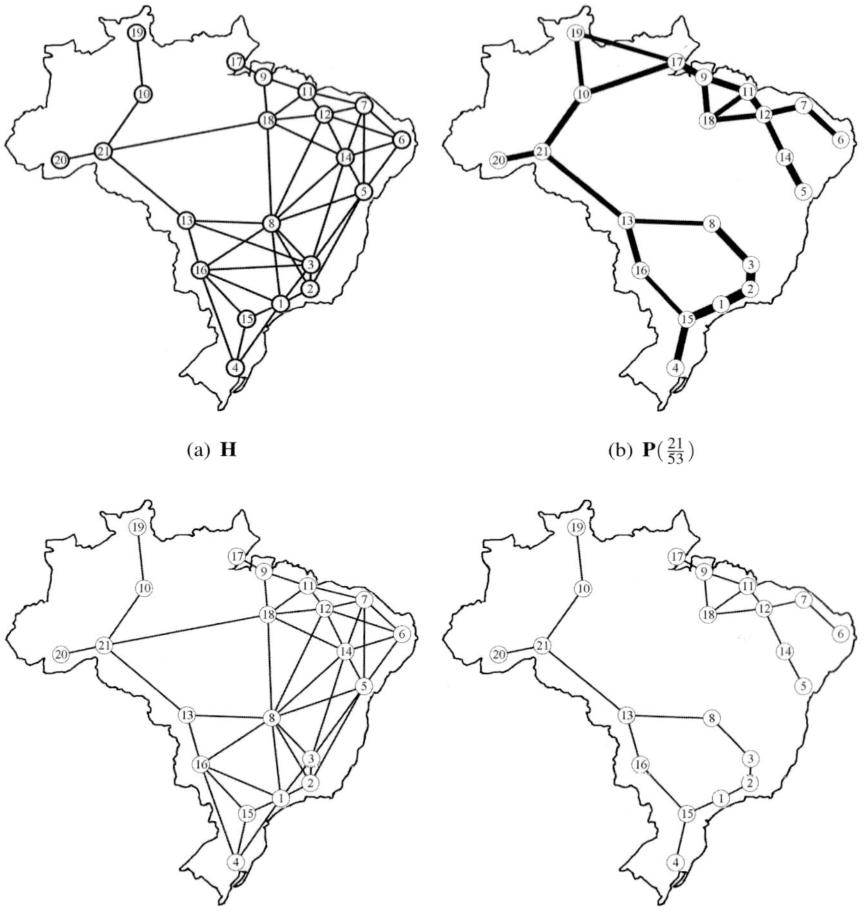

(a) **H** (b) $\mathbf{P}(\frac{21}{53})$

(c) $\mathbf{P}(0) \bigcap \mathbf{H}$ (d) $\mathbf{P}(\frac{21}{53}) \bigcap \mathbf{H}$

Fig. 6.3 Graph **H** of Brazilian highway network is shown in (a), and the highest θ connected Physarum graph is shown in (b). Intersections $\mathbf{P}(\theta) \bigcap \mathbf{H}$ of Physarum **P** and highway **H** graphs are shown in (c) for $\theta = 0$ and (d) for $\theta = \frac{21}{53}$.

6.1 Slime mould makes more highways

How well do Physarum graphs approximate the highway network? By inspection of the most well-known highway map of the country, freely available in [Guia Quatro Roda (2010)], we constructed the graph **H** of Brazilian highways. But, due to the high imbalance in quality of the Brazilian roads, depending on the region concerned, the highway network we indeed rely upon is a mix of different kinds, from standard motorways (as typical in the São Paulo state) down to roads with precarious conditions, as often happens, especially in the North Region.

The highway graph is non-planar (Fig. 6.3a). Let us compare $\mathbf{H}$ with the non-planar Physarum graph $\mathbf{P}(0)$, already shown as Fig. 6.2a, and the highest θ connected Physarum graph $\mathbf{P}(\frac{21}{53})$, introduced as Fig. 6.3b.

Finding 33. *Compared with the existing Brazilian highway network, slime mould* Physarum polycephalum *overdoes the number of transport links in all urban areas of Brazil, with an emphasis on those in the North Region and its borders.*

The highway and Physarum graphs have the same labelling, so that we can compare them straightforwardly. As Fig. 6.4a makes it evident, the distributions of degrees along nodes are almost matching in $\mathbf{H}$ and $\mathbf{P}(0)$, with all nodes of the highway graph having almost equally fewer neighbours than nodes of the Physarum graph. The node-degree distribution in $\mathbf{P}(\frac{21}{53})$ does not perfectly match that of $\mathbf{H}$, with the best matches happening in the South Region and the worst occurring in Manaus and Macapá, in the extreme north.

Spatial distributions of mismatches are shown in Fig. 6.4b and c. Let $d_i^{\mathbf{P}}$ and $d_i^{\mathbf{H}}$ be degrees of node i in graphs $\mathbf{P}$ and $\mathbf{H}$. A node i in Fig. 6.4b and c has a green (light grey) band if $d_i^{\mathbf{P}} > d_i^{\mathbf{H}}$ and a red (dark grey) band if $d_i^{\mathbf{P}} < d_i^{\mathbf{H}}$. The width of the band at i is proportional to $|d_i^{\mathbf{P}} - d_i^{\mathbf{H}}|$.

Mismatch between node degrees of $\mathbf{H}$ and $\mathbf{P}(0)$ is pronounced in urban areas of the entire Brazilian North Region and its borders with the Central-west and North-east Regions. The highest mismatch is observed in Manaus, Macapá and Imperatriz (Fig. 6.4b). In these parts the Physarum graph shows higher connectivity than the highway graph. The north-eastern part of Brazil is most characteristic of a mismatch between $\mathbf{H}$ and $\mathbf{P}(\frac{21}{53})$ (Fig. 6.4c). Urban areas lying in the eastern parts of Brazil, starting in the Central-west Region, have fewer neighbours in $\mathbf{P}$ than in $\mathbf{H}$. The Brasília urban area shows maximal mismatch.

It is also worth noticing that only Curitiba (area 15) has the same connectivity in all three situations depicted in Fig. 6.4. This represents a robustness of this southern area for ensuring full connectivity of the country, regardless of how it is conceived.

Finding 34. *Physarum matches Brazilian highways almost perfectly, by creating almost all existing connections:*

$$\mathbf{H} = \mathbf{P}(0) \cap \mathbf{H} \cup \textit{Cuiabá– Belo Horizonte} \cup \textit{Campo Grande–Belo Horizonte}.\}$$

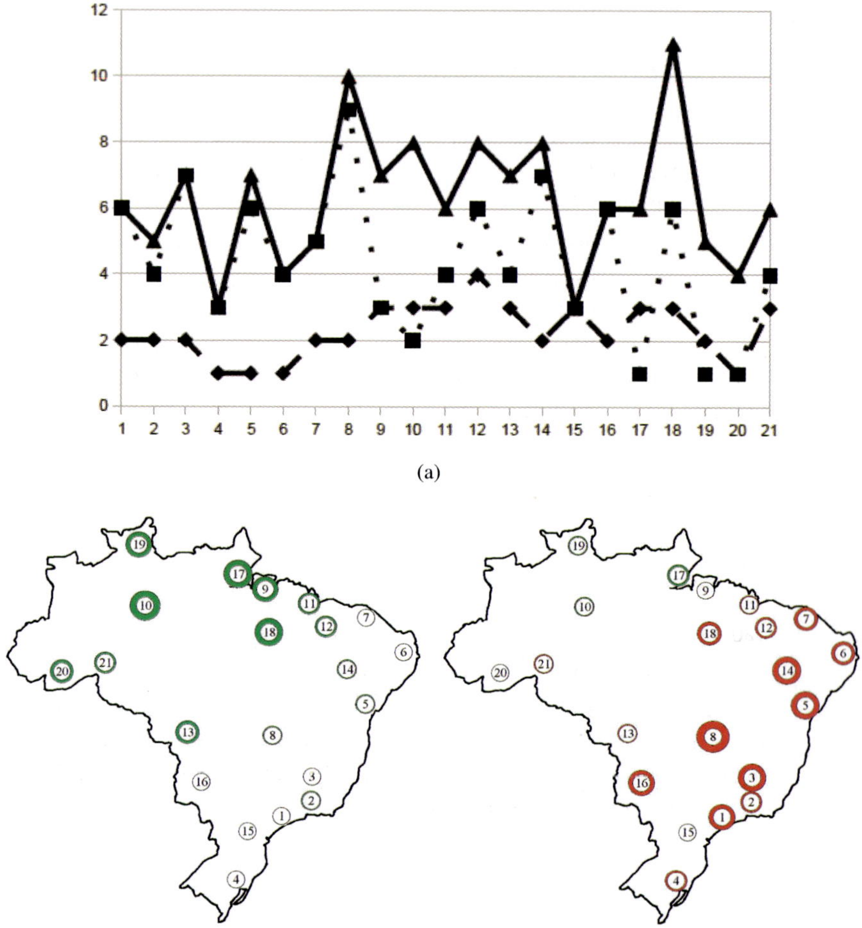

(a)

(b) (c)

Fig. 6.4 Node-degree-based comparison of Physarum and highway graphs: (a) degrees of nodes in highway graph **H** (dotted line and square markers), Physarum graphs **P**(0) (solid line and triangle markers) and **P**$(\frac{21}{53})$ (dashed line and rhomboid markers); the horizontal axis represents the node labels and the vertical axis their corresponding degrees, (b) mismatch between degrees of **H** and **P**(0) nodes and (c) **H** and **P**$(\frac{21}{53})$.

This comes as a direct observation of Fig. 6.3a and c. In contrast to the somewhat redundant connectivity degree of **P**(0), missing the links from Belo Horizonte to Cuiabá and Campo Grande may indicate that such links do not conform to slime mould's representation of local optimal connectivity.

Finding 35.

$$\mathbf{P}(\frac{21}{53}) = \mathbf{P}(\frac{21}{53}) \cap \mathbf{H} \cup \textit{Boa Vista–Macapá} \cup \textit{Manaus–Macapá}.$$

In the pruned Physarum graph $\mathbf{P}(\frac{21}{53})$, Macapá plays the same role as Belo Horizonte in the graph $\mathbf{P}(0)$. The graph $\mathbf{P}(\frac{21}{53}) \cap \mathbf{H}$ is disconnected and is composed of two components. The south-western component spans the states of Acre, Amazonas, Rondônia, Mato Grosso, Goiás, São Paulo, Minas Gerais, Rio de Janeiro, Paraná, Santa Catarina and Rio Grande do Sul. All states except for Pará and Tocantis are spanned by the north-eastern component (Fig. 6.3d). The south-western component has a single cycle: São Paulo–Rio de Janeiro–Rio de Janeiro–Brasília–Cuiabá–Campo Grande–Curitiba. The three-cycle north-eastern component of $\mathbf{P}(\frac{21}{53}) \cap \mathbf{H}$ is discussed as component c_2 earlier.

6.2 Comparing with proximity graph

Proximity graphs **RNG** and **GG** and the minimum spanning tree **MST** constructed on nodes of **U** are shown in Fig. 6.5. Strictly speaking, a spanning tree rooted in São Paulo (Fig. 6.5c) is not a minimum spanning tree. The minimal length trees are those rooted in Fortaleza, Recife or Teresina (Fig. 6.5d). However, differences in lengths are negligible: the tree rooted in São Paulo is just 4% longer than the **MST** rooted in Fortaleza, Recife or Teresina (Fig. 6.6). Therefore, further on we will consider the spanning tree rooted in São Paulo as the **MST**. We will also be referring to $\mathbf{P}(\frac{21}{53})$ as simply **P**.

Finding 36.

$$\mathbf{MST} + \textit{Cuiabá–Brasília} + \textit{Belém– São Luís} = \mathbf{RNG}.$$

This is a direct outcome of Fig. 6.5a and c, which indicate that the configuration of urban areas from **U** allows for an optimal, in a sense of proximity graphs, construction of transport networks. A minimum spanning tree is just two edges away from the relative neighbourhood graph.

Finding 37.

$$\mathbf{MST} - \textit{Manaus–Macapá} \subset \mathbf{H}$$

$$\mathbf{MST} - \textit{Belo Horizonte–Salvador} \subset \mathbf{P}.$$

Neither man-made nor Physarum-grown transport graphs allow for an optimal, in a sense of the shortest path, transportation. However, removing just one

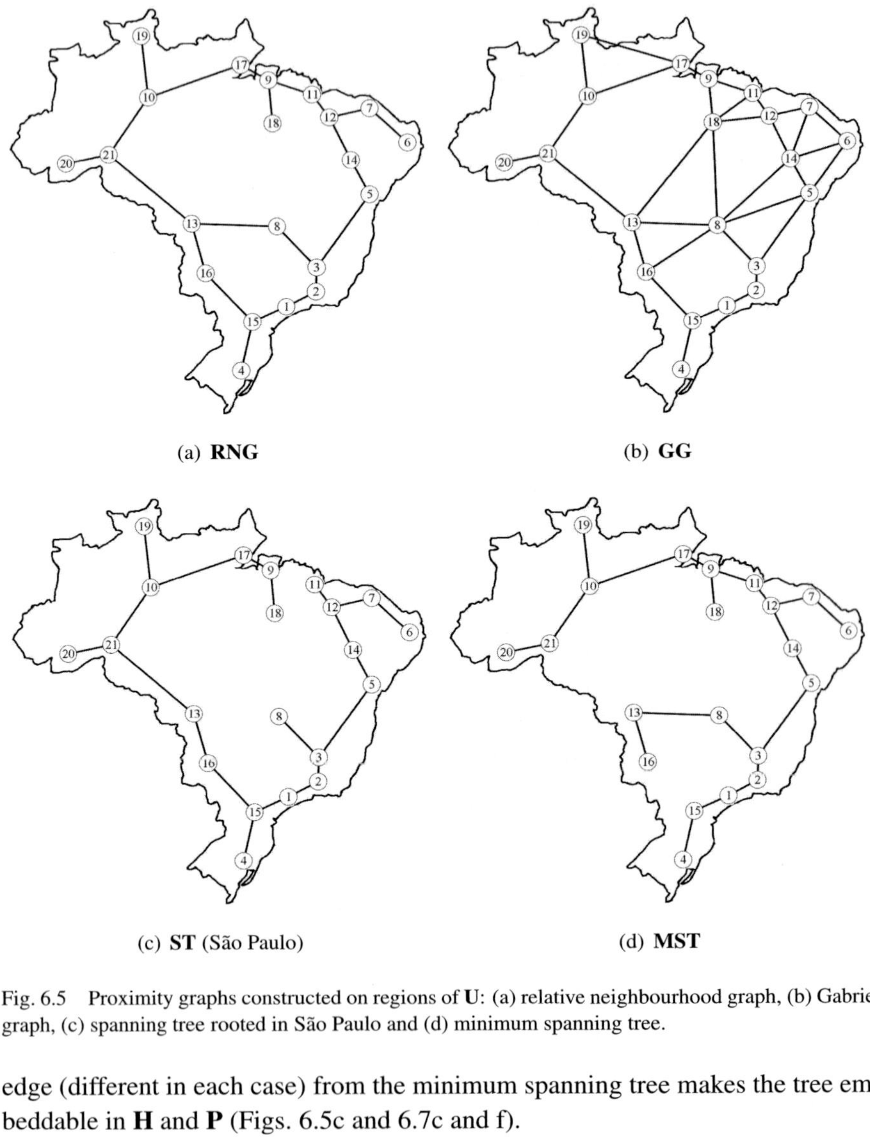

(a) **RNG** (b) **GG**

(c) **ST** (São Paulo) (d) **MST**

Fig. 6.5 Proximity graphs constructed on regions of **U**: (a) relative neighbourhood graph, (b) Gabriel graph, (c) spanning tree rooted in São Paulo and (d) minimum spanning tree.

edge (different in each case) from the minimum spanning tree makes the tree embeddable in **H** and **P** (Figs. 6.5c and 6.7c and f).

Finding 38.

$$\mathbf{RNG} - \textit{Manaus–Macapá} \subset \mathbf{H}$$

$$\mathbf{RNG} - \textit{Belo Horizonte–Salvador} \subset \mathbf{P}$$

(Fig. 6.7a and d).

Node	ST length
São Paulo	1.04
Rio de Janeiro	1.09
Belo Horizonte	1.08
Porto Alegre	1.07
Salvador	1.01
Recife	1.00
Fortaleza	1.00
Brasília	1.09
Belém	1.08
Manaus	1.10
São Luís	1.02
Teresina	1.00
Cuiabá	1.11
Petrolina-Juazeiro	1.04
Curitiba	1.07
Campo Grande	1.12
Macapá	1.05
Imperatriz	1.08
Boa Vista	1.10
Rio Branco	1.09
Porto Velho	1.09

Fig. 6.6 Ratios between the lengths of the spanning trees (ST) rooted in every node of **U** to the length of the minimum spanning tree.

Bearing in mind that **RNG** is a subgraph of **GG**, we can propose that Manaus–Macapá and Belo Horizonte–Salvador are the only transport links that prevent the relative neighbourhood graph from being embeddable in man-made and Physarum-built transport networks. The next results give us a hint of a possible position of Physarum graphs in the hierarchy of proximity graphs.

Finding 39.

$$\mathbf{GG} - \textit{Manaus–Macapá} - \textit{Boa Vista– Macapá} - \textit{Cuiabá–Imperatriz} \subset \mathbf{H}$$

$$\mathbf{P} \subset \mathbf{GG}$$

(Fig. 6.7b and e).

Thus, we can propose that a Physarum graph constructed on urban areas of **U** is a proximity graph that is neither a sub- nor a supergraph of the minimum spanning graph, but is a subgraph of the relative neighbourhood graph and Gabriel graph.

(a) **P**⋂**RNG**

(b) **P**⋂**GG**

(c) **P**⋂**MST**

(d) **H**⋂**RNG**

(e) **H**⋂**GG**

(f) **H**⋂**MST**

Fig. 6.7 Intersections of Physarum graph $\mathbf{P}(\frac{21}{53})$ (abc) and highway graph **H** (def) with proximity graphs **RNG** and **GG** and minimum spanning tree **MST**.

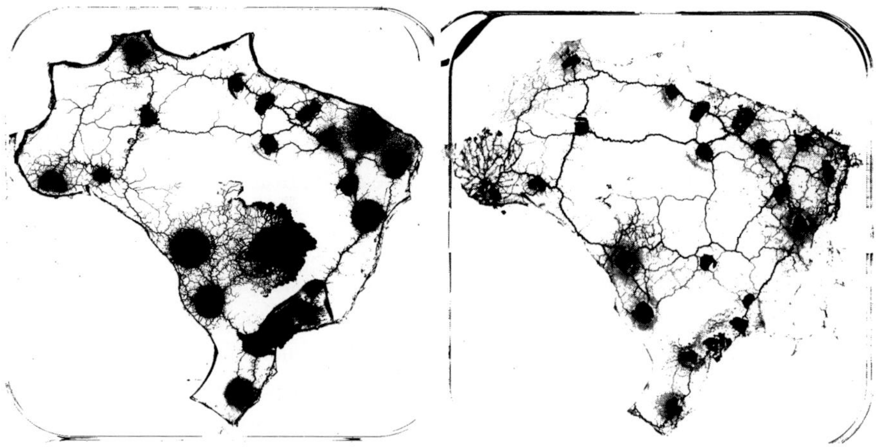

(a) $t = 0$ (b) $t = 24$ h

Fig. 6.8 Restructure of the plasmodium network in response to a crystal of sodium chloride placed in the approximate position of Angra nuclear power plant.

6.3 Physarum and Angra nuclear power plant

Finding 40. *In response to a sudden disaster, gradually spreading from its epicentre, the Physarum transport networks react by abandoning transport links affected by the disaster zone, enhancing transport links unaffected directly by the disaster, massive sprouting from the epicentre and increasing of scouting activity in the regions distant from the epicentre of the disaster.*

We imitate a hypothetical disaster in the Angra nuclear power plant by placing a crystal of sodium chloride at the approximate position of the plant (determined indeed by very low resolution of our experimental area, i.e. Petri dish). Images of the protoplasmic network were recorded before the disaster and 24 h later. A typical example of the network restructuring is shown in Fig. 6.8.

Usually, in 24 h the concentration of diffusing salt (in its repellent effect for plasmodium concentration) propagates as far as 8 and 4, and almost reaches 5. All Physarum activity in Brasília, so clearly visible in Fig. 6.8a, disappears. Also, the following transport links become destroyed and dysfunctional: Curitiba–São Paulo, Curitiba–Campo Grande, São Paulo–Campo Grande, São Paulo–Rio de Janeiro, Rio de Janeiro–Belo Horizonte, Belo Horizonte–Brasília and Belo Horizonte–Salvador. Plasmodium abandons this protoplasmic tube. The parts of the transport network not affected directly by a high concentration of salt increase their functionality (reflected in higher concentration of protoplasm and additional branching), so as to compensate for the lost transport links. Namely, transport arteries connecting Manaus and Imperatriz, Teresina and Petrolina-Juazeiro and Cuiabá and Porto Velho become particularly pronounced (Fig. 6.8b). At the same time, plasmodium activates its scouting behaviour and explores options of escape and recolonisation. Typically, it tries to relocate in the western parts of the North Region (such as Rio Branco) and the North-east Region (such as the Salvador, Fortaleza and Petrolina-Juazeiro areas).

The imitation of a disaster with its epicentre in Guararapes–Gilberto Freyre International Airport is illustrated in Fig. 6.9, with a sample of two experiments. In 24 h after the disaster its effect reaches Teresina in the North-east and Belo Horizonte in the South-east Regions. All transport links inside the disaster zone vanish. The north-eastern urban areas of Salvador, Recife, Fortaleza, Teresina and Petrolina-Juazeiro become isolated and disturbed in functioning. Slightly affected or unaffected transport arteries show signs of hypertrophy, such as in the roads connecting Brasília to Cuiabá and Cuiabá to Manaus in Fig. 6.9. Plasmodium shows exploratory behaviour and ventures onto domains not covered by growth substrates in the eastern-most (around Rio Branco) and southern-most (around Porto Alegre) parts of Brazil. In some cases we observed a 'panic' response of major transport routes, where plasmodium exhibits disorganised multisource routing. In Fig. 6.9c and d, mini wave fronts of plasmodial activity can be seen sprouting simultaneously from many sites of the main transport link that goes from Porto Alegre up to Rio Branco.

Another validation of Finding 40 was demonstrated in experiments with sodium chloride placed at the approximate position of Brasília International Airport (Fig. 6.10). A typical reaction includes hypertrophy of and 'indiscriminate' sprouting from major transport links unaffected directly by the disaster: most remarkably, a very long link connecting all areas of northern Brazil, stretching from Recife up to Rio Branco, then going to the southern parts of the Central-west Region; and also a short link between São Paulo and Porto Alegre.

6.4 Summary

In laboratory experiments with slime mould *Physarum polycephalum*, we represented the major urban areas of Brazil with oat flakes, inoculated slime mould in São Paulo, recorded the protoplasmic networks built and compared them with man-made highways and proximity graphs. We found that the plasmodium of *P. polycephalum* develops a minimal approximation of a transport network spanning urban areas of **U**. The Physarum-developed network matches the man-made high-

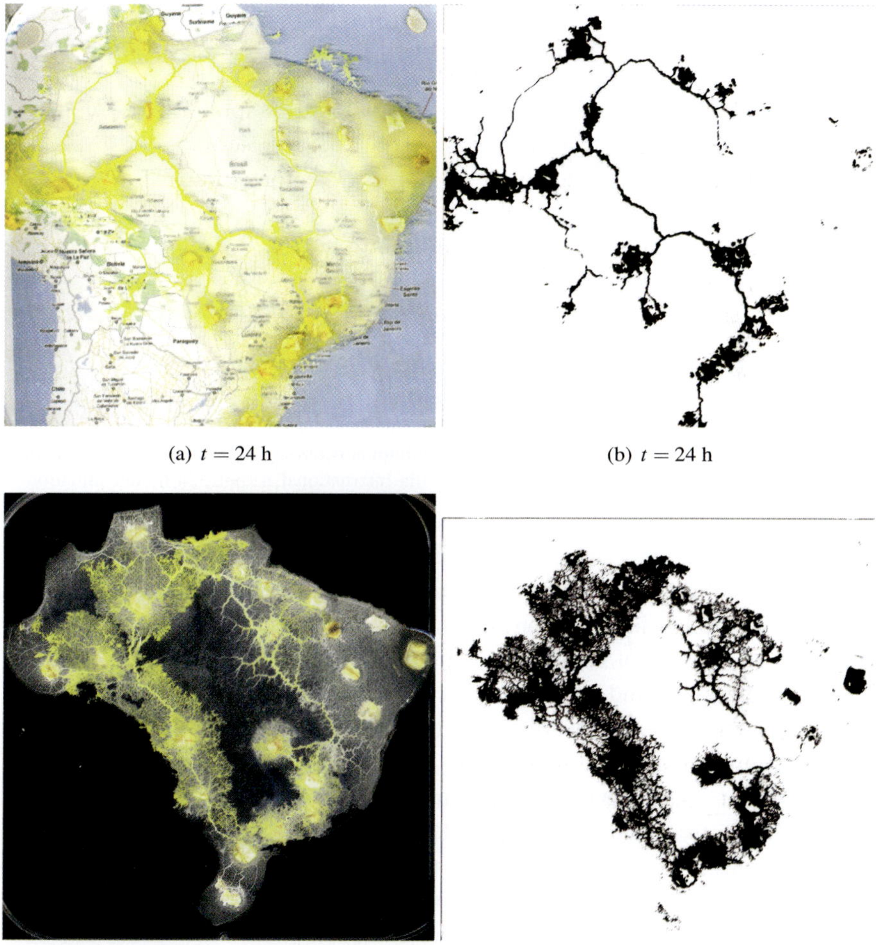

(a) $t = 24$ h (b) $t = 24$ h

(c) $t = 24$ h (d) $t = 24$ h

Fig. 6.9 Examples of two experiments on restructuring of the plasmodium network in response to a crystal of sodium chloride placed at the approximate position of the Guararapes–Gilberto Freyre International Airport, in Recife: (a and c) scanned images of experimental Petri dish 24 h after the imitated disaster, (b and d) binarised images of the protoplasmic network.

way network very well. The high degree of similarity is preserved even when we place high-demand constraints on repeatability of links in the experiments (θ threshold). Physarum approximates almost all major transport links, apart from Cuiabá–Belo Horizonte and Campo Grande–Belo Horizonte. These nice results are somewhat surprising when recalling the multitude of issues associated with

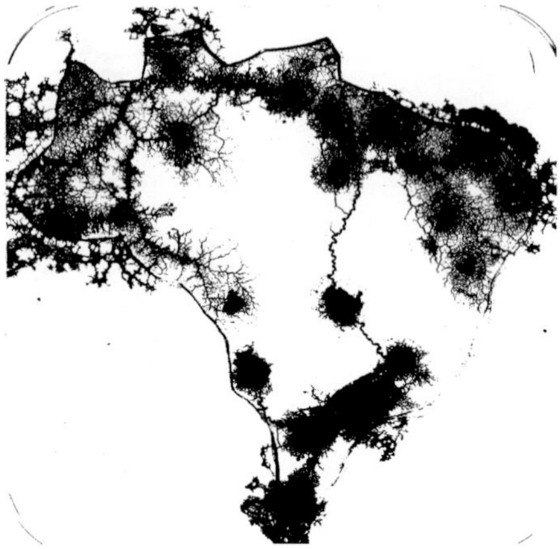

Fig. 6.10 Experiment on restructuring of the plasmodium network in response to a crystal of sodium chloride placed at the approximate position of Brasília International Airport, 24 h after initiation of contamination.

the implementation of the highway network in a country like Brazil. All in all, it seems that the Physarum graph sorts of 'integrates', so to speak, all those aspects into its own set of developmental variables, cancelling out specificities of the original problem, and leading to the empirical observations we could make.

In response to a sudden disaster, gradually spreading from its epicentre, the Physarum transport networks react by abandoning transport links affected by the disaster zone, enhancing those unaffected directly by the disaster, massive sprouting from the epicentre and increasing of scouting activity in the regions distant from the epicentre of the disaster. The space–time dynamics of slime mould responding to propagating contamination of the substrate by a repellent complements well existing computer models of spatial and economic impacts of disasters [Okuyama and Chang (2004); Andersson (2009); Takada (2010)].

Neither the minimum spanning tree constructed on urban areas nor the relative neighbourhood graph are subgraphs of the Physarum graph. However, by removing just one edge/link from the spanning tree or relative neighbourhood graph, both graphs become embeddable in the Physarum graph. Moreover, the Physarum graph is a subgraph of the Gabriel graph. We can thus claim that Physarum slime mould develops better transport links than those in the existing man-made highway network. In this light, we can reword our first finding as "according to slime mould, the Brazilian highway network is excessively redundant in the country's North Region and its borders".

This finding indicates that there are various connectivity options in the north of Brazil, with no particular preference; it is interesting that this is in tune with the true fact that the main connections between many areas of this part of Brazil lead to no particular preference, since all of them may be considered of equally bad quality.

Chapter 7

Trans-Canada slimeways: from coast to coast to coast

Andrew Adamatzky and Selim G. Akl

What are the unique properties of the Canadian transport system? The Canadian highway system gives us a good example of a logically designed transportation system whose key goal is to connect all the provinces together by highways. The highway network was built as a federal–provincial territorial cooperative project with great effort taken in coordinating work on different parts. Another attractive property of the highway system is that it was designed to provide access to remote areas where no regions of high population density exist.

We consider the 11 most populated urban areas **U** of Canada (Fig. 7.1a) and five transport nodes:

(1) Toronto area (including Hamilton, London, St. Catharines–Niagara, Windsor, Oshawa, Barrie, Guelph and Kingston),
(2) Montreal area (including Ottawa–Gatineau, Quebec City, Sherbrooke and Trois-Rivieres),
(3) Vancouver area (including Victoria, Abbotsford and Kelowna),
(4) Calgary,
(5) Edmonton,
(6) Winnipeg,
(7) Halifax–Moncton,
(8) Saskatoon–Regina,
(9) St. John's,
(10) Sudbury,
(11) Thunder Bay,
(12) Inuvik,

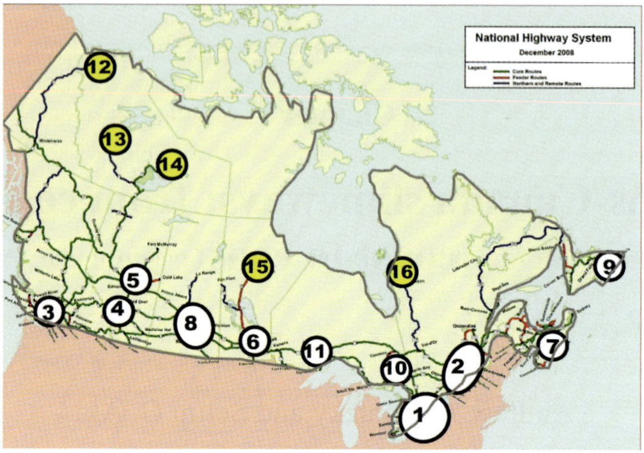

(a)

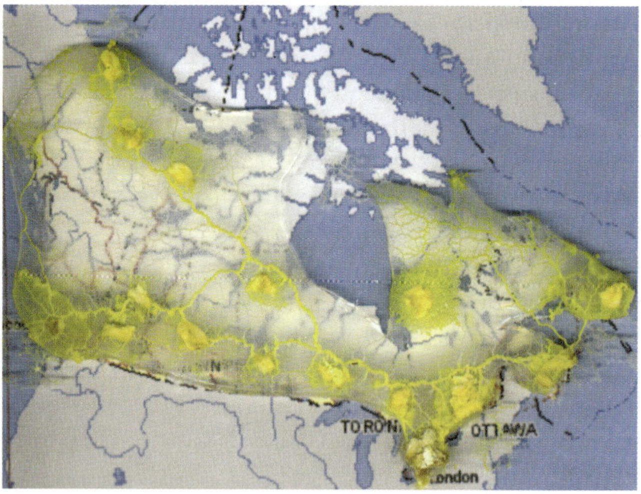

(b)

Fig. 7.1 Experimental setup: (a) urban areas and transport nodes to be represented by oat flakes, from [Canada's National Highway System (2009)], (b) snapshot of protoplasmic transport network developed by *P. polycephalum*; the snapshot is made on a highway map.

(13) Wrigley,
(14) Yellowknife,
(15) Thompson,
(16) Radisson.

The last five entries from Inuvik to Radisson are not highly populated urban areas. They are transport nodes added for completeness, i.e. to present slime mould with the same number of principal transport nodes as the man-made highway system (Fig. 7.1a). Some transport nodes such as Fort McMurray, La Ronge, Flin Flon and so on are not included in the list due to their proximity to already chosen major urban areas.

At the beginning of each experiment a piece of plasmodium, usually already attached to an oat flake in the cultivation box, was placed in the Toronto area (region 1 in Fig. 7.1a). We undertook 23 experiments.

7.1 Foraging from Toronto

It usually takes the plasmodium of *P. polycephalum* 2–5 days to span all urban areas. How fast the plasmodium colonises the space depends on many unknown factors, including seasonal variations, plasmodium's age etc. 'Younger' plasmodia, which were just recently 'woken up' from the sclerotium phase, do usually colonise the experimental arena quicker than old plasmodia, which were replanted several times in culture boxes. Images of protoplasmic networks are taken when all oat flakes, representing **U**, were colonised by plasmodium. Examples of the protoplasmic networks are shown in Fig. 7.2.

A Physarum graph $\mathbf{P}(0)$ is a non-planar graph due to the presence of a protoplasmic tube connecting Saskatoon–Regina (8) with Yellowknife (14) (Fig. 7.3a). It also exhibits two cross-Canadian transport links Inuvik (12) to Radisson (16) and Yellowknife (14) to Radisson (16). Nevertheless, these are links that might be considered as senseless from a geographical point of view because they are crossing massive mountains and forests.

All three links disappear when we increase θ to $\frac{8}{23}$ (Fig. 7.3b). Four more links become trimmed off when $\theta = \frac{8}{23}$: Sudbury (10)–Thompson (15), Montreal area (2)–Sudbury (10), Sudbury (10)–Radisson (16) and Montreal area (2)–St. John's (9).

$\mathbf{P}(\frac{8}{23})$ is the last connected graph in a series of Physarum graphs, $\mathbf{P}(\theta)$, $0 \leq \theta \leq 1$. The urban areas St. John's (9) and Radisson (16) become isolated in $\mathbf{P}(\frac{9}{23})$ due to the disappearance of transport links from the Montreal area (2) to Radisson (16) and from Radisson (16) to St. John's (9) (Fig. 7.3c).

The Physarum graph splits into a tree and two isolated nodes for $\theta = \frac{17}{23}$ (Fig. 7.3d). Five long-distance routes from urban areas to transport nodes are removed: Vancouver area (3)–Inuvik (12), Vancouver area (3)–Wrigley (13), Edmonton (5)–Wrigley (13), Saskatoon–Regina (8)–Thompson (15) and Winnipeg

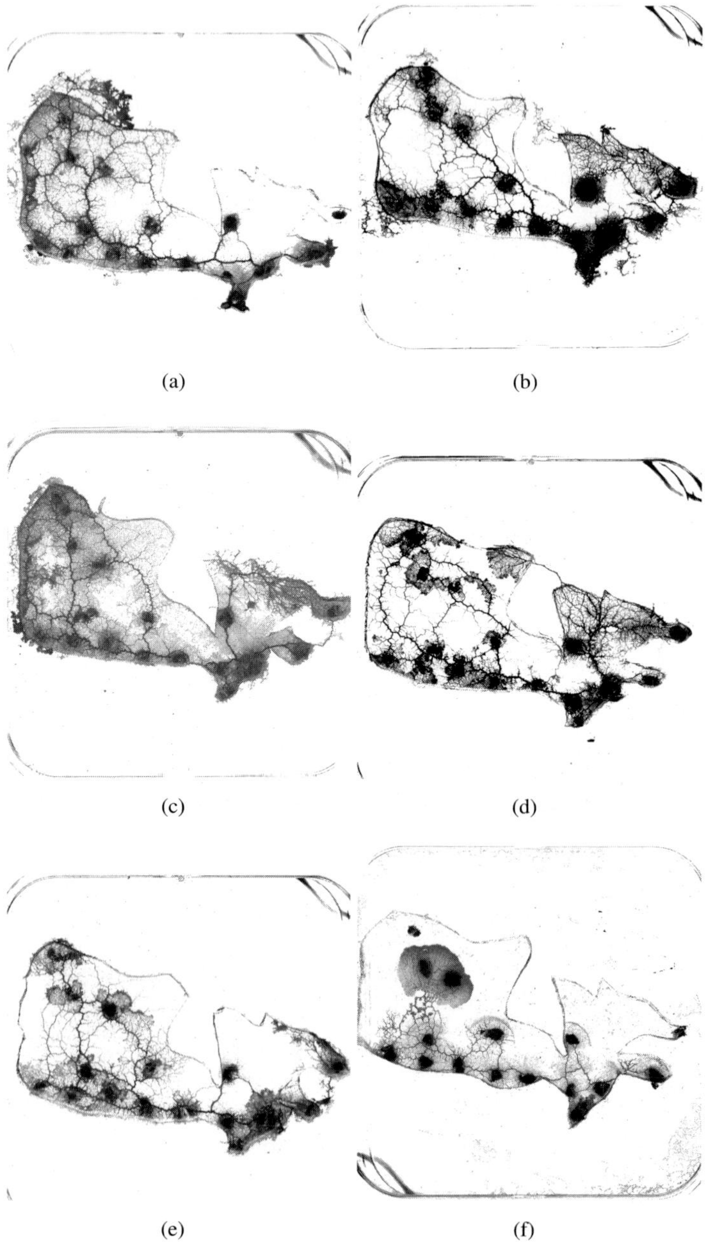

Fig. 7.2 Examples of protoplasmic networks developed by *P. polycephalum* on major urban areas and transport nodes **U**. Grey-scale images.

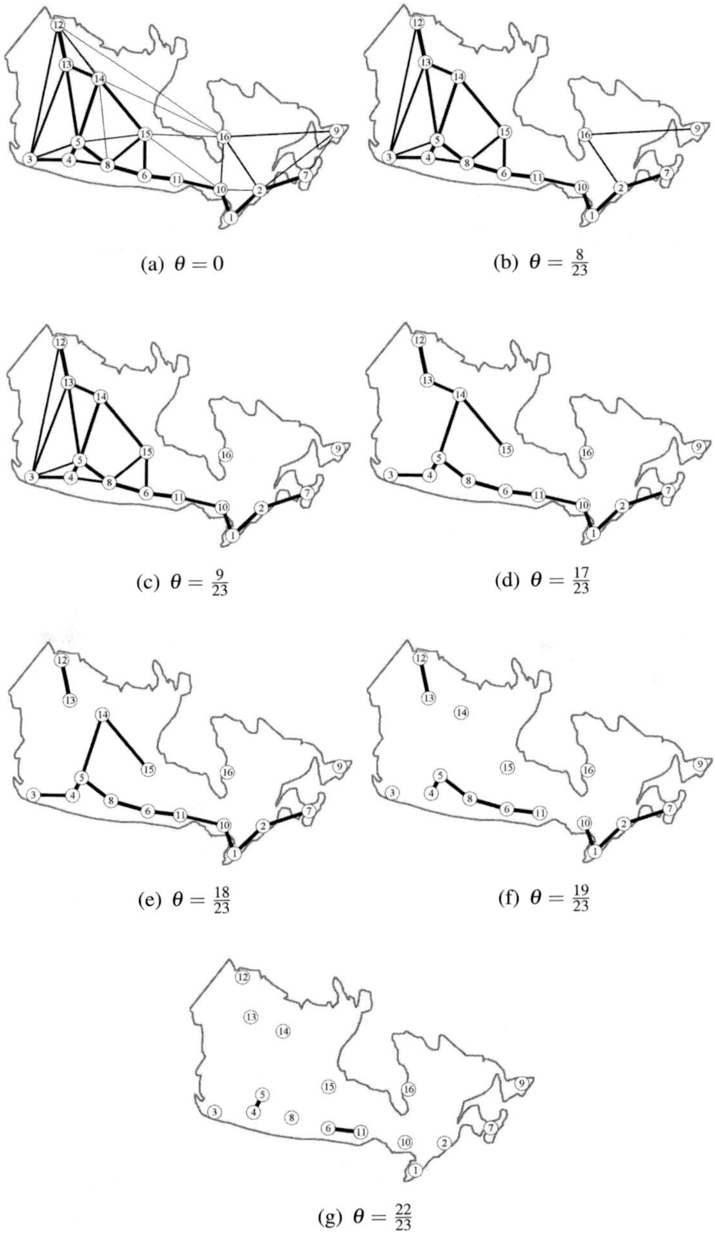

(a) $\theta = 0$ (b) $\theta = \frac{8}{23}$

(c) $\theta = \frac{9}{23}$ (d) $\theta = \frac{17}{23}$

(e) $\theta = \frac{18}{23}$ (f) $\theta = \frac{19}{23}$

(g) $\theta = \frac{22}{23}$

Fig. 7.3 Configurations of Physarum graph $\mathbf{P}(\theta)$ for various cutoff values of θ. Thickness of each edge is proportional to the edge's weight.

(6)–Thompson (15); and two short-distance routes between major urban areas: Vancouver area (3)–Edmonton (5) and Calgary (4)–Saskatoon–Regina (8).

The graph $\mathbf{P}(\frac{18}{23})$ consists of several disconnected components: two isolated nodes, St. John's (9) and Radisson (16), a two-node segment Inuvik (12)–Wrigley (13) and a tree spanning the rest of the urban areas (Fig. 7.3e). A further increase of θ to $\frac{19}{23}$ (Fig. 7.3f) leads to the formation of:

- five isolated nodes: Vancouver area (3), St. John's (9), Yellowknife (14), Thompson (15) and Radisson (16),
- a segment: Inuvik (12)–Wrigley (13),
- a chain: Calgary (4)–Edmonton (5)–Saskatoon–Regina (8)–Winnipeg (6)–Thunder Bay (11),
- a chain: Sudbury (10)–Toronto area (1)–Montreal area (2)–Halifax–Moncton (7).

Only segment routes Calgary (4) to Edmonton (5) and Winnipeg (6) to Thunder Bay (11) are represented by protoplasmic tubes in almost all experiments (Fig. 7.3g).

7.2 Physarum almost approximates Canadian highways

The highway graph **H** shown in Fig. 7.4b is extracted from a scheme of the Canadian transport network (Fig. 7.4a).

Finding 41. *Physarum almost approximates the Canadian highway network.*

The physarum graph $\mathbf{P}(0)$ approximates 21 of 22 edges of the highway graph **H**. Only one edge, Vancouver area (3) to Yellowknife (14), of the highway graph **H** is not represented by protoplasmic tubes in any of the 23 experiments undertaken (Fig. 7.4c). The graph $\mathbf{P}(0)$ gives us a rather relaxed approximation because it even includes links which occurred just once in a set of experiments. Let us look at $\mathbf{P}(\frac{8}{23})$, which represents links which occurred in over 35% of experiments. The physarum graph $\mathbf{P}(\frac{8}{23})$ approximates 18 of 22 edges of **H** (Fig. 7.4d). The only edges of **H** not represented in $\mathbf{P}(\frac{8}{23})$ are Vancouver area to Yellowknife, Sudbury to Radisson, Sudbury to Montreal area and St. John's to Montreal area.

Finding 42. *A core component of the Physarum transport network and the Canadian highway network consists of a chain passing along the southern border from the Halifax–Moncton area to Edmonton, and a fork attached to Edmonton; the southern branch of the fork is the Edmonton–Calgary–Vancouver area and the northern branch is Edmonton–Yellowknife–Wrigley.*

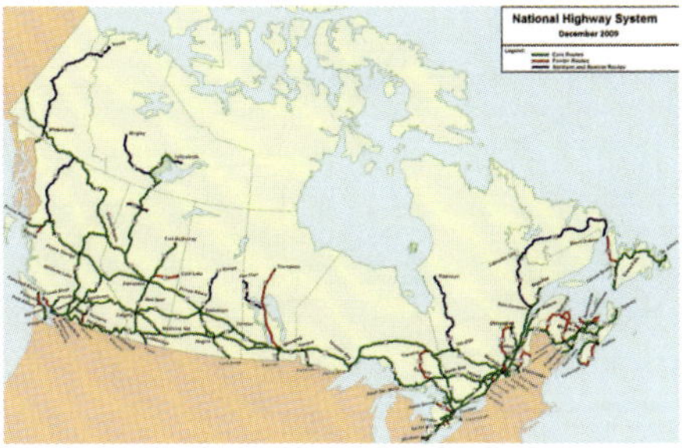

(a)

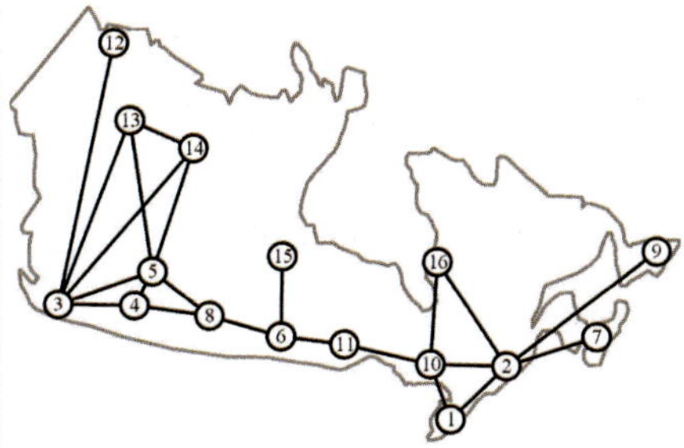

(b) **H**

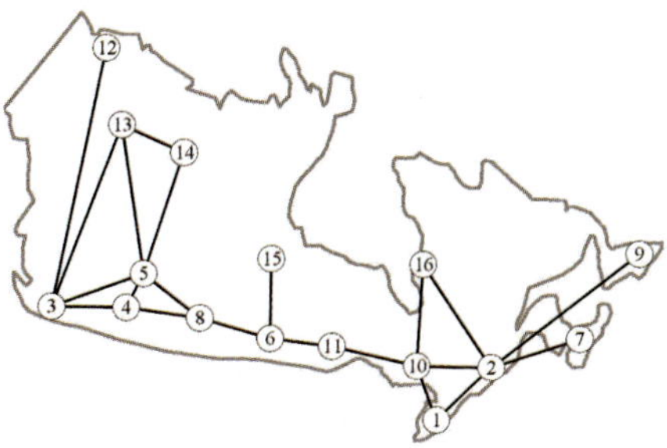

(c) $\mathbf{P}(0) \bigcap \mathbf{H}$

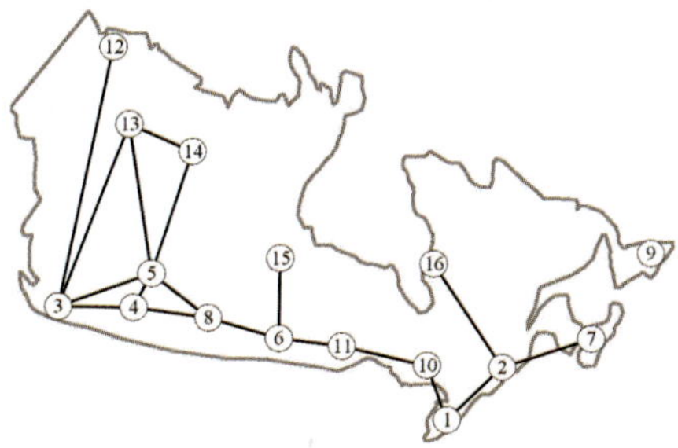

(d) $\mathbf{P}(\frac{8}{23}) \bigcap \mathbf{H}$

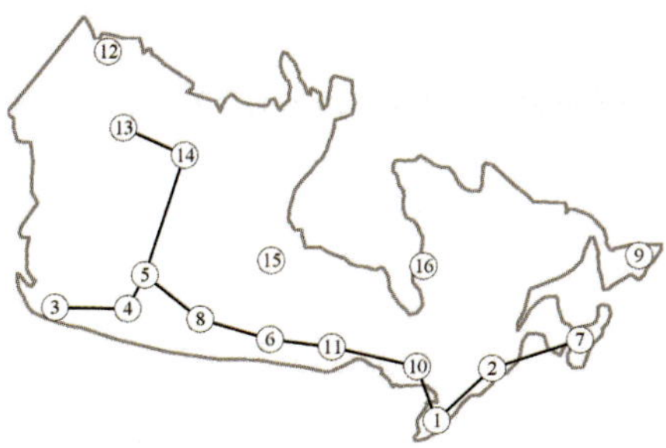

(e) $\mathbf{P}(\frac{17}{23}) \bigcap \mathbf{H}$

Fig. 7.4 Physarum vs highway network: (a) highway network in Canada [Canada's National Highway System (2009)], (b) highway graph **H**, (c, d, e) intersections of Physarum graph with highway graph for $\theta = 0, \frac{8}{23}, \frac{17}{23}$.

The component above is the only connected component in the intersection of $\mathbf{P}(\frac{17}{23})$ with **H** (Fig. 7.4e).

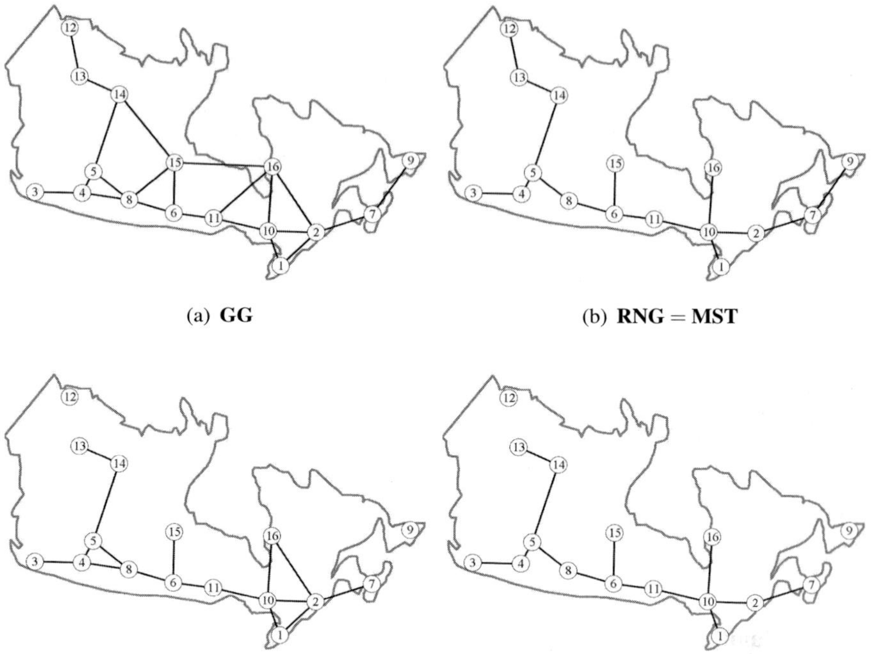

(a) **GG** (b) **RNG** = **MST**

(c) **GG**$\bigcap$**H** (d) **RNG**$\bigcap$**H**

Fig. 7.5 Proximity graphs and their intersections with highway graph **H**: (a) Gabriel graph, (b) relative neighbourhood graph and minimum spanning tree, (c) intersection of Gabriel graph with highway graph, (d) intersection of relative neighbourhood graph with highway graph.

7.3 On optimality of Canadian highways

Finding 43. *For a given configuration of nodes of* **U**, **RNG** = **MST**.

The finding implies that the configuration of urban areas of **U** is 'spanning friendly'.

Finding 44. *Let* $\mathbf{T}_i$ *be a minimum spanning tree rooted in node* $i \in \mathbf{U}$*; then* $\mathbf{T}_i = \mathbf{T}_j$ *for any* $i, j \in \mathbf{U}$.

We demonstrated this by direct computation of all possible spanning trees on **U**.

Finding 45. *The St. John's and Inuvik urban areas are isolated in* **GG**$\bigcap$**H** *and* **RNG**$\bigcap$**H**.

Finding 46. **MST**/$\{$*(Inuvik–Wrigley, Halifax–Moncton–St. John's)*$\}$ $\subset$ **H**.

This means that the Canadian highway network is almost optimal (Fig. 7.5d and e).

Intersections of Physarum graphs for principal values of the threshold θ with the Gabriel graph and the minimum spanning tree are shown in Fig. 7.6.

Finding 47. **MST** $\subset$ **P**(0).

Physarum graphs for low values of θ include an 'ideal' acyclic spanning network **MST**. This somehow characterises a good quality of a slime mould approximation of a transport network. The minimum spanning tree is not included in the high-threshold Physarum graph $\mathbf{P}(\frac{17}{23})$. However, there is a 'strong' component of **MST** which is included in the high-threshold Physarum graph (Fig. 7.6). The strong component is a tree rooted in the Toronto area. The tree's stem is Toronto–Sudbury–Thunder Bay–Winnipeg–Saskatoon–Regina–Edmonton. It has two end branches: Edmonton–Calgary–Vancouver area and Edmonton–Yellowknife–Wrigley–Inuvik.

7.4 Contamination from Bruce nuclear power station

To imitate propagating contamination, we placed a crystal of sea salt in the site of the Bruce nuclear power station (eastern shore of Lake Huron, near Tiverton, Ontario). We studied plasmodium's response approx. 24 h after initiation of contamination. During 24 h the contamination zone expands as far as Winnipeg and Thompson in the west and St. John's in the east. In some cases the contamination spreads to cover the area Saskatoon–Moncton. In a few experiments the plasmodium colony occupying St. John's remains unaffected.

Finding 48. *In response to contamination propagating from the Bruce nuclear power station, the plasmodium of* P. polycephalum *takes one or more of the following actions: migrates outside Canada, enhances the transport network outside the contaminated zone and sprouts indiscriminately from urban areas and transport links.*

The plasmodium's reactions are illustrated in Figs. 7.7 and 7.8. Four types of responses are observed in laboratory experiments.

- Plasmodium migrates outside Canada (Fig. 7.7a). Typical waves of migration are from Nunavut towards Baffin Bay and Greenland and from British Columbia towards Washington and Oregon in the USA. Due to the growth

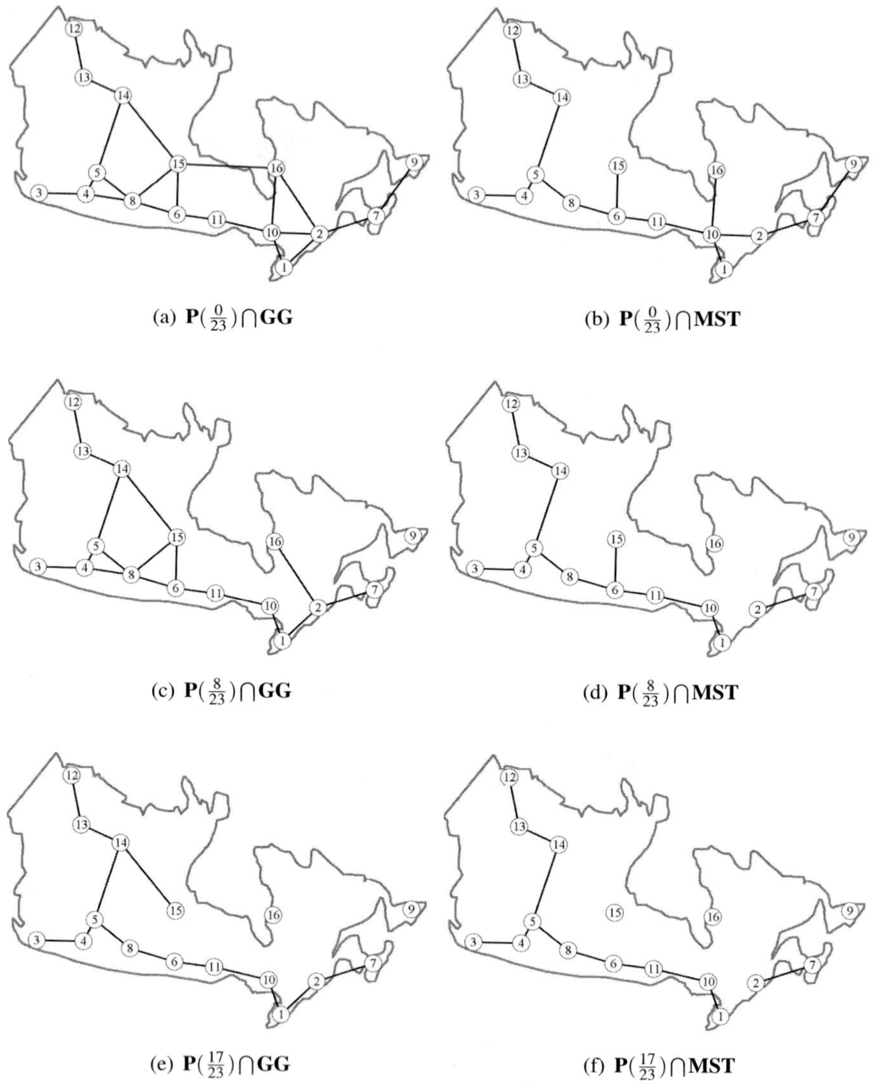

(a) $\mathbf{P}(\frac{0}{23}) \cap \mathbf{GG}$ (b) $\mathbf{P}(\frac{0}{23}) \cap \mathbf{MST}$

(c) $\mathbf{P}(\frac{8}{23}) \cap \mathbf{GG}$ (d) $\mathbf{P}(\frac{8}{23}) \cap \mathbf{MST}$

(e) $\mathbf{P}(\frac{17}{23}) \cap \mathbf{GG}$ (f) $\mathbf{P}(\frac{17}{23}) \cap \mathbf{MST}$

Fig. 7.6 Intersections of (a, c, e) Gabriel graph **GG** and (b, d, f) minimum spanning tree **MST** with Physarum graph $\mathbf{P}(\theta)$ for (a, b) $\theta = 0$, (c, d) $\theta = \frac{8}{23}$, (e, f) $\theta = \frac{17}{23}$.

substrate being an agar plate cut in the shape of Canada, the plasmodium ends up on the bare plastic bottom of the Petri dish; therefore, it does not migrate far away from Canada.

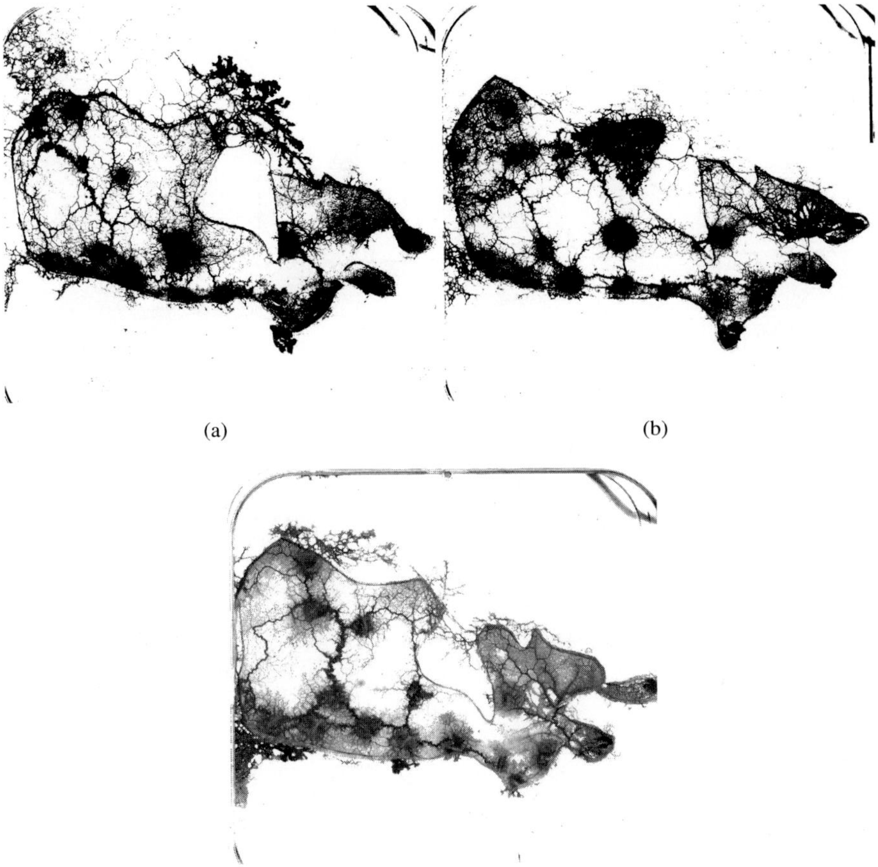

(a) (b)

(c)

Fig. 7.7 Physarum response to contamination: (a) migration outside Canada, (b, c) compensating activation of transport network unaffected by contamination.

- Plasmodium enhances its foraging and colonisation activity in the parts unaffected by contamination (Fig. 7.7b and c), mainly in Alberta, British Columbia, Northern Territories, Yukon and Nunavut. Protoplasmic tubes themselves are often increased in size and intensity of their colours, which reflect increased propagation of cytoplasm inside the tubes. For example, in Fig. 7.7c and d the plasmodium clearly shows hyperactivity in the Nunavut area, with the whole territory covered by spreading plasmodium. Figure 7.7c and d illustrate hyperactivation of transport routes, particu-

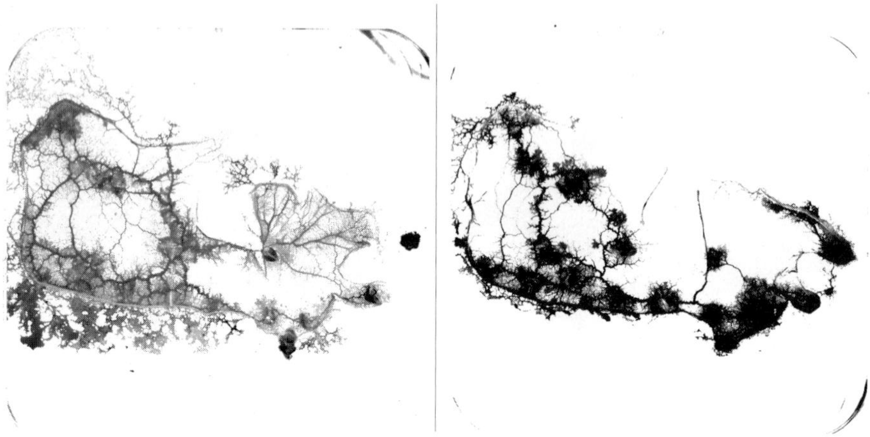

(a) (b)

Fig. 7.8 Physarum sprouting in response to contamination: (a) sprouting from edges, (b) sprouting from nodes.

larly the links Inuvik–Wrigley, Yellowknife–Wrigley, Edmonton–Wrigley, Edmonton–Vancouver area, Saskatoon–Regina–Edmonton and Winnipeg–Saskatoon.

- Plasmodium expands outside oat flakes and also produces processes protruding from the protoplasmic tubes (Fig. 7.8). This a common reaction of plasmodium in response to mechanical damage, e.g. cutting of protoplasmic tubes [Adamatzky (2010)b].

7.5 Summary

To imitate transport networks in Canada, we represented major urban areas and transport nodes with oat flakes, inoculated plasmodium of *P. polycephalum*, allowed the plasmodium to span all oat flakes with its network of protoplasmic tubes and analysed the structure of the protoplasmic network. We found that in over 75% of experiments *P. polycephalum* formed an acyclic transport network consisting of a chain spanning urban areas along the southern boundary of Canada, from Halifax–Moncton to the Vancouver area, with branches Edmonton to Yellowknife to Wrigley to Inuvik and Yellowknife to Thompson. In all experiments slime mould approximates all but the Vancouver to Calgary links of the Canadian highway networks. Both the slime mould and the Canadian highway networks have a strong spanning tree component and thus can be thought of as optimal transport networks. In laboratory experiments with slime mould, we also detailed possible scenarios of transport network restructuring in response to a spreading contamination.

Chapter 8

Slime mould imitates highways in China

Andrew Adamatzky, Xin-She Yang and Yu-Xin Zhao

The expressway network in China, known as the national trunk highway system, has been developed very recently and it is a high-standard transport system planned by the central government. The total length of the expressways has grown rapidly from virtually zero in 1988 to about 74,000 km in 2011. It mainly consists of five longitudinal and seven latitudinal high-standard trunk roads, interlinking with regional highway systems, and its naming system was standardised as the 7918 network [National Bureau of Statistics of China (2012); Li and Shum (2001); Wang et al. (2011)].

It was intended as an optimal system, considering many factors including terrains and landscape. However, a Physarum system usually does not consider any landscape variations. Therefore, a network study using Physarum simulation can show insight into how significant terrestrial variations can affect the routing and optimality of the network system.

We selected most populated major urban areas listed below (see configuration of the areas in Fig. 8.1a, which roughly corresponds to the distribution of population densities by 2010 [National Bureau of Statistics of China (2012)]):

(1) Beijing,
(2) Tianjin,
(3) Shijiazhuang (capital city of Hebei Province),
(4) Taiyuan (capital city of Shanxi Province),
(5) Hohhot (capital city of Inner Mongolian Autonomous Region),
(6) Shenyang (capital city of Liaoning Province),
(7) Changchun (capital city of Jilin Province),

Fig. 8.1 Outline map of continental China with major urban areas **U** shown by encircled numbers.

(8) Harbin (capital city of Heilongjiang Province),
(9) Shanghai,
(10) Nanjing (capital city of Jiangsu Province),
(11) Hangzhou (capital city of Zhejiang Province),
(12) Hefei (capital city of Anhui Province),
(13) Fuzhou (capital city of Fujian Province),
(14) Nanchang (capital city of Jiangxi Province),
(15) Jinan (capital city of Shandong Province),
(16) Zhengzhou (capital city of Henan Province),
(17) Wuhan (capital city of Hubei Province),
(18) Changsha (capital city of Hunan Province),
(19) Guangzhou (capital city of Guangdong Province),
(20) Nanning (capital city of Guangxi Zhuang Autonomous Region),
(21) Haikou (capital city of Hainan Province),
(22) Chongqing,
(23) Chengdu (capital city of Sichuan Province),
(24) Guiyang (capital city of Guizhou Province),
(25) Kunming (capital city of Yunnan Province),
(26) Lhasa (administrative city of Tibet Autonomous Region),
(27) Xi'an (capital city of Shanxi Province),
(28) Lanzhou (capital city of Gansu Province),

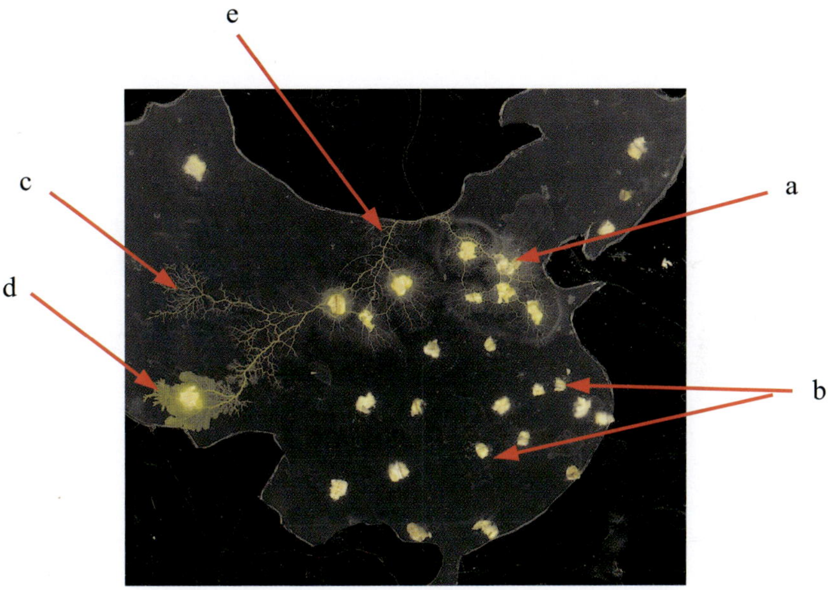

Fig. 8.2 A typical image of slime mould *P. polycephalum* growing on a non-nutrient substrate and connecting oat flakes, which represent major urban areas **U**, by a network of protoplasmic tubes. (a) Site of inoculation, (b) virgin oat flakes, (c) active zone, propagating part of plasmodium in a search for nutrients, (d) oat flake occupied by plasmodium's active zone, (e) protoplasmic tube.

(29) Xining (capital city of Qinghai Province),
(30) Yinchuan (capital city of Ningxia Hui Autonomous Region),
(31) Urumqi (capital city of Xinjiang Uyghur Autonomous Region).

At the beginning of each experiment an oat flake colonised by plasmodium was placed in the Beijing area. We undertook 22 experiments. A typical image of an experimental Petri dish is shown in Fig. 8.2

8.1 From Beijing to Urumqi

In a few hours after inoculation in Beijing, plasmodium recovers from the initial shock, starts exploring its substrate, detects gradients of chemoattractants emitted by virgin oat flakes and propagates towards them. Examples of typical scenarios of plasmodium development are shown in Figs. 8.3, 8.4 and 8.5. Urban areas Tianjin, Shijiazhuang, Taiyuan and Jinan are usually occupied by slime mould in one go, because they are close to Beijing and to each other in the Petri dish scale.

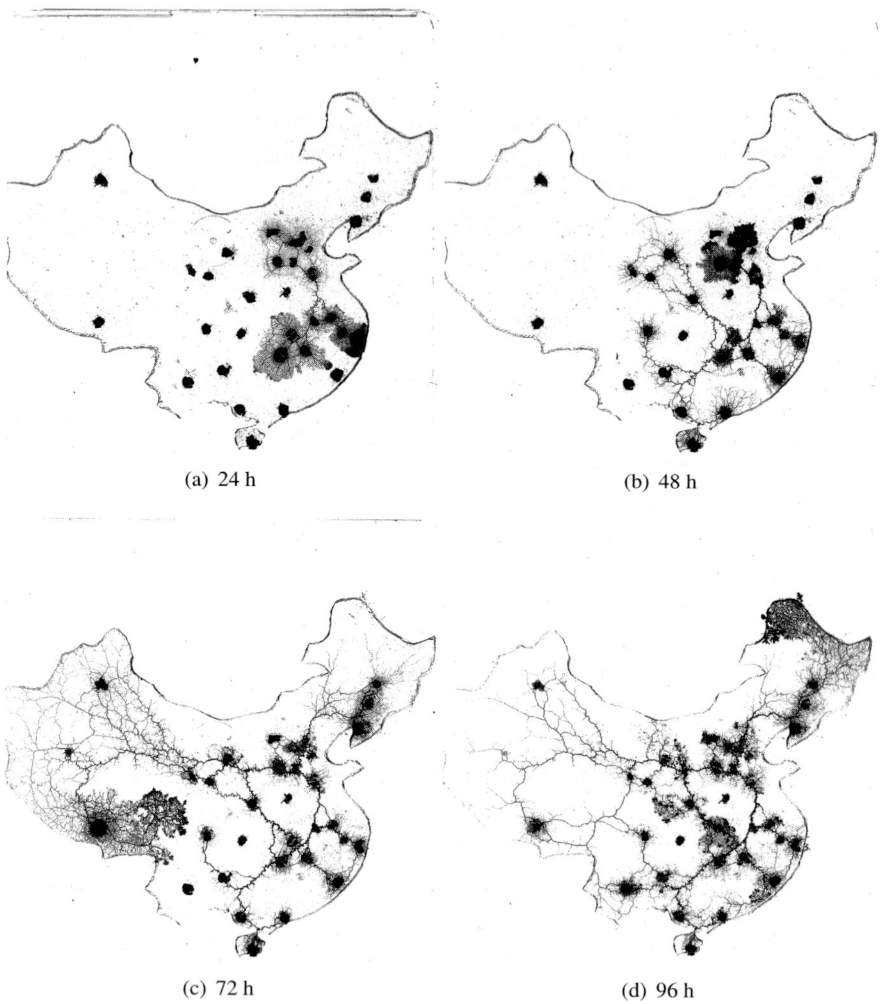

Fig. 8.3 Experimental laboratory example of south-eastern and then north-western propagation of the slime mould. Petri dishes with slime mould were scanned every 24 h.

Further development may differ from one experiment to another.

In the experiment, illustrated in Fig. 8.3, the plasmodium occupies the cluster Beijing, Tianjin, Shijiazhuang, Taiyuan and Jinan, and then colonises Hefei and Nanjing. From Nanjing it propagates to Hangzhou and Shanghai. From Hefei the slime mould propagates to Wuhan and Nanchang. After the urban areas Wuhan and Nanchang are colonised, the plasmodium colonises Changsha (Fig. 8.3a).

Thus, in the first 24 h of the laboratory experiment the plasmodium propagates from Beijing as far as Changsha and connects the urban areas in between with protoplasmic networks.

In the next 24 h, two directions of the colonisation could be observed: south-west and north-west. In the south-western colonisation the plasmodium propagates from Changsha to Guiyang and from Guiyang the slime mould branches to Chengdu and Nanning. It propagates from Nanning to Guangzhou and Haikou afterwards (Fig. 8.3b). The first 48 h of colonisation in the north-western direction goes as follows. The plasmodium propagates from Wuhan to Xi'an. It then links Xi'an with Lanzhou and Yinchuan with its protoplasmic tubes, and propagates from Lanzhou to Xining (Fig. 8.3b).

By the third day of the laboratory experiment, the plasmodium grows from Xining to Urumqi and from Urumqi to Lhasa (Fig. 8.3c). The slime mould propagates from Beijing and Tianjin to Shenyang, Changchun and Harbin. Thus, by 72 h from inoculation the plasmodium occupies almost all major urban areas in China. The urban area Kunming becomes colonised by the plasmodium on the 4th day of the experiment, while Chongqing and Zhengzhou remain uncolonised (Fig. 8.3d).

In the experiment illustrated in Fig. 8.4, the plasmodium colonises the cluster Beijing, Tianjin, Shijiazhuang and Hohhot in the first 24 h (Fig. 8.4a) and then propagates towards and colonises Taiyuan and Hohhot and Shenyang, Changchun and Harbin in the next 24 h (Fig. 8.4b). Only then does it propagate along the east coast towards the south and occupies urban areas as far as Kunming and Chongqing in the west and Haikou in the south (Fig. 8.4c). The slime mould then moves north and colonises Chengdu, Xi'an, Lanzhou, Xining, Yinchuan and finally Urumqi and Lhasa (Fig. 8.4d).

Often the urban areas Shenyang, Changchun and Harbin remain untouched by slime mould, as e.g. in the example shown in Fig. 8.5. In the first 24 h the plasmodium colonises and connects with its protoplasmic tubes the cluster Beijing, Tianjin, Hohhot , Shijiazhuang, Taiyuan and Jinan with Nanjing and Hefei. The plasmodium also occupies Hangzhou and Shanghai, and propagates from Nanjing to Nanchang and Wuhan and then to Changsha. It propagates from Changsha to Chongqing and Chengdu (Fig. 8.5a). In the next 24 h the plasmodium propagates from Chengdu to Lhasa and Kunming and from Changsha to Guangzhou. It propagates from Guangzhou to Haikou and Nanning (Fig. 8.5b). In the last stages of colonisation the plasmodium propagates from Xi'an to Lanzhou, Xining and Yinchuan and then from Xining to Urumqi (Fig. 8.5c). After that the plasmodium ceases to explore any more substrate and remains static for a few days (Fig. 8.5d). The urban areas Shenyang, Changchun and Harbin remain unoccupied.

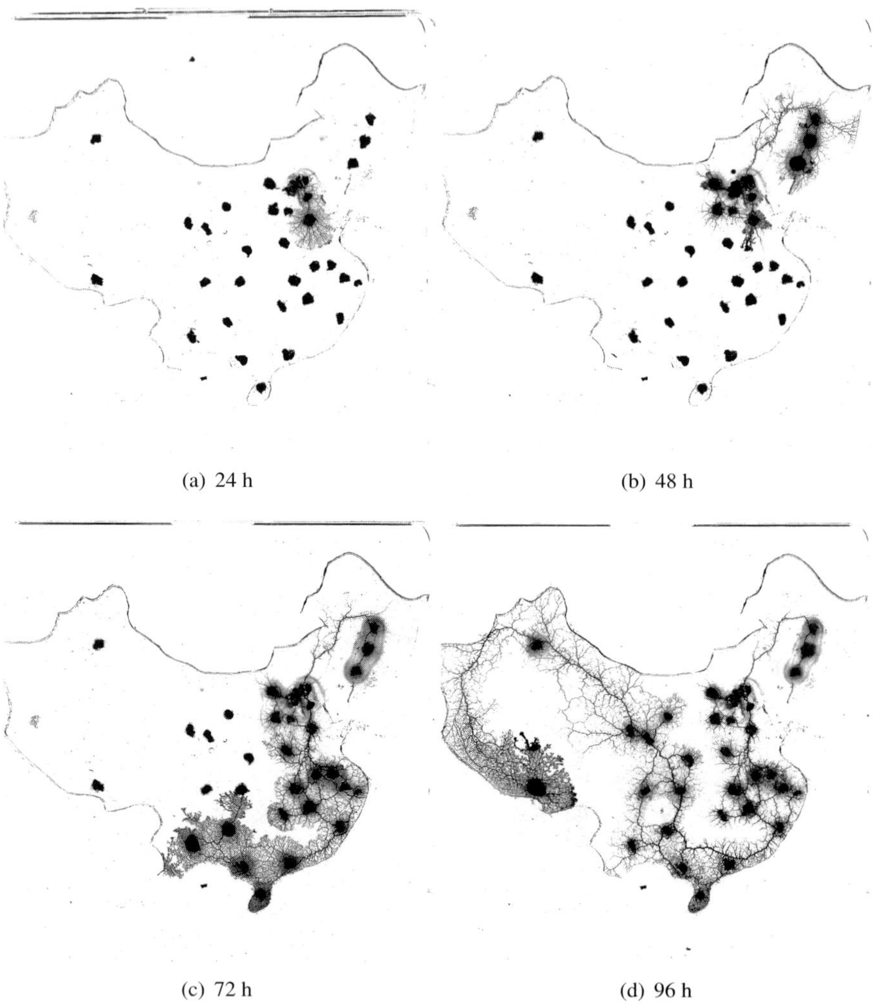

(a) 24 h

(b) 48 h

(c) 72 h

(d) 96 h

Fig. 8.4 Experimental laboratory example of north-eastern then south-eastern and then north-western propagation of the slime mould. Petri dishes with slime mould were scanned every 24 h.

Examples of Physarum graphs for various values of θ are shown in Fig. 8.6 and special cases, mostly for critical values of θ, are shown in Fig. 8.7.

Finding 49. *The Physarum graph parameterised by θ loses its connectivity before becoming planar.*

A Physarum graph $\mathbf{P}(\frac{1}{22})$ is connected and non-planar (Fig. 8.7a). The graph

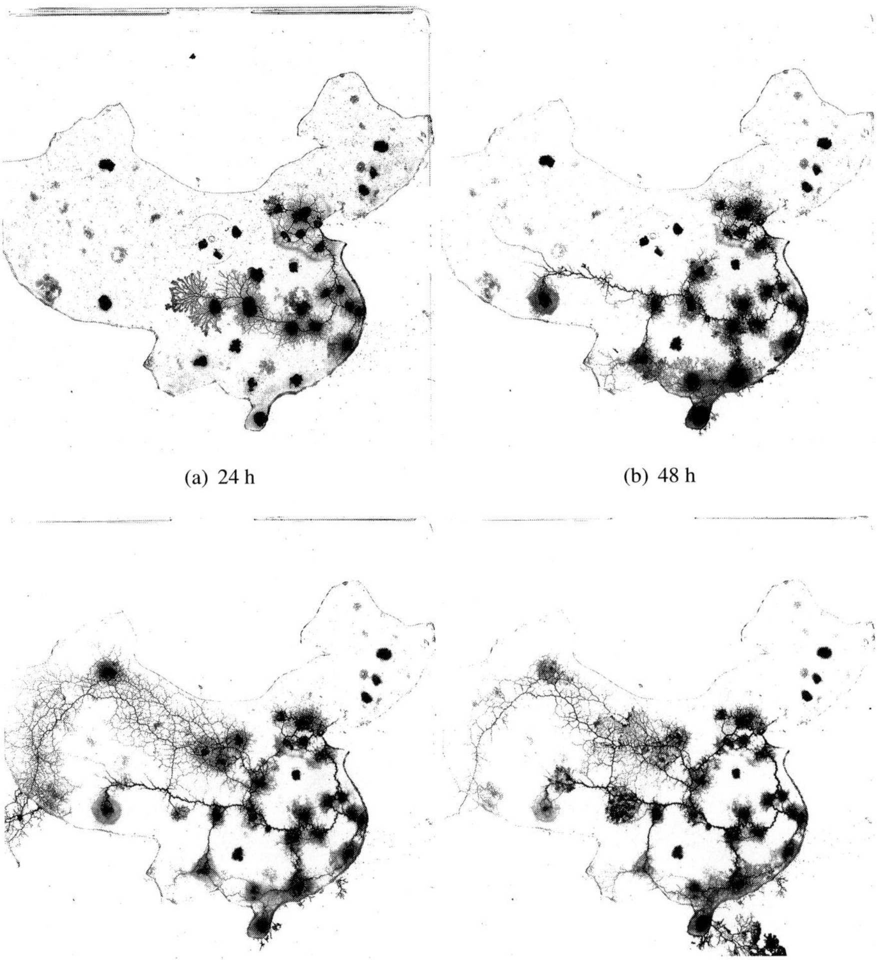

(a) 24 h (b) 48 h

(c) 72 h (d) 96 h

Fig. 8.5 Experimental laboratory example of south-eastern then western, then southern and then north-western propagation of the slime mould. Petri dishes with slime mould were scanned every 24 h.

remains connected until $\theta = \frac{4}{22}$ (Fig. 8.7b). The Physarum graph becomes planar when $\theta = \frac{6}{22}$ and splits into two disconnected components: the chain Shenyang–Changchun–Harbin and the rest of the graph (Fig. 8.7c). The urban areas Shenyang, Changchun and Harbin become isolated (Fig. 8.7d) when θ increases to $\frac{8}{22}$.

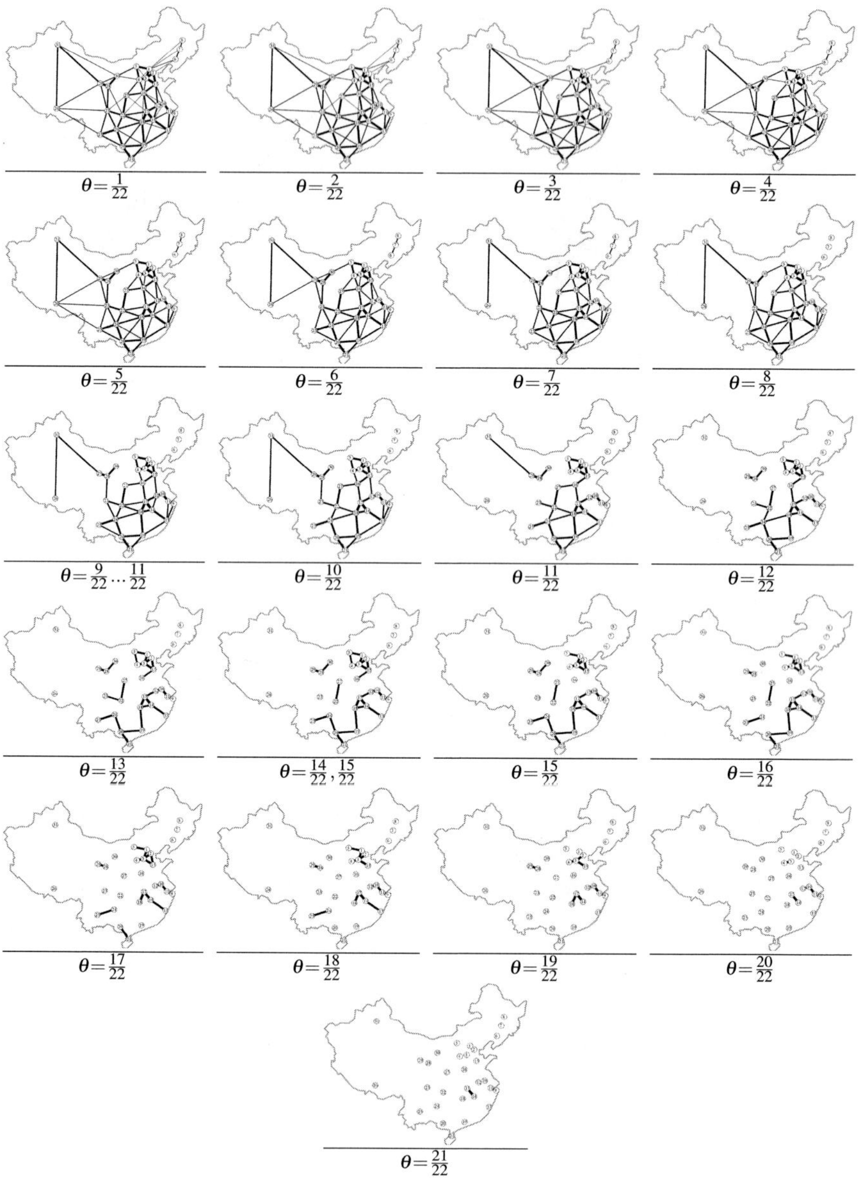

Fig. 8.6 Physarum graphs $\mathbf{P}(\theta)$ for $\theta = \frac{1}{22},\dots,\frac{21}{22}$.

Finding 50. *The urban areas Shenyang, Changchun and Harbin are rarely colonised by the slime mould and in over 80% of experiments remain isolated.*

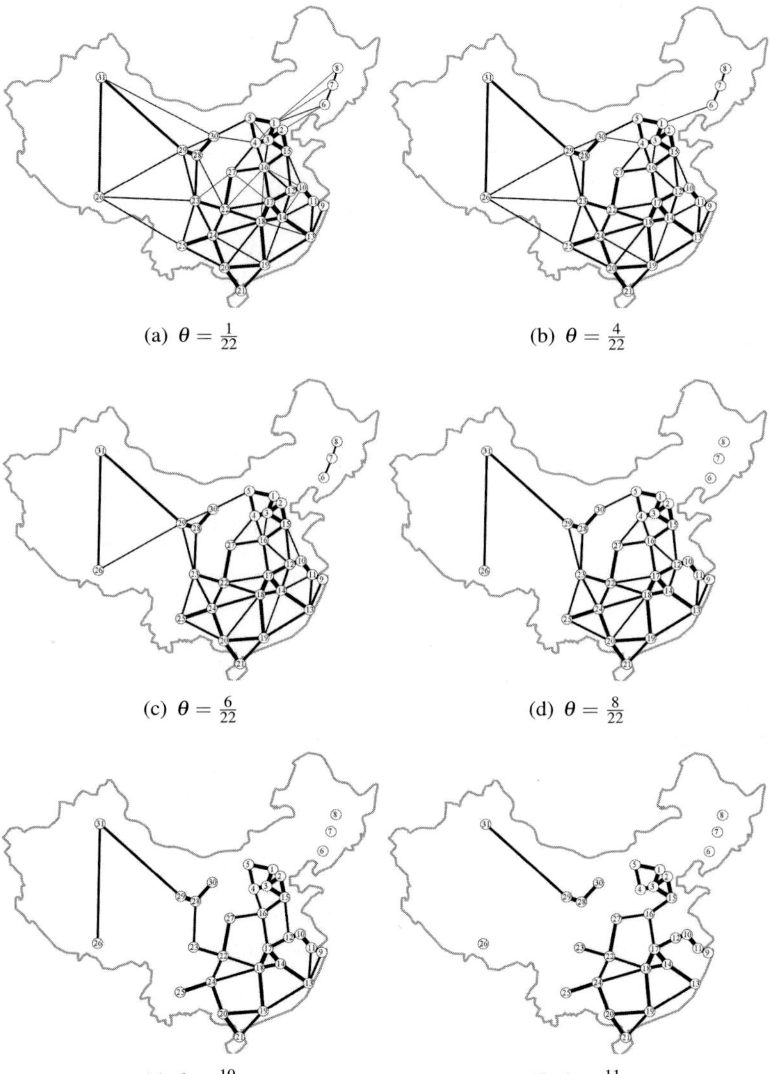

(a) $\theta = \frac{1}{22}$ (b) $\theta = \frac{4}{22}$

(c) $\theta = \frac{6}{22}$ (d) $\theta = \frac{8}{22}$

(e) $\theta = \frac{10}{22}$ (f) $\theta = \frac{11}{22}$

Fig. 8.7 Physarum graphs $\mathbf{P}(\theta)$ for selected values of θ.

The value $\theta = \frac{10}{22}$ is the highest value of the parameter θ before the Physarum graph disintegrates into more than two disconnected components (Fig. 8.7e). The graph $\mathbf{P}(\frac{11}{22})$ consists of

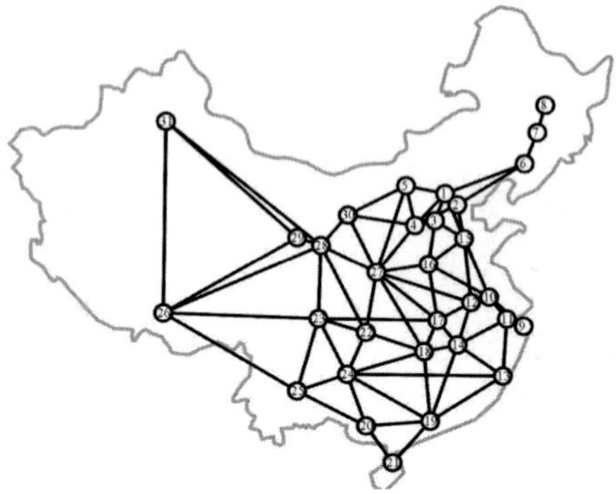

Fig. 8.8 Motorway graph **H** of China.

- the isolated urban areas Lhasa, Shenyang, Changchun and Harbin,
- the chain Yinchuan–Lanzhou–Xining–Urumqi,
- the connected subgraph of south-eastern urban areas (Fig. 8.7f).

We call the graph $\mathbf{P}(\frac{11}{22})$ a strong component of the protoplasmic transport system connecting sites of **U** because edges of this graph are represented in over half of all laboratory trials with the slime mould. Isolated vertices are not included in the strong component.

Finding 51. *The strong component of the transport system built by slime mould* P. polycephalum *on major urban areas of China consists of one chain of four nodes and one planar graph with three leaves and eight cycles.*

Based on the above, we can summarise a list of weakest links.

Finding 52. *The transport links less likely to be represented by the slime mould in laboratory experiments are as follows: Beijing (Tianjin) to Shenyang, Chengdu to Lanzhou, Urumqi to Lhasa.*

8.2 Physarum graph belongs to motorway graph

The Chinese motorway graph is shown in Fig. 8.8.

Finding 53. *Slime mould represents 90% of man-made motorways in China by its protoplasmic tubes at least once.*

As we see in Fig. 8.9a, the only man-made transport links *never* represented by protoplasmic tubes are the following: Lhasa–Lanzhou, Xi'an–Yinchuan, Xi'an–

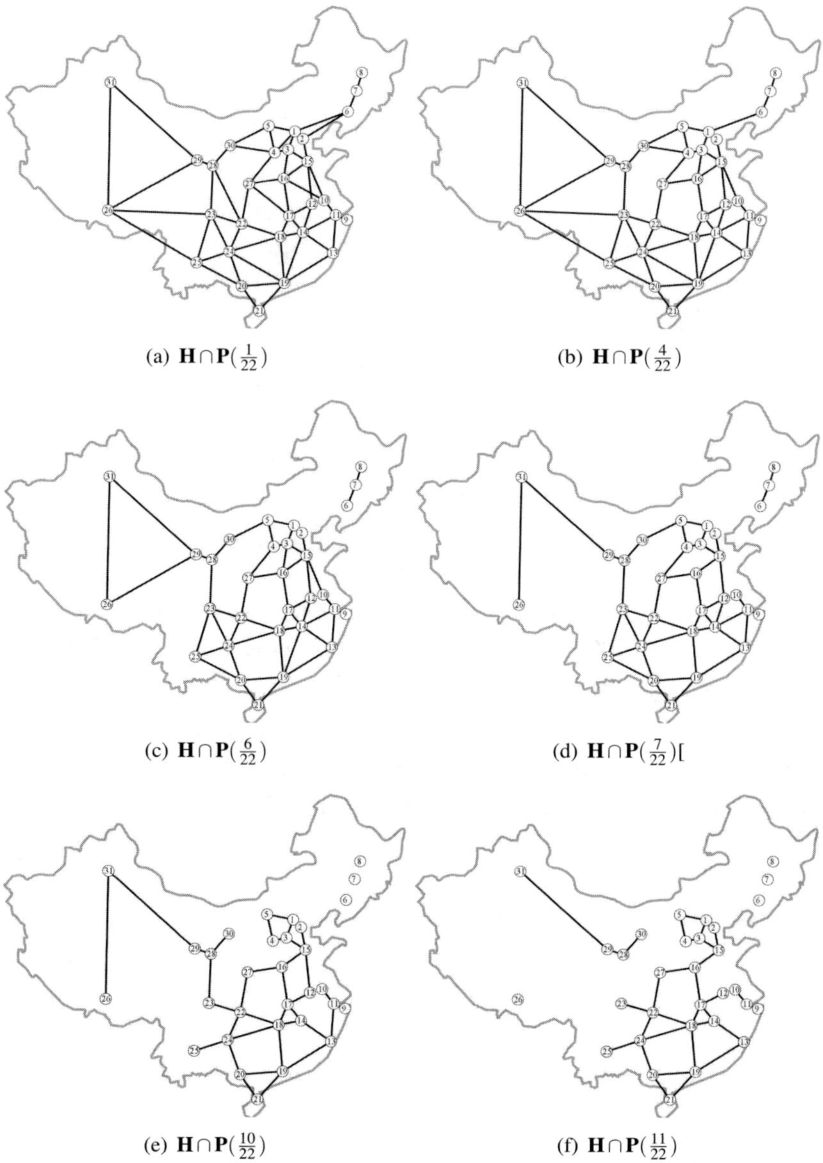

(a) $\mathbf{H} \cap \mathbf{P}(\frac{1}{22})$ (b) $\mathbf{H} \cap \mathbf{P}(\frac{4}{22})$

(c) $\mathbf{H} \cap \mathbf{P}(\frac{6}{22})$ (d) $\mathbf{H} \cap \mathbf{P}(\frac{7}{22})$[

(e) $\mathbf{H} \cap \mathbf{P}(\frac{10}{22})$ (f) $\mathbf{H} \cap \mathbf{P}(\frac{11}{22})$

Fig. 8.9 Intersections of motorway graph $\mathbf{H}$ and Physarum graph $\mathbf{P}(\theta)$ for (a) $\theta = \frac{1}{22}$, (b) $\theta = \frac{4}{22}$, (c) $\theta = \frac{6}{22}$, (d) $\theta = \frac{7}{22}$, (e) $\theta = \frac{10}{22}$, (f) $\theta = \frac{11}{22}$.

Lanzhou, Tianjin–Shijiazhuang, Xi'an–Hefei, Chengdu–Wuhan and Guiyang–Fuzhou.

Finding 54. *The strong component of the slime mould network is a subgraph of the Chinese motorway graph.*

Intersections of the Physarum graph and the motorway graph for selected values of θ are shown in Fig. 8.9. Several transport links built by the slime mould are not represented by the man-made motorway links: Xining–Yinchuan, Xining–Chengdu, Taiyuan–Zhengzhou, Zhengzhou–Hefei, Chongqing–Wuhan, Changsha–Nanning and Tianjin–Shijiazhuang when we consider the intersection of the highest-θ connected Physarum graph and **H** (Fig. 8.9b).

The number of slime-mould links not represented in the intersections decreases with increase of θ; see e.g. Fig. 8.9c and d. For high values of θ, all slime-mould links have their man-made motorway matches (Fig. 8.9e and f).

8.3 Slime and man-made networks vs proximity graphs

Finding 55. *The Chinese motorway network almost includes a minimum spanning tree as its core transport structure.*

A minimum spanning tree built on urban areas **U** of China would be a subgraph of the Chinese motorway graph **H** if there were motorway links connecting Taiyuan to Zhengzhou, Shijiazhuang to Tianjin and Shanghai to Fuzhou (Fig. 8.11c and d). The finding above demonstrates that, in principle, the Chinese motorway transport network has an optimal structure and can be reduced to a minimum spanning core.

Finding 56. *The relative neighbourhood graph and the Gabriel graph are almost subgraphs of the Chinese motorway graph.*

Less than 14% of edges of **RNG** (Fig. 8.11a) and 10% of edges of **GG** are not represented by edges of **H** (Fig. 8.11a). We can conclude that the Chinese motorway network is optimal in the senses of both minimum spanning and cyclic proximity graphs.

Finding 57. *A minimum spanning tree built on* **U** *is almost a subgraph of the Physarum graph* $\mathbf{P}(\frac{4}{22})$.

We are comparing $\mathbf{P}(\frac{4}{22})$ with proximity graphs because this is the Physarum graph with the highest value of θ yet connected (without disconnected components and isolated vertices). Two edges of the minimum spanning tree, rooted in

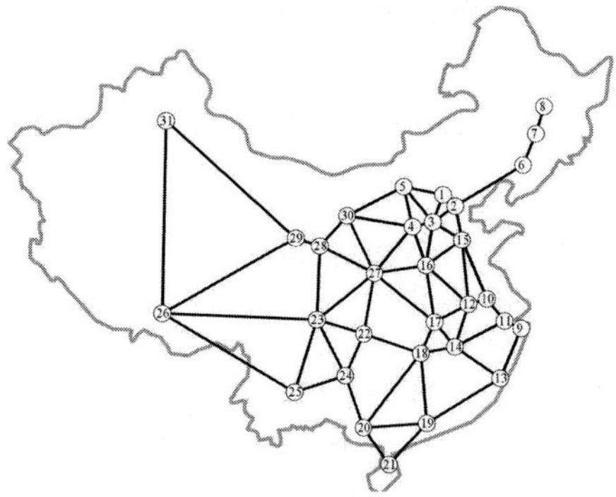

(a) **GG**

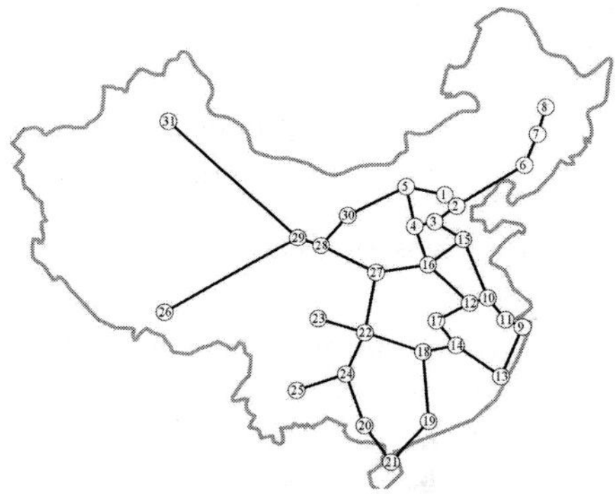

(b) **RNG**

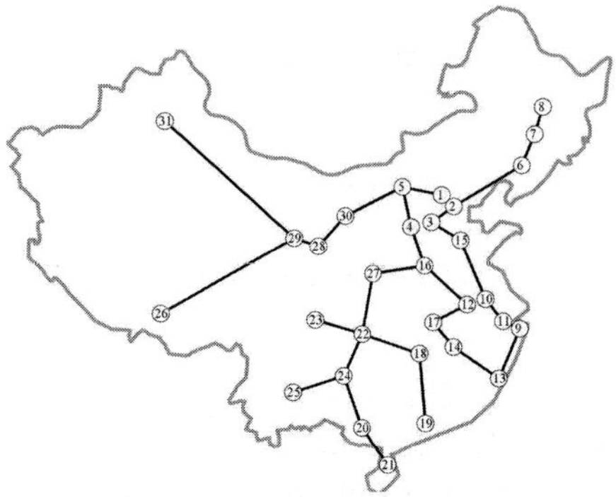

(c) **ST**(Beijing)

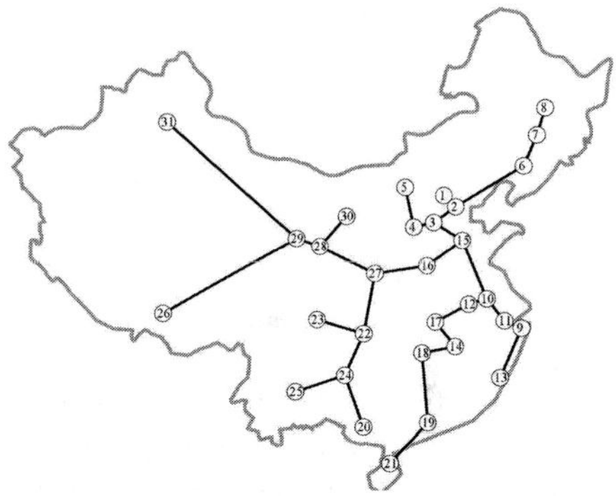

(d) **MST**(Nanjing)

Fig. 8.10 Proximity graphs constructed on sites of **U**: (a) Gabriel graph, (b) relative neighbourhood graph, (c) spanning tree rooted in Beijing, (d) minimum spanning tree rooted in Nanjing.

Nanjing, are not represented by slime mould: Tianjin–Shenyang and Lanzhou–Xi'an (Fig. 8.12d). Only one edge of the spanning tree rooted in Beijing is not represented by an edge of $\mathbf{P}(\frac{4}{22})$: Tianjin–Shenyang (Fig. 8.12c). The graph $\mathbf{P}(\frac{4}{22})$ has an edge Beijing–Shenyang which is just slightly longer than Tianjin–Shenyang; thus, we can assume that edges are interchangeable, and therefore the spanning tree rooted in Beijing is a subgraph of $\mathbf{P}(\frac{4}{22})$.

Finding 58. *If the Physarum graph* $\mathbf{P}(\frac{4}{22})$ *had edges Lanzhou–Xi'an and Tianjin–Shenyang, then it would be a subgraph of the relative neighbourhood graph and the Gabriel graph constructed on* **U**.

This is illustrated in Fig. 8.12a and b.

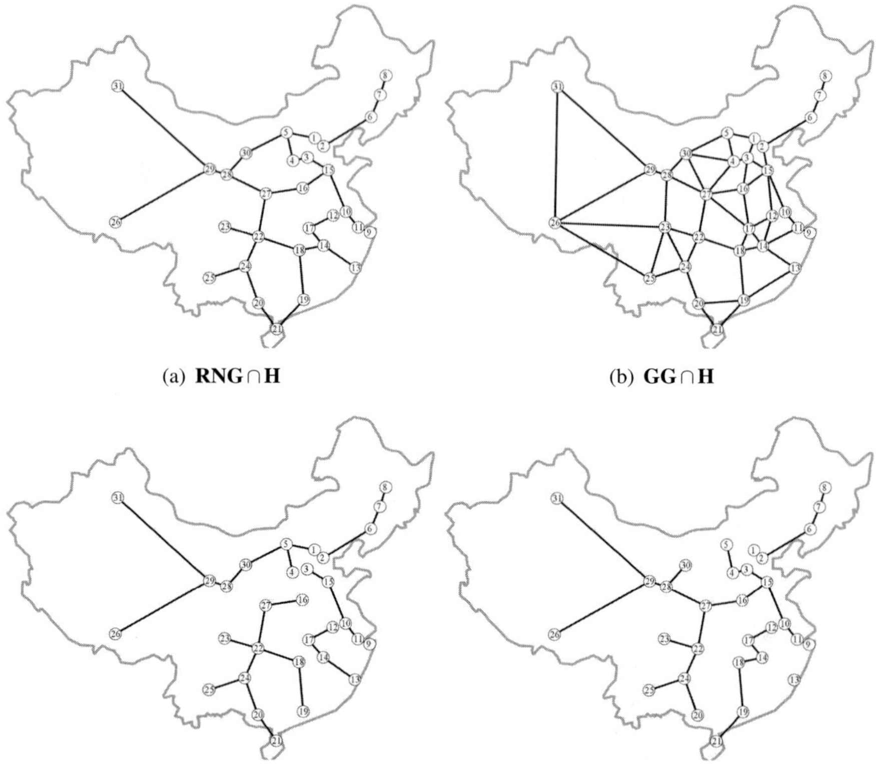

(a) **RNG**∩**H** (b) **GG**∩**H**

(c) **MST**(Beijing)∩**H** (d) **MST**(Nanjing)∩**H**

Fig. 8.11 Intersections of highway graph **H** with (a) relative neighbourhood graph, (b) Gabriel graph, (c) spanning tree rooted in Beijing, (d) spanning tree rooted in Nanjing.

8.4 Summary

Using slime mould *P. polycephalum*, we imitated development of the transport network of China. We found that the slime mould represents 90% of man-made motorways in China by its protoplasmic tubes at least once in laboratory experiments. The strong component of the slime mould network is a subgraph of the Chinese motorway graph. The transport links less likely to be represented by the slime mould in laboratory experiments are as follows: Beijing (Tianjin) to Shenyang, Chengdu to Lanzhou and Urumqi to Lhasa. The urban areas Shenyang, Changchun and Harbin are rarely colonised by the slime mould and in over 80% of experiments remain isolated. We demonstrated that the strong component of the transport system built by slime mould *P. polycephalum* on major urban areas of

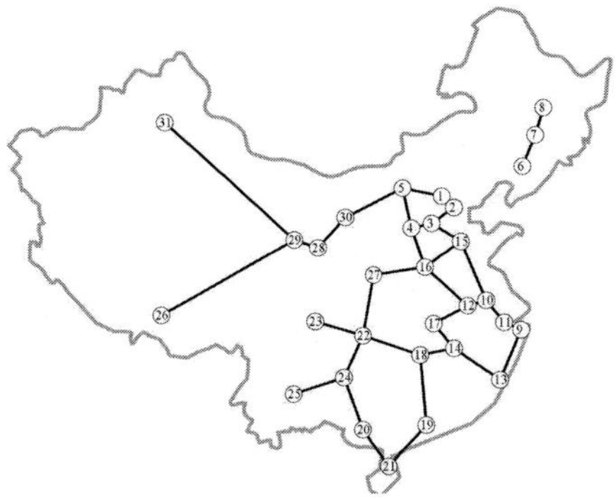

(a) $\mathbf{RNG} \cap \mathbf{P}(\frac{4}{22})$

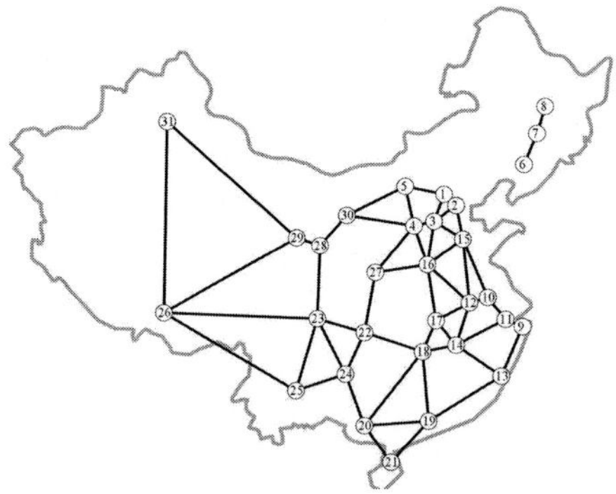

(b) $\mathbf{GG} \cap \mathbf{P}(\frac{4}{22})$

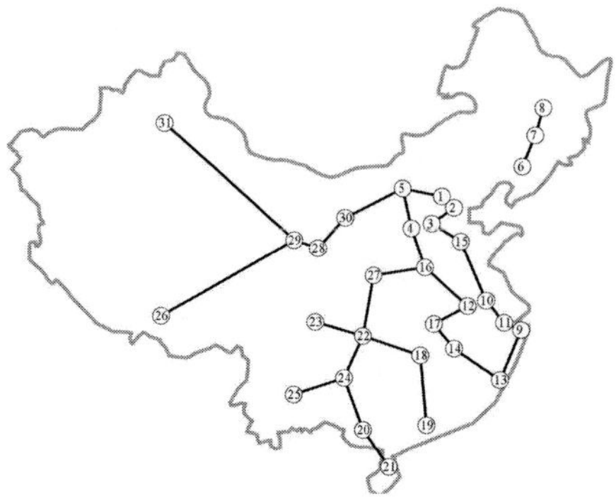

(c) $\mathbf{ST}$(Beijing)$\cap \mathbf{P}(\frac{4}{22})$

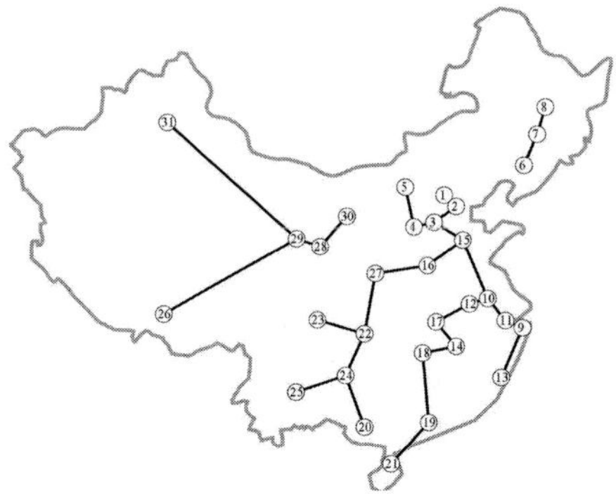

(d) $\mathbf{MST}$(Nanjing)$\cap \mathbf{P}(\frac{4}{22})$

Fig. 8.12 Intersections of Physarum graph $\mathbf{P}(\frac{4}{22})$ with (a) relative neighbourhood graph, (b) Gabriel graph, (c) spanning tree rooted in Beijing, (d) spanning tree rooted in Nanjing.

China consists of one chain of four nodes and one planar graph with three leaves and eight cycles; the planar graph resides on the urban areas in the south-eastern part of China.

With regard to planar proximity graphs, we discovered that the Chinese motorway network is almost a subgraph of a minimum spanning tree, relative neighbourhood graph and Gabriel graph. The same can be said about the Physarum graph, which is a subgraph of these three proximity graphs. These findings demonstrate that not only Chinese motorways are underpinned by engineering logic of proximity but also the motorway topology is supported by the biologic of the slime mould *P. polycephalum*.

Chapter 9

Schlauschleimer auf Autobahnen: the case of Germany

Andrew Adamatzky and Theresa Schubert

In contrast to other countries, motorways in Germany are not just means of transportation but a pivotal part of citizens' mentality. The history of the German motorway network dates back to 1926 when the so-called HaFraBa association was formed to build a motorway linking Hamburg with Basel via Frankfurt [Rothengatter (2005)]. Only a very small part of the project was built until Hitler came to power and reassessed the planning. The whole idea of free-flowing vehicular transport networks — *Reichsautobahnen*, motorways of the empire — became a revolutionary driving force of the miraculous economical and technological progress of prewar Germany and an important instrument for propaganda. After the Second World War the 'Hitler Autobahns' were reinterpreted as a means of reconstruction of West German democracy [Zeller (2007)]. The 'Green Nazi' environmentally oriented approach towards the construction of autobahns and the aesthetics and design of surrounding landscapes was — at different stages of history — propagandised, mythologised, refuted, dismissed [Zeller (2007)] but then partially 'resurrected' in the 1990s and now actively considered by specialists in the context of transport network integrative development in Germany.

German motorways, or 'autobahns', are a unique road system because it is amongst the earliest state-planned transport networks; it was planned precisely and meticulously yet possesses a wide range of quality; and it has the highest traffic load in Europe. Peculiar features of German motorways are based on long-standing traditions of car ownership, the wide range of road quality and the absence of speed limits in some parts of the motorways [Brilon (1994)]. Germany occupies a central geographical position in Europe. This leads to a dramatic increase of traffic on autobahns, which in turns leads to an increase in the number

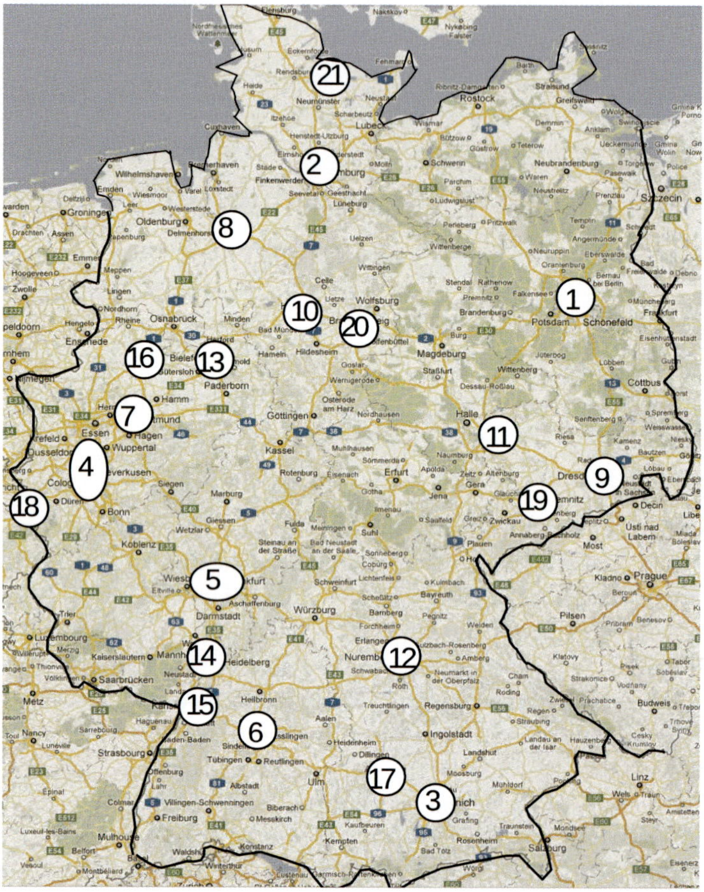

Fig. 9.1 A map of Germany with major urban areas **U** shown by encircled numbers.

of traffic accidents. Expansion of the autobahn network is amongst many possibilities of solving the traffic problem [Garnowski and Manner (2011)]. However, the possibilities for adding new autobahn routes are very limited.

How would the autobahn network develop from scratch under the current configuration of urban areas in Germany? Are autobahns optimal from primitive living creatures' point of view? Does the topology of autobahns satisfy any principles of natural foraging behaviour and fault tolerance? Are there any matches between German transport networks and basic planar proximity graphs? All these are — at least partially — answered by physically imitating the autobahn development in laboratory experiments with slime mould *Physarum polycephalum*.

We selected 21 most populated major urban areas listed below (see configuration of the areas in Fig. 9.1a, which roughly corresponds to the distribution of population densities by 2009 [Statistische (2012)]):

(1) Berlin,
(2) Hamburg,
(3) Munich,
(4) Cologne, including Düsseldorf and Bonn,
(5) Frankfurt, including Wiesbaden,
(6) Stuttgart,
(7) Dortmund area,
(8) Bremen,
(9) Dresden,
(10) Hanover,
(11) Leipzig,
(12) Nuremberg,
(13) Bielefeld,
(14) Mannheim,
(15) Karlsruhe,
(16) Münster,
(17) Augsburg,
(18) Aachen,
(19) Chemnitz,
(20) Braunschweig,
(21) Kiel.

Essen, Duisburg, Bochum, Wuppertal, Gelsenkirchen and Mönchengladbach are not included in the list because they are very close to Cologne and/or Dortmund.

At the beginning of each experiment an oat flake colonised by plasmodium (25–30 mg plasmodial weight) was placed in the Berlin area. We undertook 22 experiments. A typical image of the experimental Petri dish with a Germany-shaped gel plate colonised by *P. polycephalum* is shown in Fig. 9.2.

The autobahn graph is planar (Fig. 9.3).

9.1 Germany colonised

Being inoculated in Berlin, slime mould *P. polycephalum* develops by one of two scenarios: anticlockwise propagation along boundaries of Germany, in the direc-

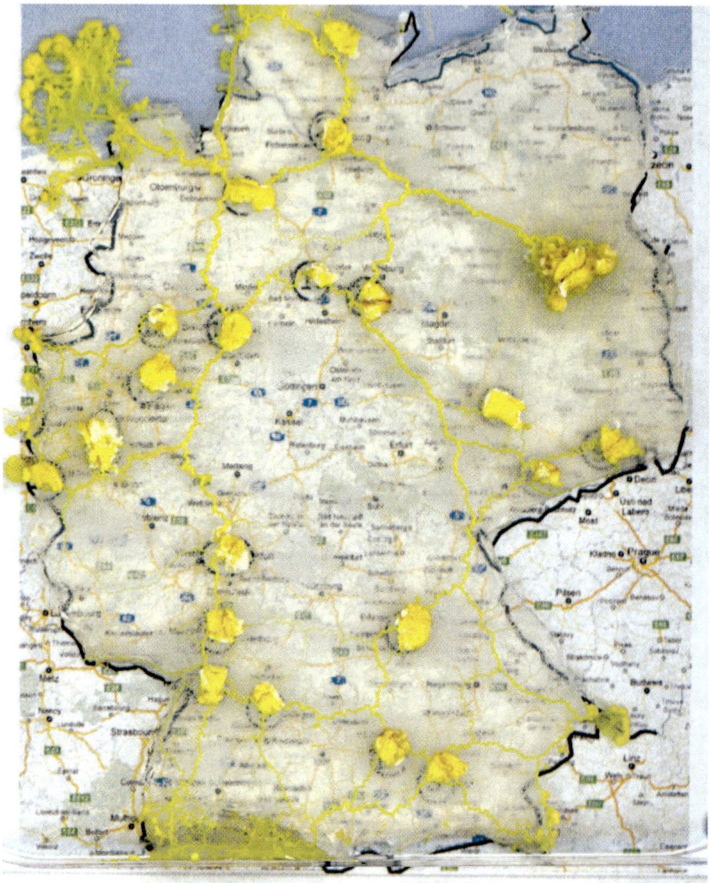

Fig. 9.2 A typical image of slime mould *P. polycephalum* growing on a non-nutrient substrate and connecting oat flakes, which represent major urban areas **U**, by a network of protoplasmic tubes.

tions west–south–east–north; and clockwise propagation, in the directions south–east–north–west. In 13 of 22 experiments plasmodium propagates anticlockwise, in nine experiments clockwise.

The anticlockwise scenario of colonisation is shown in Fig. 9.4. The plasmodium is inoculated in Berlin. It propagates to and colonises Leipzig and Dresden simultaneously. Then it builds protoplasmic tubes connecting Dresden and Leipzig with Chemnitz (Fig. 9.4a). At the same time, the plasmodium grows from Leipzig to Braunschweig, and then occupies almost all urban areas from Bremen in the north to Augsburg in the south (Fig. 9.4a). By the third day after inoculation the

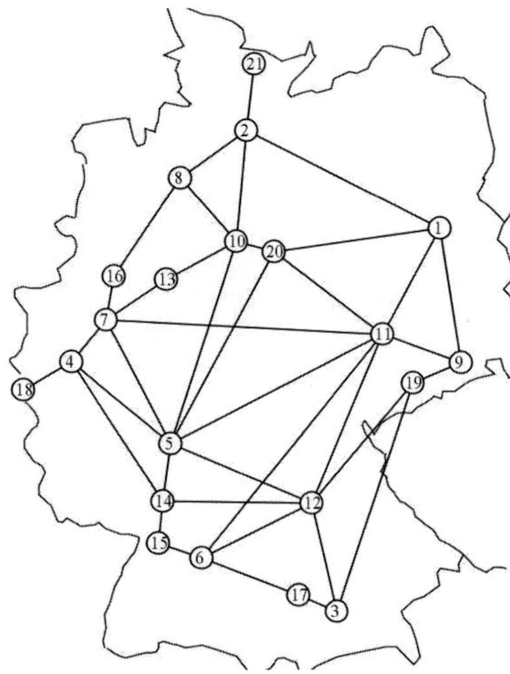

Fig. 9.3 Autobahn graph **H** of Germany.

plasmodium propagates from Augsburg to Munich and Nuremberg, from Bremen to Hamburg and from Hamburg to Kiel (Fig. 9.4b).

The clockwise (south–east–north) scenario is illustrated in Fig. 9.5. The slime mould propagates from Berlin to Leipzig and Dresden simultaneously. Then it colonises Chemnitz, forming protoplasmic tubes Leipzig–Chemnitz and Dresden–Chemnitz. It moves further south and occupies the oat flake corresponding to the urban area Nuremberg (Fig. 9.5a). In its subsequent development, by the 48th hour after inoculation in Berlin, the plasmodium develops a chain of protoplasmic tubes linking Nuremberg, Augsburg and Munich in the south, Stuttgart, Karlsruhe, Mannheim and Frankfurt in the south-west and Aachen, Cologne, Dortmund, Münster and Bielefeld in the west (Fig. 9.5a). The colonisation of the urban areas **U** is completed by the slime mould by 72 h after inoculation, when it develops protoplasmic tubes connecting Bremen and Hanover with Hamburg and Kiel (Fig. 9.5b).

A basic decision-making process implemented by *P. polycephalum* is shown in Fig. 9.6. In the first two days after being inoculated in Berlin, the plasmodium

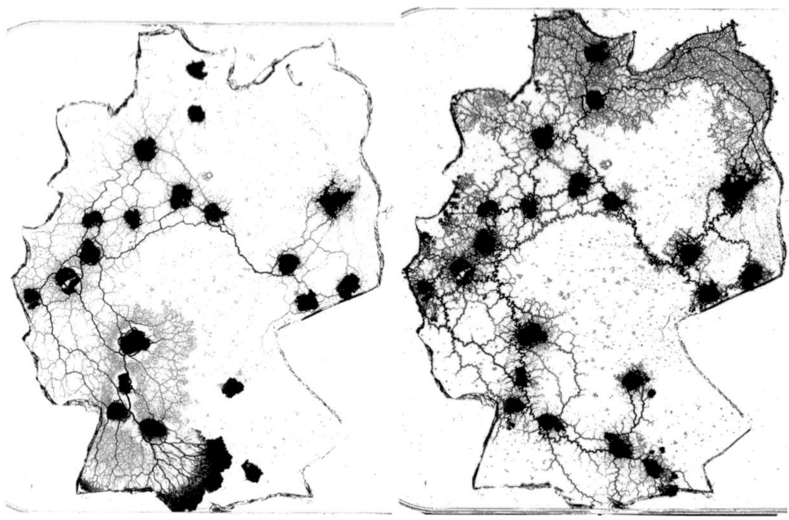

(a) 48 hr (b) 72 hr

Fig. 9.4 Illustration of anticlockwise propagation. Experimental laboratory snapshots of *P. polycephalum* colonising urban areas **U**. The snapshots are taken (a) 48 h and (b) 72 h after inoculating the plasmodium in Berlin.

propagates to Leipzig. It then branches from Leipzig to Braunschweig, Chemnitz and Dresden simultaneously (Fig. 9.6a). At this moment the plasmodium has three options:

(1) propagate from Chemnitz to Nuremberg, Augsburg and Munich and then westward and towards the north,
(2) propagate from Braunschweig westward and then southward,
(3) implement options 1 and 2 in parallel.

In this particular experimental laboratory example, the plasmodium chooses option 2, because the concentration of nutrients (detected via chemoattractants) is higher in the region of Braunschweig than around Chemnitz. See the scheme of propagation in Fig. 9.6d. It propagates from Braunschweig to Hanover and Hamburg; from Hamburg to Kiel; from Hamburg to Bremen; from Bremen to Münster and Bielefeld; and from Münster and Bielefeld to Dortmund (Fig. 9.6b). On the fourth day after inoculation, the plasmodium colonises Cologne and Aachen and builds protoplasmic veins linking Cologne to Frankfurt and Mannheim; Mannheim to Karlsruhe; and Karlsruhe to Stuttgart. The snapshot in Fig. 9.6c shows the 'moment' when the plasmodium has just colonised Munich, Nuremberg

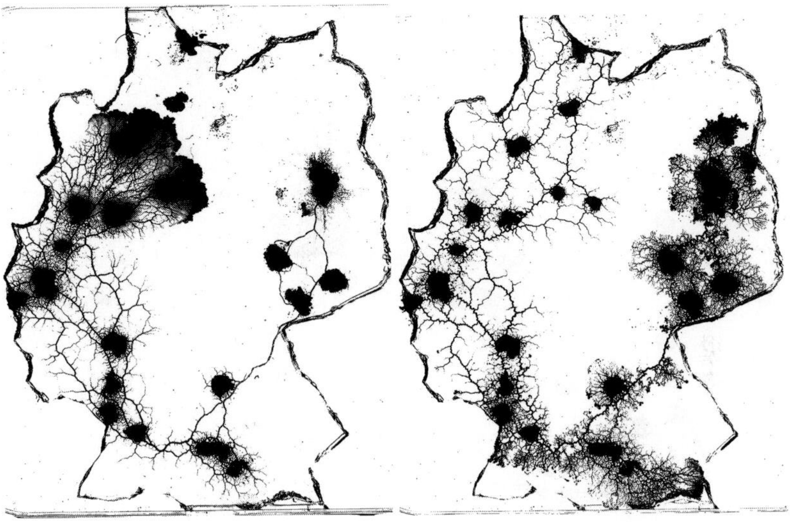

(a) 48 hr (b) 72 hr

Fig. 9.5 Illustration of clockwise propagation. Experimental laboratory snapshots of *P. polycephalum* colonising urban areas **U**. The snapshots are taken (a) 48 h and (b) 72 h after inoculating the plasmodium in Berlin.

and Augsburg and starts to develop growing zones to explore the space around the newly occupied areas. The plasmodium also is detecting shortest ways, and evaluating feasibility of propagation, towards Chemnitz.

9.2 More connections in the west

Examples of Physarum graphs for some values of θ are shown in Fig. 9.7. The Physarum graph $\mathbf{P}(\frac{1}{22})$ is non-planar (Fig. 9.7). With increase of θ — the higher the value of θ in $\mathbf{P}(\theta)$, the more often edges of $\mathbf{P}(\theta)$ appear in laboratory experiments — the Physarum graphs undergo the following transformations.

- $\frac{1}{22} \leq \theta \leq \frac{4}{22}$: graph $\mathbf{P}(\theta)$ is non-planar and connected (Fig. 9.7a).
- $\frac{5}{22} \leq \theta \leq \frac{8}{22}$: graph $\mathbf{P}(\theta)$ is planar and connected (Fig. 9.7b).
- $\theta = \frac{9}{22}$: graph $\mathbf{P}(\theta)$ splits into two disconnected components, one consists of urban areas Berlin, Dresden, Leipzig and Chemnitz, the other includes the remaining areas (Fig. 9.7c).
- $\theta = \frac{11}{22}$: Berlin becomes an isolated vertex (Fig. 9.7d).

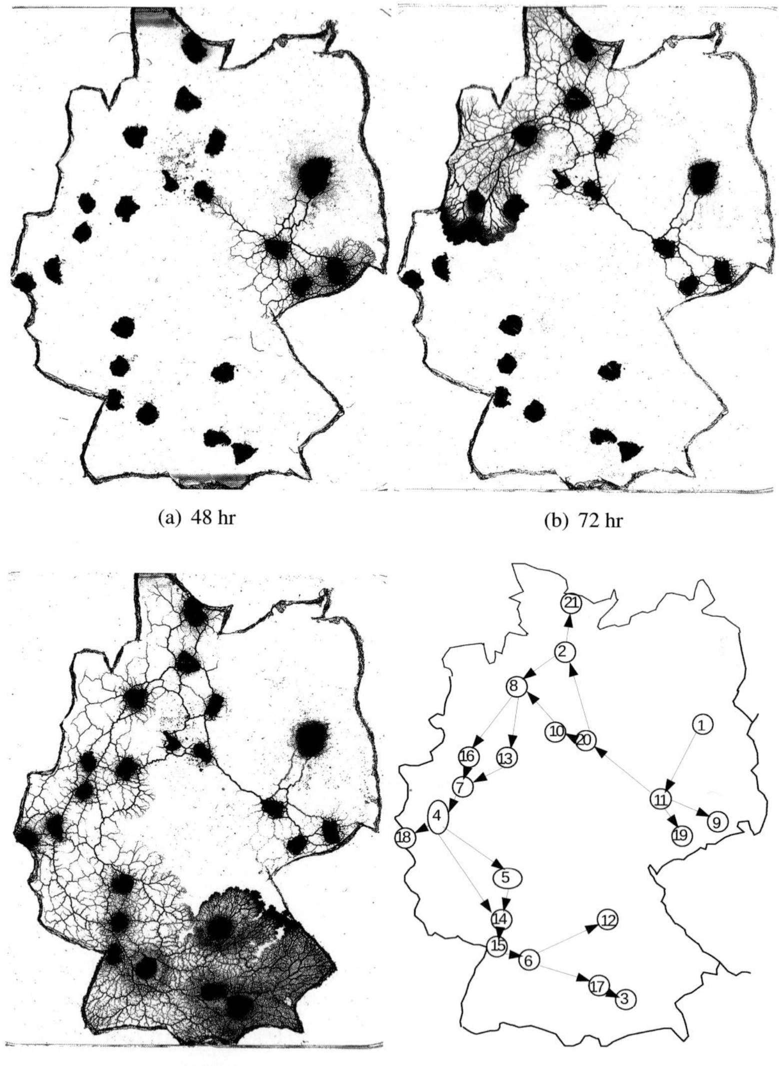

(a) 48 hr (b) 72 hr

(c) 96 hr (d) Scheme

Fig. 9.6 Illustration of decision making by plasmodium. The snapshots are taken (a) 48 h, (b) 72 h and (c) 96 h after inoculating the plasmodium in Berlin.

- $\theta = \frac{14}{22}$: Berlin, Dresden, Leipzig and Chemnitz become isolated vertices (Fig. 9.7e).
- $\theta = \frac{15}{22}$: graph $\mathbf{P}(\theta)$ becomes acyclic (Fig. 9.7f).

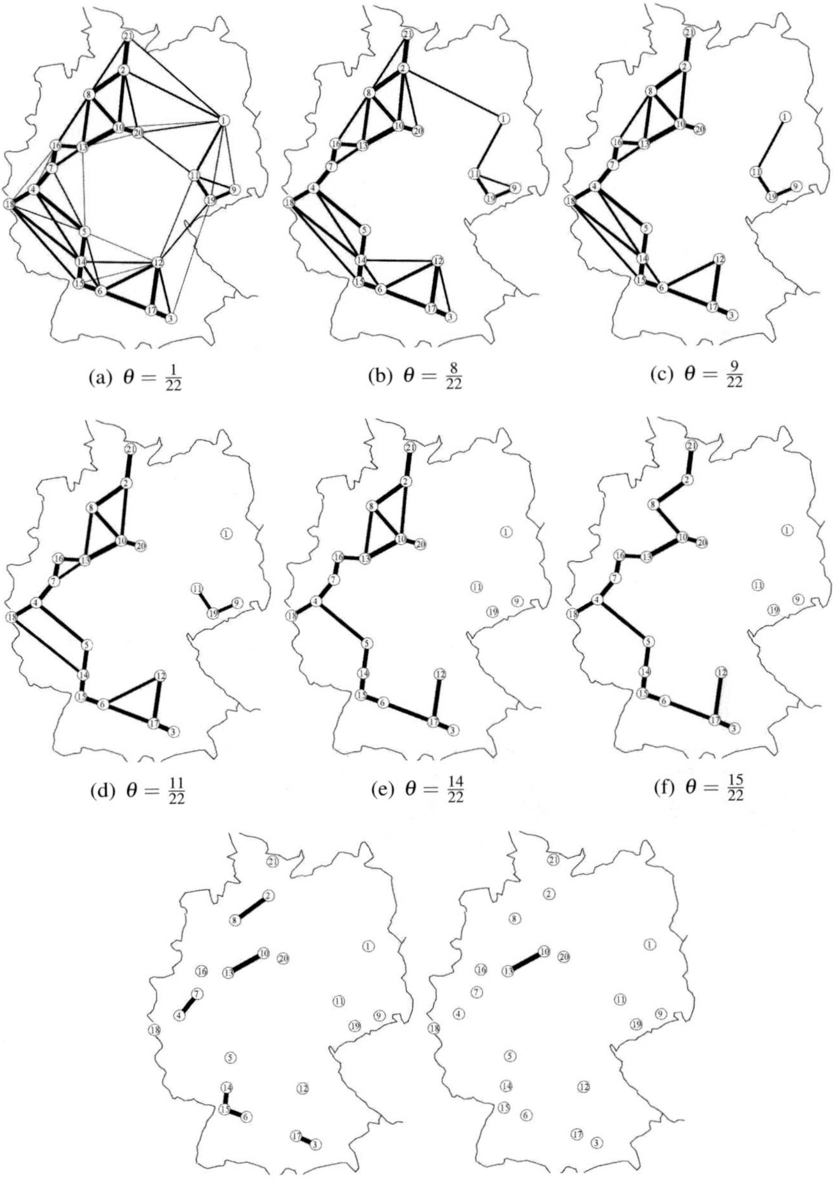

(a) $\theta = \frac{1}{22}$ (b) $\theta = \frac{8}{22}$ (c) $\theta = \frac{9}{22}$

(d) $\theta = \frac{11}{22}$ (e) $\theta = \frac{14}{22}$ (f) $\theta = \frac{15}{22}$

(g) $\theta = \frac{20}{22}$ (h) $\theta = \frac{21}{22}$

Fig. 9.7 Physarum graphs $\mathbf{P}(\theta)$ for selected values of θ.

Note. The slime mould *P. polycephalum* imitates the 1947 separation of Germany into East Germany and West Germany.

See Fig. 9.7c. When we consider only edges represented by protoplasmic tubes in over 41% of laboratory experiments, the Physarum graph becomes split into two disconnected components. The eastern component is a chain of urban areas Berlin–Leipzig–Chemnitz–Dresden lying exactly in the territory of former Eastern Germany (Fig. 9.7c). The construction of the German transport network was interrupted by the Second World War and resumed only in 1953. Over half of the existing autobahns in the West German network had been constructed by 1975 [Rothengatter (2005)], well before the fall of the Berlin Wall. The 'weak' fragments, i.e. those which are removed from Physarum graphs with increase of θ, of the motorway network between West and East Germany are the ones which now link the territories of East and West Germany. Another contributing factor to the weak links exposed in laboratory experiments with *P. polycephalum* could be that car ownership and roads were symbols of freedom and democracy in West Germany, while East Germany was characterised with a low level of car ownership and poorly maintained roads.

Further increase of θ to $\frac{16}{22}$ leads to separation of the graph into the following disconnected components: isolated vertices Berlin, Dresden, Leipzig and Chemnitz, two three-link chains Kiel–Hamburg–Bremen and Braunschweig–Hanover–Bielefeld, one four-link chain Münster–Dortmund–Cologne–Aachen, the six-link chain Frankfurt–Mannheim–Karlsruhe–Stuttgart–Augsburg–Munich and the branch Augsburg–Nuremberg.

Finding 59. *Transport links represented in over 68% of laboratory experiments form a connected component consisting of a chain Kiel–Hamburg–Bremen–Hanover–Bielefeld–Münster–Dortmund–Cologne–Frankfurt–Mannheim–Karlsruhe–Stuttgart–Augsburg–Munich with three branches: Hanover–Braunschweig, Cologne–Aachen and Augsburg–Nuremberg.*

See Fig. 9.7f. The slime mould representation thus reflects an ever-existing imbalance. Historically the west of Germany has always been richer than the east; the highest concentration of steel and coal-mining industry and engineering was situated in the Ruhrgebiet (Essen, Dortmund etc.).

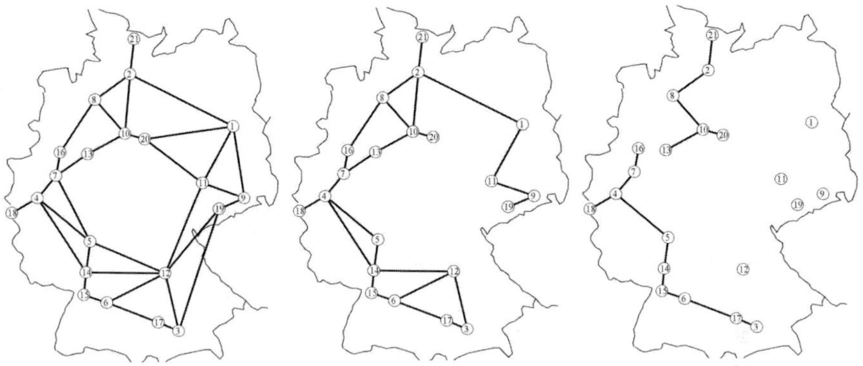

(a) $\mathbf{H} \cap \mathbf{P}(\frac{1}{22})$ (b) $\mathbf{H} \cap \mathbf{P}(\frac{7}{22})$ (c) $\mathbf{H} \cap \mathbf{P}(\frac{15}{22})$

Fig. 9.8 Intersections of autobahn graph $\mathbf{H}$ with Physarum graphs: (a) $\mathbf{P}(\frac{1}{22})$, (b) $\mathbf{P}(\frac{7}{22})$, (c) $\mathbf{P}(\frac{15}{22})$.

9.3 Reichsautobahn rediscovered

Finding 60. *The only autobahn links represented by* P. polycephalum *in over 90% of laboratory trials are Hamburg–Bremen, Hanover–Bielefeld, Cologne–Dortmund, Mannheim–Karlsruhe–Münster and Augsburg–Munich.*

All these links are parts of the so-called Reichsautobahn built by 1940 during Hitler's Germany [Autobahns (1940)]. The only transport link connecting Hanover to Bielefeld is represented by protoplasmic tubes in 95% of all laboratory experiments (Fig. 9.7h and j).

Finding 61. *Autobahn links Frankfurt–Braunschweig, Frankfurt–Hanover, Frankfurt–Leipzig and Stuttgart–Leipzig are never represented by protoplasmic tubes of* P. polycephalum.

As we can see in Fig. 9.8a, the Physarum graph, consisting of edges, which are represented by protoplasmic tubes in at least two experiments, is almost a subgraph of the autobahn graph apart from edges Frankfurt–Braunschweig, Frankfurt–Hanover, Frankfurt–Leipzig and Stuttgart–Leipzig. The intersection of the Physarum graph, whose edges appear in at least eight of 22 experiments, and the autobahn graph remains a connected graph (Fig. 9.8b). The intersection $\mathbf{H} \cap \mathbf{P}(\frac{15}{22})$ consists of:

- five isolated vertices: Berlin, Dresden, Leipzig, Nuremberg and Chemnitz;
- the chain Kiel–Hamburg–Bremen–Hanover–Bielefeld with branch Hanover–Braunschweig;

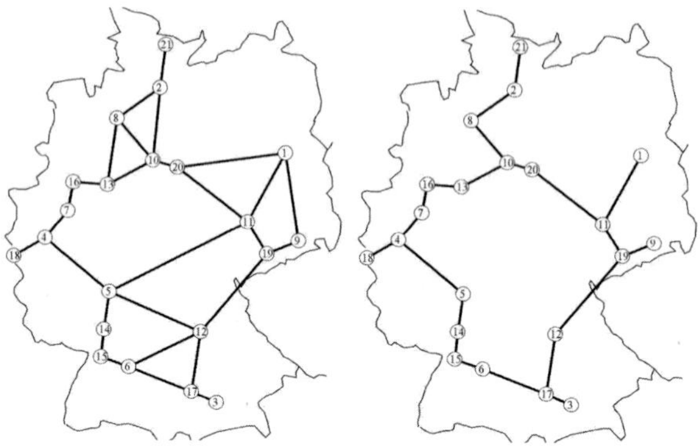

(a) **GG** (b) **RNG**

Fig. 9.9 Proximity graphs constructed on sites of **U**: (a) Gabriel graph, (b) relative neighbourhood graph.

- the chain Münster–Dortmund–Cologne–Frankfurt–Mannheim– Karlsruhe–Stuttgart–Augsburg–Munich with branch Cologne– Aachen (Fig. 9.8c).

9.4 Slimy proximity graphs

The most common proximity graphs constructed on sites of **U** are shown in Fig. 9.9. Seven topologies of spanning trees on **U** are shown in Fig. 9.10. The tree rooted in Berlin is not exactly the minimum tree but only 1.02 times longer.

Finding 62. *Let* $E(\mathbf{G})$ *be the number of edges in a graph* **G** *and* **ST** *be a spanning tree rooted from any node of* **U***; then* $E(\mathbf{H} \cap \mathbf{ST}) = E(\mathbf{ST}) - 3$.

See the illustration in Fig. 9.11.

Finding 63. *If* **RNG** *was a subgraph of* **H** *and if* **H** *had edges Bielefeld–Münster and Nuremberg–Augsburg, then* **RNG** *would have an edge Leipzig–Chemnitz.*

Finding 64. $\mathbf{H} \cap \mathbf{RNG} = \mathbf{H} \cap \mathbf{ST}_{12}$.

The fact that just two edges Bielefeld–Münster and Nuremberg–Augsburg must be removed from **RNG** and only one edge removed from **H** to make the relative neighbourhood graph a subgraph of the autobahn graph demonstrates that the autobahn graph fits well into existing concepts of near-optimal planar proximity graphs.

Finding 65. $\mathbf{ST}(a) \subset \mathbf{P}(\frac{1}{22})$ *for any* $a \in \mathbf{U}$.

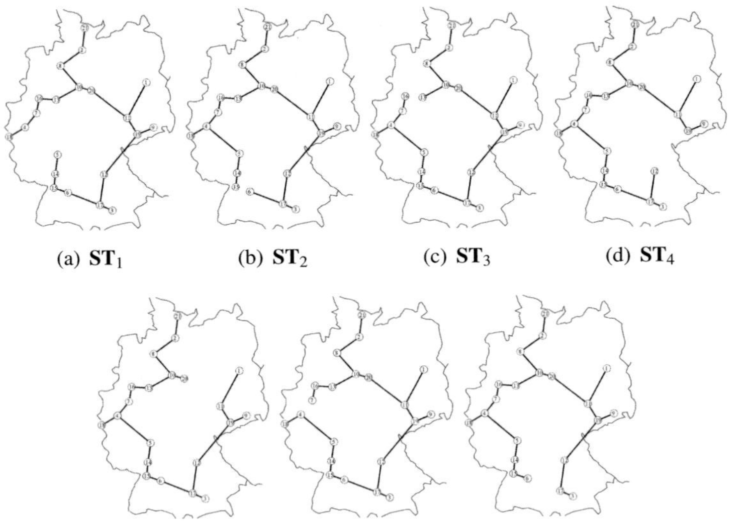

(a) $\mathbf{ST}_1$ (b) $\mathbf{ST}_2$ (c) $\mathbf{ST}_3$ (d) $\mathbf{ST}_4$

(e) $\mathbf{ST}_5$ (f) $\mathbf{ST}_{12}$ (g) $\mathbf{ST}_{13}$

Fig. 9.10 Topological classes of spanning trees, rooted in (a) $\mathbf{ST}_1$: Berlin, Leipzig, Chemnitz, Dresden, $l = 1.02$, (b) $\mathbf{ST}_2$: Hamburg, Kiel, Bremen, Hanover, $l = 1.07$, (c) $\mathbf{ST}_3$: Munich, Augsburg, $l = 1.07$, (d) $\mathbf{ST}_4$: Cologne, Aachen, Dortmund, $l = 1$, (e) $\mathbf{ST}_5$: Frankfurt, Stuttgart, Mannheim, Karlsruhe, $l = 1.01$, (f) $\mathbf{ST}_{12}$: Nuremberg, $l = 1.07$, (g) $\mathbf{ST}_{13}$: Bielefeld, Münster, $l = 1.04$. Lengths l of trees are normalised to a length of the minimum spanning tree (d).

Namely, in laboratory experiments protoplasmic networks almost always maintain a minimum spanning core as an underlying structure of their topology. This can be demonstrated by direct comparison of graphs presented in Figs. 9.7 and 9.10.

Finding 66. $\mathbf{GG} \subset \mathbf{P}(\frac{1}{22})$.

See Fig. 9.13. The Gabriel graph is a supergraph of a relative neighbourhood graph and a minimum spanning tree. Thus, the slime mould *P. polycephalum* imitates, in its foraging patterns, all three basic planar proximity graphs.

9.5 Mass migration due to contamination

To study the effect of a large-scale contamination on the dynamic and structure of transport networks, imitated by *P. polycephalum* protoplasmic networks, we placed a grain of sea salt in the site of the agar plate corresponding to the approximate position of Emsland nuclear power plant [Emsland Nuclear Power Plant

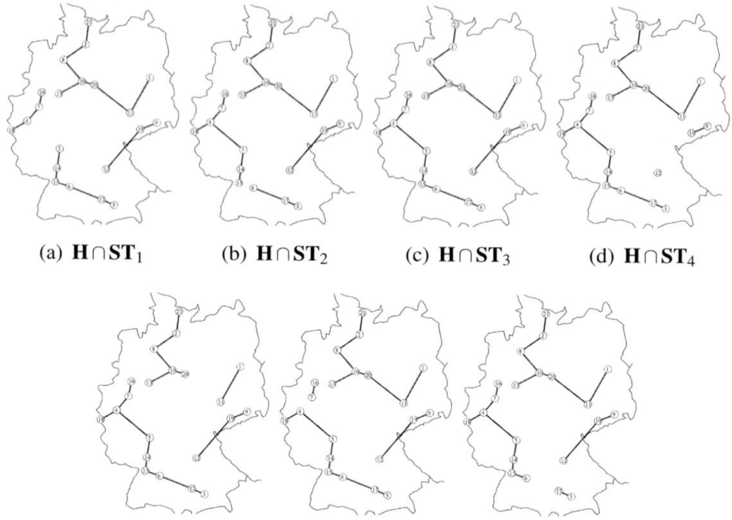

(a) $\mathbf{H} \cap \mathbf{ST}_1$ (b) $\mathbf{H} \cap \mathbf{ST}_2$ (c) $\mathbf{H} \cap \mathbf{ST}_3$ (d) $\mathbf{H} \cap \mathbf{ST}_4$

(e) $\mathbf{H} \cap \mathbf{ST}_5$ (f) $\mathbf{H} \cap \mathbf{ST}_{12}$ (g) $\mathbf{H} \cap \mathbf{ST}_{13}$

Fig. 9.11 Intersection of autobahn graph **H** with spanning trees: (a) $\mathbf{ST}_1$, (b) $\mathbf{ST}_2$, (c) $\mathbf{ST}_3$, (d) $\mathbf{ST}_4$, (e) $\mathbf{ST}_5$, (f) $\mathbf{ST}_{12}$, (ag) $\mathbf{ST}_{13}$.

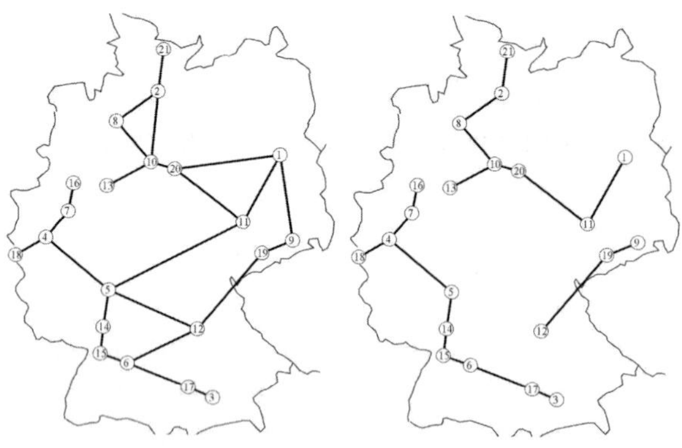

(a) $\mathbf{H} \cap \mathbf{GG}$ (b) $\mathbf{H} \cap \mathbf{RNG}$

Fig. 9.12 Intersections of autobahn graph with (a) Gabriel graph and (b) relative neighbourhood graph.

Identifier (2012)]. The position is marked by a star in Fig. 9.14. As an immediate response to diffusing chloride the plasmodium withdraws from the zone immediate to the epicentre of the contamination. Further response usually fits into two

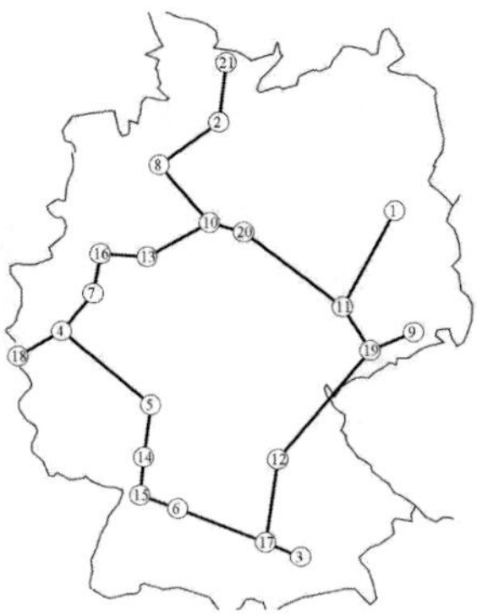

(a) $\mathbf{RNG} \cap \mathbf{P}(\frac{1}{22})$

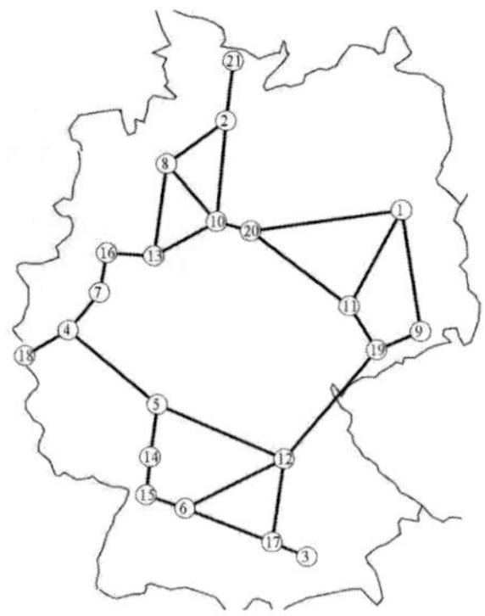

(b) $\mathbf{GG} \cap \mathbf{P}(\frac{1}{22})$

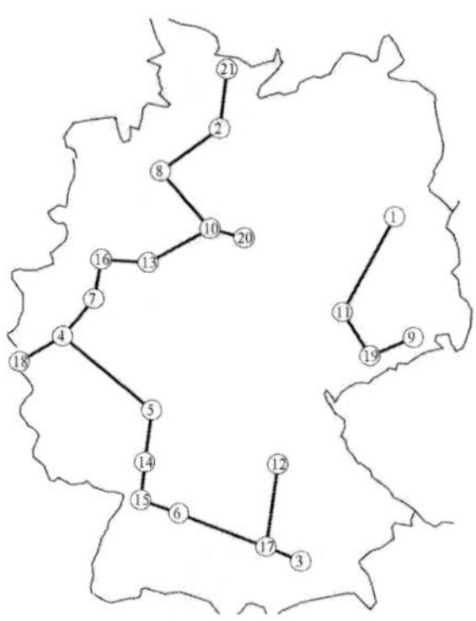

(c) $\mathbf{RNG} \cap \mathbf{P}(\frac{7}{22})$

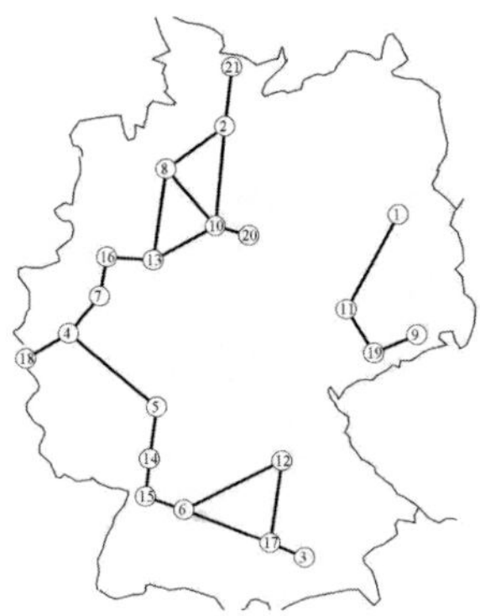

(d) $\mathbf{GG} \cap \mathbf{P}(\frac{7}{22})$

Fig. 9.13 Intersections of Physarum graphs (a, b) $\mathbf{P}(\frac{1}{22})$ and (c, d) $\mathbf{P}(\frac{7}{22})$ with (a, c) relative neighbourhood graph and (b, d) Gabriel graph.

types: hyperactivation of transport as an attempt to deal with the situation and migration away from the contamination, sometimes even beyond the agar plate, as an attempt to completely avoid the contaminated area.

Typical scenarios of slime mould's response to contamination are illustrated in Fig. 9.14. Thus, we observe a substantial increase of foraging activity in Schleswig–Holstein, around Berlin and at the boundary between Thuringia and Upper Franconia.

In the example of Fig. 9.14a, we witness that transport links are substantially enhanced between Berlin and Hamburg, Braunschweig and Hanover and Berlin and Leipzig and Dresden. Also, an auxiliary transport link from Hamburg and Kiel to Berlin emerges along the north-eastern boundary of the country (Fig. 9.14a);

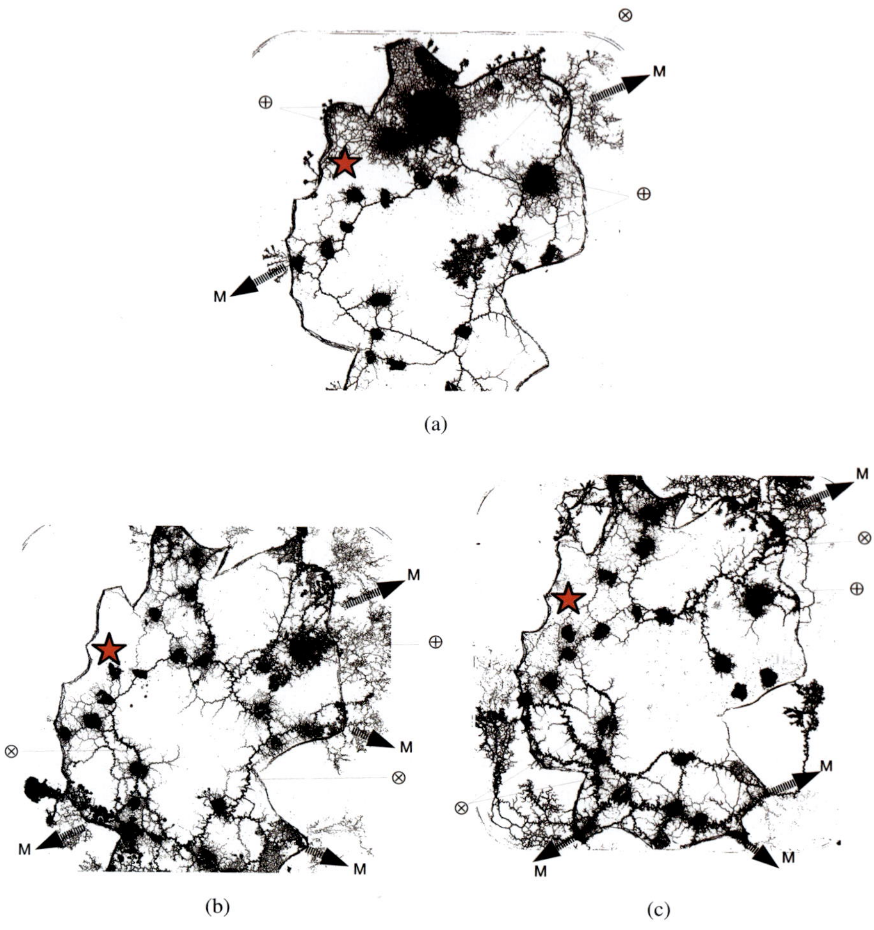

Fig. 9.14 Exemplar snapshots of laboratory experiments on reconfiguration of protoplasmic network in response to propagating contamination. The snapshots are taken 24 h after initiation of contamination. Mass-escape routes are marked by 'M', increase of activity in certain urban areas is labelled by $\oplus$ and increase in traffic along certain routes by $\otimes$.

this link may play a key role in preparation of mass migration from Germany to north-western Poland. Hyperactivity of transport links connecting Chemnitz and Munich, Nuremberg and Stuttgart, Karlsruhe and Mannheim and Cologne and Mannheim is recorded in examples shown in Fig. 9.14b and c.

Mass migration is observed from Aachen to Limburg in Belgium, from Mecklenburg, western Pomerania inwards of north-western Poland (Fig. 9.14a), from the Dessau, Leipzig and Chemnitz areas towards Wroclaw in Poland (Fig. 9.14b) and from the Baden–Württemberg area inwards of south-eastern France and Switzerland (Fig. 9.14b and c).

9.6 Summary

In our experimental laboratory research we represented major urban areas of Germany with oat flakes and inoculated plasmodium of *Physarum polycephalum* in Berlin. The plasmodium propagated from Berlin to nearby urban areas and then to fertile urban areas close to already colonised urban areas. Eventually all urban areas were colonised by the plasmodium. We conducted 22 experiments. We found that autobahn and protoplasmic networks match each other satisfactorily in many integral characteristics and topological indices, especially connectivity, average link length and Randić index. With regard to exact matching between edges of autobahn and Physarum graphs, in 40% of laboratory experiments almost 60% of the autobahn segments are represented by the slime mould's protoplasmic tubes.

We found that only the autobahn links Frankfurt–Braunschweig and Frankfurt–Leipzig are never represented by protoplasmic tubes of *P. polycephalum* in laboratory experiments. The following transport links are imitated by the slime mould in almost over 70% of laboratory experiments: Kiel–Hamburg–Bremen–Hanover–Bielefeld–Münster–Dortmund–Cologne–Frankfurt–Mannheim–Karlsruhe–Stuttgart–Augsburg–Munich, Hanover–Braunschweig, Cologne–Aachen and Augsburg–Nuremberg. The transport links represented by the plasmodium network in over 90% of laboratory experiments are Hamburg–Bremen, Hanover–Bielefeld, Cologne–Dortmund, Mannheim–Karlsruhe–Münster and Augsburg–Munich. The man-made autobahn links represented in over half of laboratory experiments are the chain Kiel–Hamburg–Bremen–Hanover–Bielefeld with the branch Hanover–Braunschweig and the chain Münster–Dortmund–Cologne–Frankfurt–Mannheim–Karlsruhe–Stuttgart–Augsburg–Munich with the branch Cologne–Aachen.

We did not aim to give any conclusive answer of whether autobahns are mathematically optimal and environmentally friendly or not. We attempted to find how 'good' are autobahns from slime mould's point of view, and how the transport network in Germany would develop if it was developed by foraging principles and biomechanics of the slime mould from scratch. We demonstrated that *P. polycephalum* approximates autobahns satisfactorily and thus can be considered as a valuable and user-friendly experimental laboratory tool for imitation of man-made transport networks with amorphous living creatures.

Chapter 10

Vie Physarale: Roman roads with slime mould

Emanuele Strano, Andrew Adamatzky and Jeff Jones

Developing physical, chemical and biological analogies of socioeconomic processes is becoming increasingly popular nowadays because they give rise to new metaphors and uncover unique similarities. Successful examples of such cross-disciplinary fertilisation include the theory of fractal cities [Batty and Longley (1994)], leaf-inspired simulation of street network growth [Runions et al. (2005); Barthelemy and Flammini (2008)], urban theories by Alexander [Alexander (1964)] and Salingaros [Salingaros (2005)], approaches relating urban morphology to biological morphogenesis [Mouson (1997)] and indeed the whole branch of sociophysics [Galam (2012)].

Despite the overwhelming success of the bioinspired simulation and sociophysics, the prevailing majority of publications deal with purely theoretical works and computer simulations. Almost no attempts have been made to undertake experimental laboratory comparisons between very-large-scale socioeconomic developments and spatiotemporal dynamics of chemical or biological systems. This could be explained by difficulties in finding a suitable experimental substrate which does not require sophisticated laboratory equipment and expensive support. A breakthrough came in 2009 when the first experimental results on imitating road networks in the United Kingdom with plasmodium of slime mould *Physarum polycephalum* were published [Adamatzky and Jones (2009)] followed by imitation of rail networks in Japan [Tero et al. (2010)].

We focus on the well-studied and historically understood *Vie Consolari* street system (VC) in the Italian peninsula. The VC is one of the first modern human transportation systems linking cities and certainly the first in Europe. The VC is also a modern regional transportation network at the first stage of its development and it is also possible to trace back its evolution to observe that Rome linked all the important primordial principal cities in the Italian peninsula. This fact is important given that the VC is a transportation network that is not affected by the complex amount of changes derived by technological evolutions in transportation systems and — more generally — by the stratification of historical facts like wars, demography dynamics and natural hazards. It is derived by a top-down and self-organised local organisation given the complete absence of bottom-up technologies in regional planning.

Experimental simulations confirm that *P. polycephalum* matches the street network to a certain degree but we find also that there is an interesting similarity between the colonisation processes in which principal streets have been built; this can suggest very relevant questions regarding self organisation in the urbanisation processes. It is now known that the VC is now mostly overlapped by modern motorways [Davies (2006)] and there is a body of study dedicated to tracking the original street system in all of the Roman Empire [Snead et al. (2009)]. For this study, we are choosing the reconstruction proposed in [Villa (1995)], choosing the street systems it supposed to link the major cities in Italy at the Imperial Age (I Cent. A.D.).

The Vie Consolari is amongst few, if any other, ancient road networks which survived and are still in use in modern times. The Vie Consolari was the regional transportation network between major cities in the Italian peninsula at the time of the Roman Empire, and evolution of the Italian transport system can be traced back and hopefully replayed using the configuration of 'primordial' cities in the Italian peninsula.

We selected 11 cities **U** for laboratory experiments (see configuration of the areas in Fig. 10.1):

(1) Genua/Genova,
(2) Placentia/Piacenza,
(3) Aquileia/Venezia,
(4) Bononia/Bologna,
(5) Florenzia/Firenza,
(6) Ariminum/Rimini,
(7) Roma/Roma,
(8) Capua/Capua,
(9) Venusia/Venusia,
(10) Brundisium/Brindisi,
(11) Rheghium Reggio/Calabria.

To represent the Apennine mountains, impassable for vehicular traffic, we re-

(a)

(b) **H**

Fig. 10.1 Experimental setup: (a) outline map of Italy (from [Villa (1995)]) with Roman cities **U** shown by discs, (b) example of experimental setup.

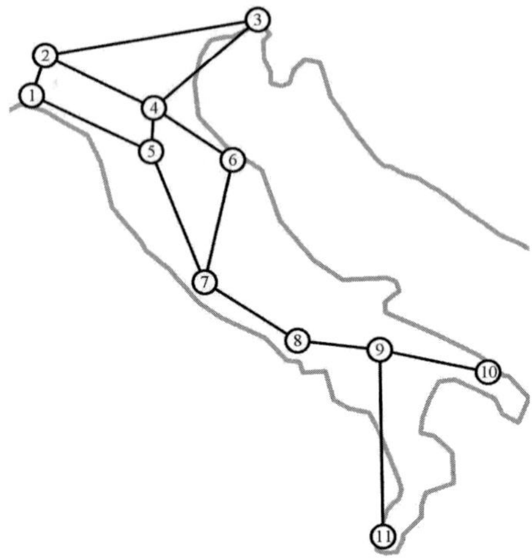

Fig. 10.2 Motorway graph **H**.

moved corresponding parts of the agar plate. At the beginning of each experiment an oat flake colonised by plasmodium was placed in the Rome area. We undertook 28 experiments.

The Roman road graph is planar (Fig. 10.2).

10.1 From Piacentia to Bononia and from Genua to Florenzia are missing

Exemplars of Physarum graphs $\mathbf{P}(\theta)$ extracted from laboratory experiments are shown in Fig. 10.3. We call the Physarum graph $\mathbf{P}(\frac{1}{28})$ weak and the Physarum graph $\mathbf{P}(\frac{9}{28})$ strong. This is because $\mathbf{P}(\frac{1}{28})$ represents protoplasmic links developed in at least one experiment. The graph $\mathbf{P}(\frac{9}{28})$ gives us a more reliable representation because its edges are represented by slime mould in a third of all experiments.

One weak Physarum graph $\mathbf{P}(\frac{1}{28})$ is non-planar due to the edge Bononia to Roma (Fig. 10.3a). The graphs become planar for $\theta \geq 2$ (Fig. 10.3b) and remain connected until $\theta = 9$ (Fig. 10.3c). The Physarum graph $\mathbf{P}(\frac{9}{28})$ is the last connected graph in the series $\mathbf{P}(\theta)$, $\theta = \frac{1}{28}, \frac{2}{28}, \ldots, 1$. When θ becomes $\frac{10}{28}$, the Physarum graphs split into a northern component: the transport link between Genua and Placentia (1 and 2) becomes separated from the rest of the network (Fig. 10.3d) and Aquileia (3) becomes an isolated vertex.

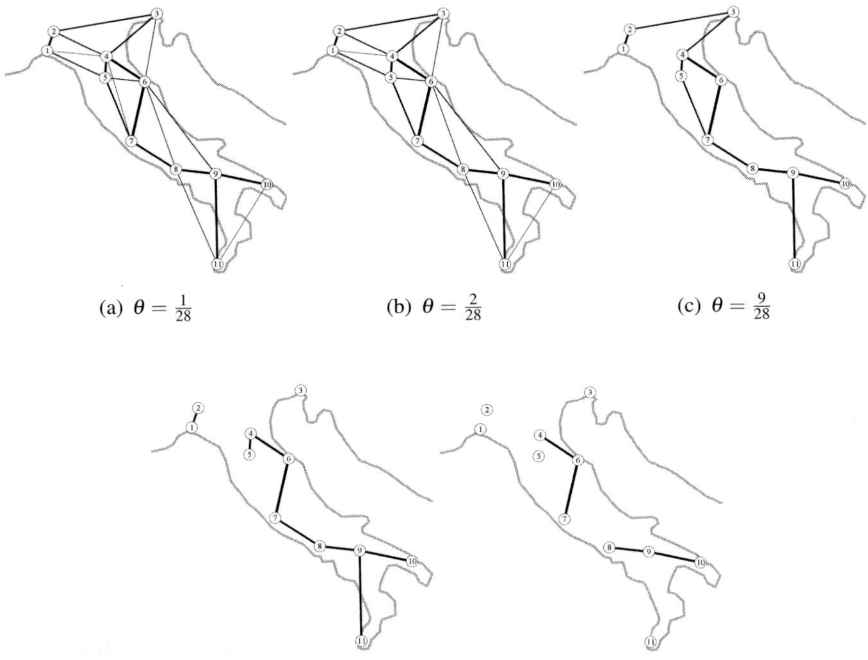

(a) $\theta = \frac{1}{28}$ (b) $\theta = \frac{2}{28}$ (c) $\theta = \frac{9}{28}$

(d) $\theta = \frac{10}{28}$ (e) $\theta = \frac{14}{28}$

Fig. 10.3 Physarum graphs $\mathbf{P}(\theta)$ for selected values of θ.

The structure of the Physarum graph $\mathbf{P}(\theta)$ remains unchanged for $\theta = \frac{10}{28}, \frac{12}{28}$ and $\frac{13}{28}$. However, when $\theta = \frac{14}{28}$, half of the cities become isolated vertices: Genua, Aquileia, Piacentia, Filorenzia and Reghium Reggio, and only two two-node transport chains are sustained (Fig. 10.3e). The transport chains Capua–Venusia–Brundisium and Bononia–Ariminu–Roma (segments of Via Aemelia and Via Flamimia) are represented in half of the experiments.

Finding 67. *The weak Physarum graph includes the road graph. The strong Physarum graph is included in the road graph.*

This is because $\mathbf{P}(\frac{1}{28}) \cap \mathbf{H} = \mathbf{H}$ (compare Figs. 10.4a and 10.2) and $\mathbf{P}(\frac{9}{28}) \cap \mathbf{H} = \mathbf{P}(\frac{9}{28})$ (compare Figs. 10.4b and 10.2). The strong Physarum graph would provide a perfect approximation of the road graph $\mathbf{H}$ if roads from Piacentia to Bononia and from Genua to Florenzia were represented by protoplasmic tubes in over half of laboratory experiments.

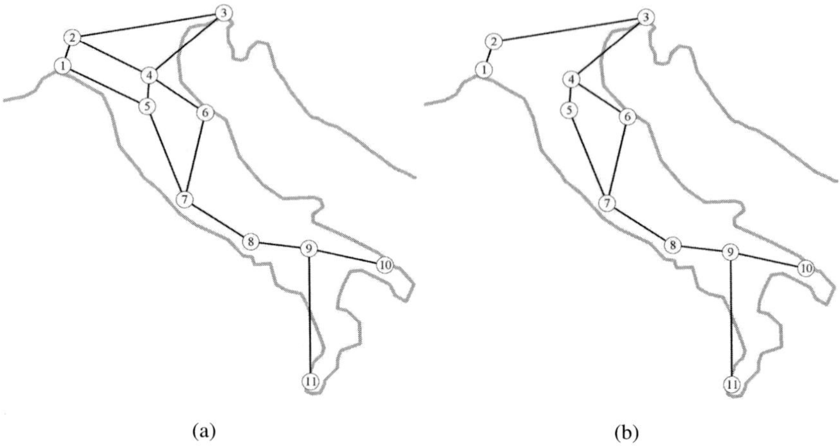

Fig. 10.4 Intersections of road graph **H** and Physarum graphs (a) $\mathbf{P}(\frac{1}{28})$ and (b) $\mathbf{P}(\frac{9}{28})$.

The road graph **H** and the Physarum graphs $\mathbf{P}(\theta)$, $\theta \geq \frac{2}{28}$, are planar graphs. Roman builders aimed to provide the shortest transport connection between neighbouring cities while slime mould followed gradients of chemoattractants emitted from oat flakes, representing the cities. Therefore, it would be reasonable to check what is the place of the graphs **H** and $\mathbf{P}(\theta)$ in the family of planar proximity graphs.

Finding 68. $\mathbf{MST}(\mathbf{U}) = \mathbf{RNG}(\mathbf{U})$.

This follows directly from Fig. 10.5a and c. Equality of spanning and relative neighbourhood graphs may be attributed to the fact that the Italian peninsula is rather narrow yet long and therefore planar proximity graphs built on **U** are heavily constrained.

Finding 69. $\mathbf{MST} \subseteq \mathbf{RNG} \subseteq \mathbf{GG} \subseteq \mathbf{P}(\frac{1}{28})$.

The weak Physarum graph is a subgraph of the three principal proximity graphs. This follows directly from Figs. 10.5 and 10.3a.

Physarum graph $\mathbf{P}(\frac{1}{28})$ does not include **MST** because this Physarum graph does not represent the transport links Placentia–Bononia and Genua–Florenzia.

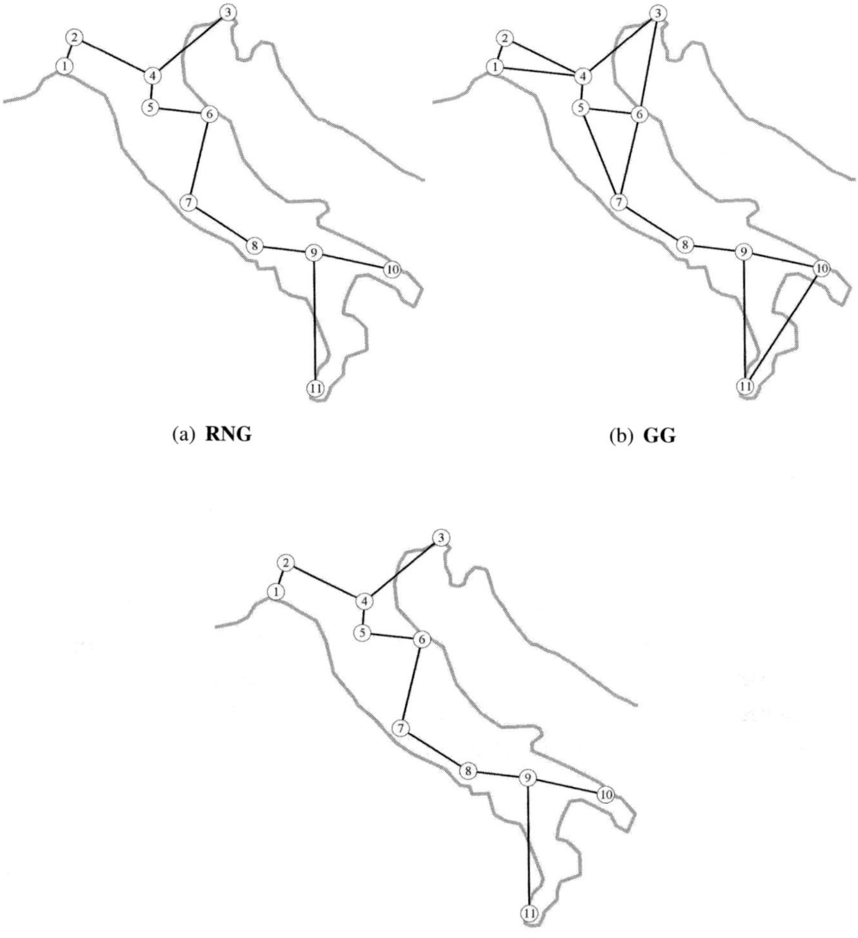

(a) **RNG** (b) **GG**

(c) **MST**(Rome)

Fig. 10.5 Proximity graphs constructed on sites of **U**: (a) relative neighbourhood graph, (b) Gabriel graph, (c) minimum spanning tree rooted in Rome.

10.2 Simulation: linking Bononia to Ariminum and Roma

The simulation model is a particle-based reaction–diffusion mechanism. Unlike classical reaction–diffusion models, which are composed of the interactions of two simulated activator/inhibitor reactants in a diffusive environment, there is only a single representative reactant — a mobile particle which senses and deposits simulated chemoattractant as it moves within a diffusive environment. The model

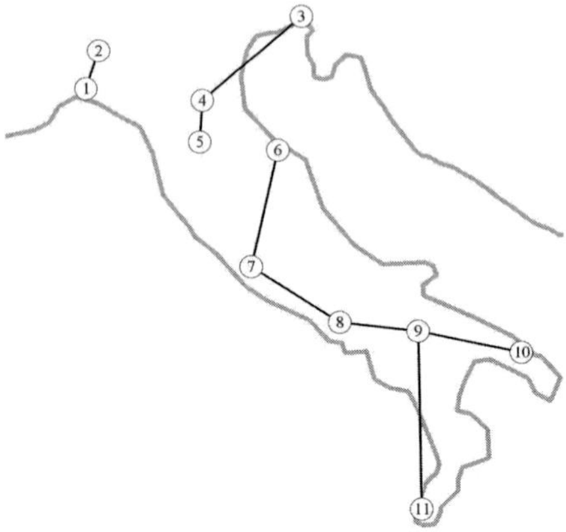

(a) $\mathbf{MST}(\text{Rome}) \cap \mathbf{P}(\frac{1}{28})$

Fig. 10.6 Intersection of Physarum graph and minimum spanning tree.

is based on the LALI (local activation, lateral inhibition) approach to reaction–diffusion pattern formation [Bonabeau (1997)] that was used in [Jones (2010)] to generate emergent dynamical transport networks.

A population of mobile particles is created and initialised on a two-dimensional lattice configured to the experimental pattern of Italian cities. The diffusive medium is represented by a discrete two-dimensional floating point lattice. Particle positions are stored on a discrete lattice isomorphic to the diffusive lattice. Particles also store internal floating point representations of position and orientation which are rounded to discrete values to compute movement updates and sensory inputs. A single particle, and an aggregation of particles, are related to the *P. polycephalum* plasmodium in the following way: the plasmodium is conceptualised as an aggregate of identical components. Each particle represents a hypothetical unit of gel/sol interaction. Gel refers to the relatively stiff sponge-like matrix composed of actin–myosin fibres and sol refers to the protoplasmic solution which flows within the matrix. The *structure* of the protoplasmic network is indicated by the particle positions and the *flux* of sol within the network is represented by the movement of the particles. The resistance of the gel matrix to protoplasmic flux of sol is generated by particle–particle movement collisions.

The morphology of each particle is shown in Fig. 10.7a. The particle 'body' occupies a single cell in the lattice, whose configuration is specified by a digitised grey-scale image. Specific grey-scale values of the image represent habitat features such as empty habitable space, boundaries and food sources. Each particle

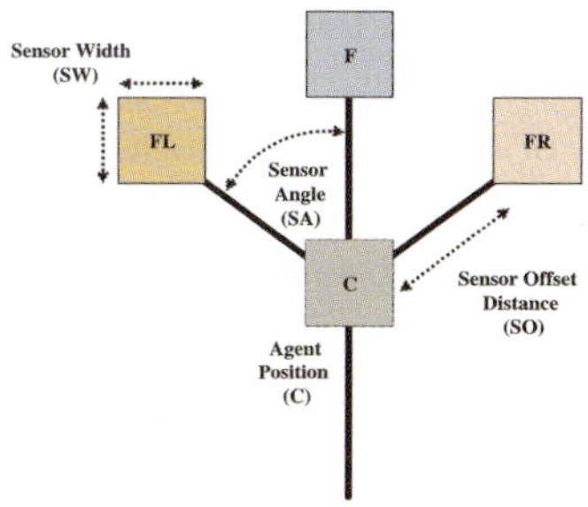

(a) **Morphology**

```
[Sensory stage]

- Sample chemoattractant map values
- if (F > FL) && (F > FR)
       - Continue facing same direction
- Else if (F < FL) && (F < FR)
       Rotate by RA towards larger of FL and FR
- Else if (FL < FR)
       Rotate right by RA
- Else if (FR < FL)
       Rotate left by RA
- Else
       Continue facing same direction
```

(b) **SensoryAlgorithm**

Fig. 10.7 Single agent particle: (a) morphology of agent showing sensors FL, F, FR and position C, (b) sensory algorithm.

has three sensors which sample the environment at some distance away (SO — sensor offset distance, in pixels) from the particle. The sensor offset positions generate local sensory coupling of the particle population. The cohesion of the aggregate population is generated by the mutual attraction to the stimuli deposited by the particles. The coupling of particle sensors and the autocatalytic nature of the particle movement result in complex collective pattern formation and pattern evolution.

The sensory stage of the algorithm is given in Fig. 10.7b. Particles sense chemoattractant concentrations at lattice sites covered by the three sensors. At each scheduler step, particles orient towards the strongest concentration of local stimuli. Particle behaviour is very simple and is explicitly forward biased — the particle does not contemplate its current position — so there is an implicit emphasis on forward movement. By adjusting the SA/RA parameters, a wide variety of reaction–diffusion patterning can be generated. The characteristics of the pattern-formation types and their parametric evaluation were discussed in [Jones (2010)].

The motor stage is executed after the sensory stage. Each particle attempts to move forward in its current direction (represented by an internal state from 0 to 360°). If the new site is vacant, the particle occupies the new site, depositing chemoattractant (5 arbitrary units) at the new location. If the site is occupied, no deposition is made and a new random orientation is selected. The particle population is iterated in a random order for both sensory and motor stages to avoid any influence from sequential updates. Each sensory and motor update, combined with environmental diffusion of chemoattractant, is considered as one scheduler step.

Chemoattractant nutrient stimuli at city locations are projected to the diffusive map at every step of the scheduler (2.55 units for low nutrient concentration and

255 units for high concentration). This level of stimulus attracts the adapting virtual plasmodium and acts to constrain the evolution of the network pattern. The chemoattractant stimuli are diffused by means of a simple 3×3 mean filter kernel. The diffusing chemoattractant values are damped by multiplying by 0.9 in order to adjust the strength of the diffusion gradient and the distance at which the plasmodium collective can sense the nutrients.

Growth and adaptation of the particle model population are currently implemented using a simple method based upon local measures of space availability (growth) and overcrowding (adaptation by population reduction). Growth and shrinkage states are iterated separately for each particle at every two scheduler steps. For growth, we assess if there are 1 to 10 particles in a 9×9 neighbourhood of a particle, and the particle has moved forwards successfully. If so, the particle attempts to divide into two if there is an empty location in the immediate neighbourhood surrounding the particle. For shrinkage, we assess if there are 0 to 24 particles in a 5×5 neighbourhood of a particle. If so, the particle survives, otherwise it is annihilated.

The topology of *P. polycephalum* transport networks is, in part, influenced by unpredictable influences on its formation, for example the initial migration direction of the plasmodium or the presence of previously laid down protoplasmic tubes. This raises the question as to whether the topology of the networks would be more regular under idealised adaptation conditions. In order to minimise this unpredictability, we initialised the simulation model with a uniform distribution of a fully grown virtual plasmodium to solely assess the effect of the spatial arrangement of nutrients (corresponding to city locations) had on the morphological adaptation of the virtual transport network. The results of an example evolution of the virtual plasmodium are shown in Fig. 10.8. Uniform coverage was attained by populating 50% of the habitable area of Italy with 14,789 particles. The simulation was started and the collective adapted to the nutrient stimulus by shrinking in size to form a transport network connecting the city locations and avoiding the mountainous regions. Twenty runs of the simulation were performed at both low and high nutrient concentrations.

We used the same method of assessing connectivity between cities as in the experimental approach and this resulted in an adjacency matrix containing the frequencies of connected nodes for the 20 experiments in both nutrient conditions. The connectivity graphs at different threshold weights are shown in Fig. 10.9 for low nutrient concentrations and Fig. 10.10 for high nutrient concentrations.

Finding 70. $\mathbf{MST} \subset \mathbf{V}(\frac{12}{20})$ *at low nutrient concentration and* $\mathbf{MST} \subset \mathbf{V}(\frac{20}{20}) \cup$ $\{$*Placentia, Bolonia*$\}$ *at high nutrient concentration.*

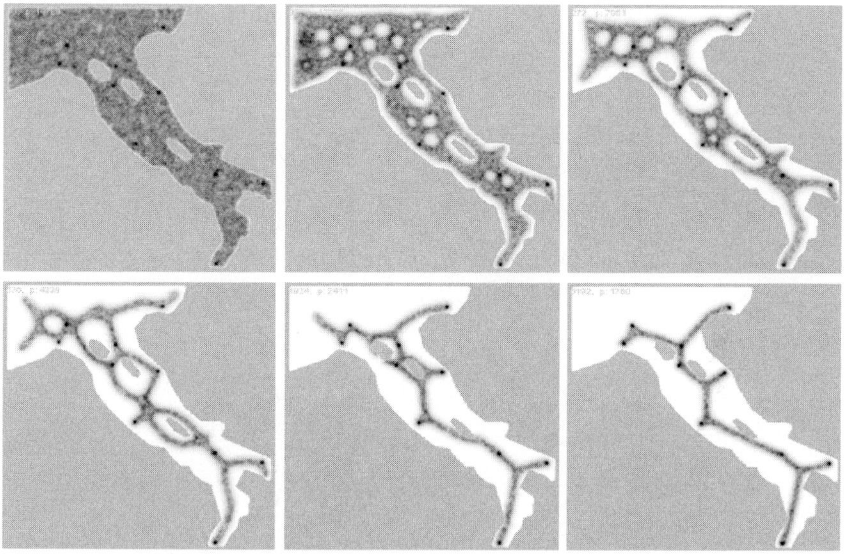

Fig. 10.8 Example of evolution of virtual plasmodium adapting its morphological pattern to form transport network connecting cities at low nutrient strengths. Image snapshots taken at 10, 91, 272, 576, 1924 and 6192 scheduler steps.

The higher the attractive value of nodes, the more likely is virtual slime mould to connect the nodes with a minimum spanning tree.

Finding 71. $\mathbf{P}(\frac{10}{20}) \subset \mathbf{V}(\frac{20}{20})$.

The results demonstrate that the idealised virtual plasmodium redundantly represents generalised Physarum. Moreover, strong components Bononia–Ariminum–Roma and Capua–Venusia–Brundisium are represented by $\mathbf{P}(\frac{14}{20})$ and $\mathbf{V}(\frac{20}{20})$.

10.3 Summary

Experimental laboratory studies and computer simulations confirmed that *P. polycephalum* matches the street network to a certain degree. Based on the laboratory experiments, we can propose the following sequence of road network development in the Italian peninsula. At the initial stage Roma is linked with Corfinium, Capua and Reate (Fig. 10.11a). The next stage is characterised by northbound development: the roads are built to Populonia, Clusium and Ariminum (Fig. 10.11b).

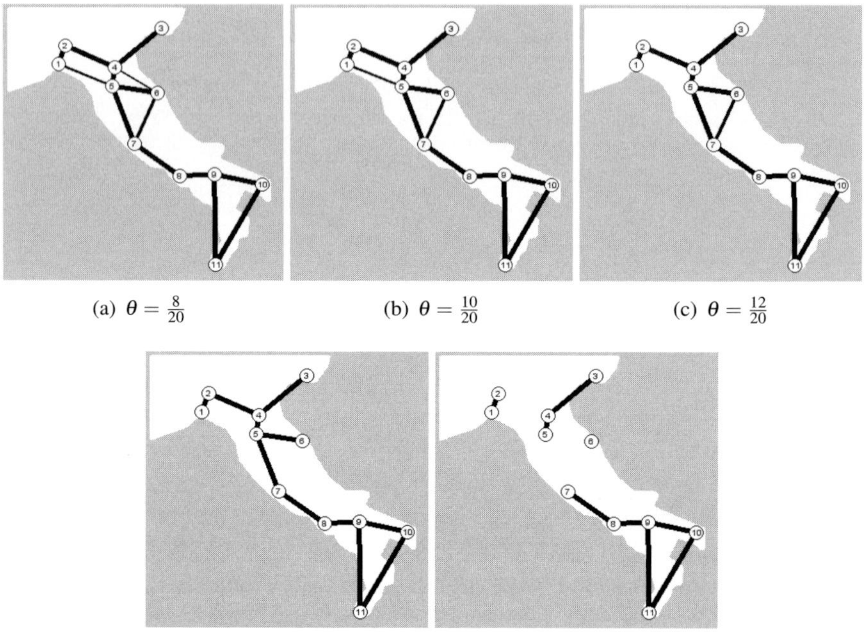

(a) $\theta = \frac{8}{20}$ (b) $\theta = \frac{10}{20}$ (c) $\theta = \frac{12}{20}$

(d) $\theta = \frac{19}{20}$ (e) $\theta = \frac{20}{20}$

Fig. 10.9 Virtual plasmodium graphs $\mathbf{V}(\theta)$ for selected values of θ at low nutrient concentration.

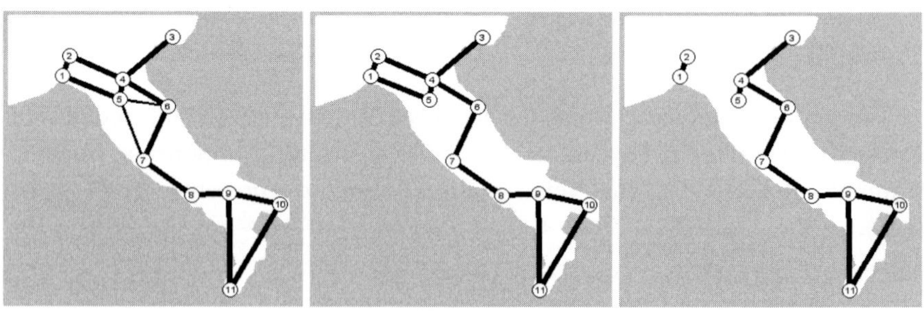

(a) $\theta = \frac{10}{20}$ (b) $\theta = \frac{19}{20}$ (c) $\theta = \frac{20}{20}$

Fig. 10.10 Virtual plasmodium graphs $\mathbf{V}(\theta)$ for selected values of θ at high nutrient concentration.

The propagation continues towards northern parts until Roma is connected with Genua, Placentia, Verona, Aquileia and Pola (Fig. 10.11c). After the northern part of the Italian peninsula is well serviced with roads, the development starts southward. The roads are built from Capua to Brundisiu and Rhegnium (Fig. 10.11d).

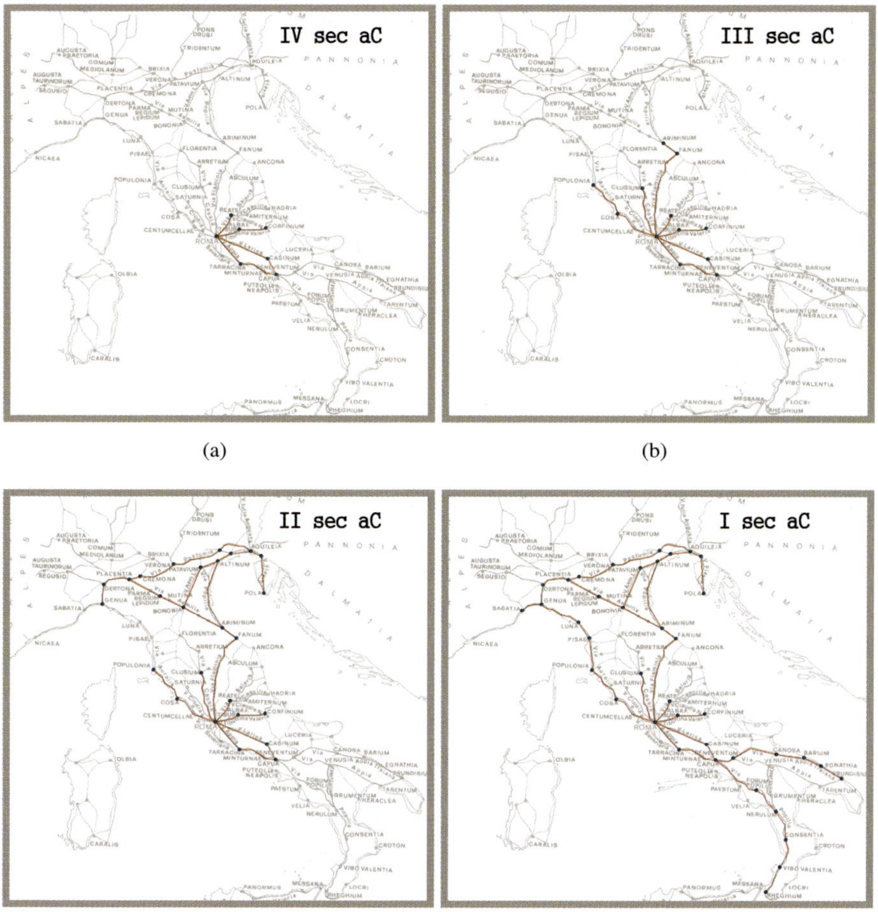

Fig. 10.11 Proposed stages of Roman street network development based on experiments with *P. polycephalum*. Routes of possible propagation are shown by red/dark-grey superimposed on the map of Iron Age Italy 1000–100 B.C. [Villa (1995)].

Such development is substantially matched by dynamics of a spanning tree growing from Roma 10.12.

By comparing the road networks and slime mould networks with proximity graphs, we found that in this particular case of the Italian peninsula the road networks and slime mould networks show a high degree of affinity to spanning trees and relative neighbourhood graphs (see also [Watanabe (2005); Watanabe (2008)] for use of proximity graphs in simulation of urban networks).

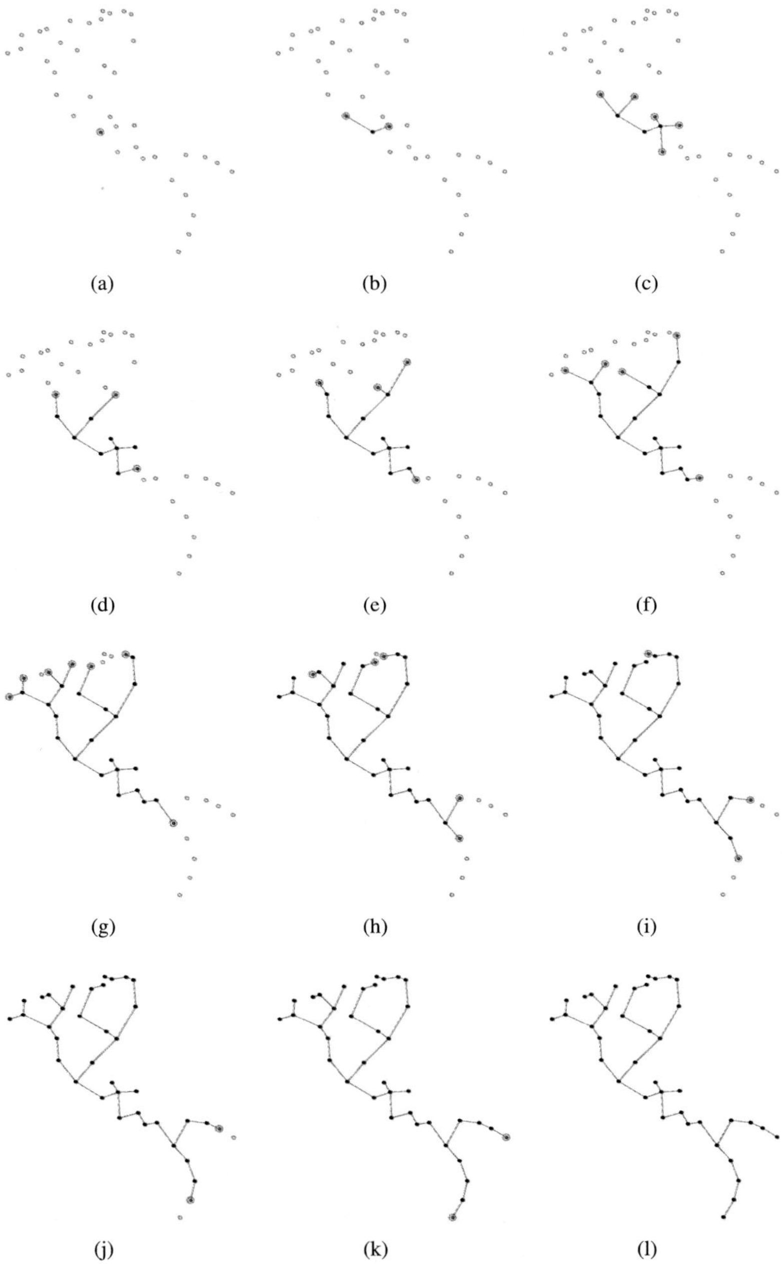

Fig. 10.12 Growth of a spanning tree rooted in Roma, on cities used as nodes in illustration of Fig. 10.11. Nodes occupied by 'active zones', or growth cones of the tree, are encircled.

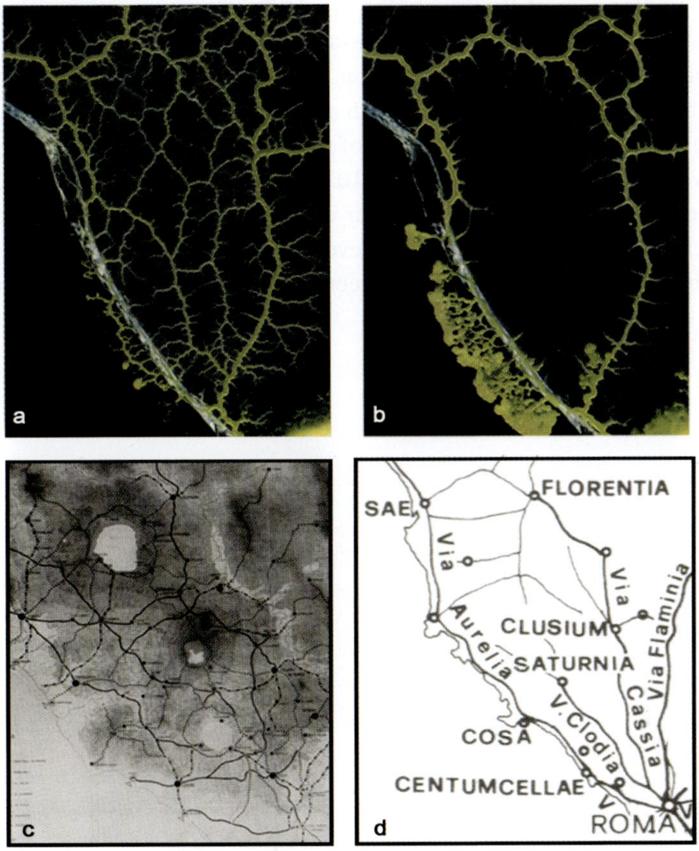

Fig. 10.13 Similarity between evolution of road networks in Italian peninsula and corresponding protoplasmic networks: (a) early protoplasmic tube network, (b) after tube network adaptation, (c) primordial Etrurian trails (approx. V Cent. B.C.) [Villa (1995)], (d) Imperial Roman street networks (I Cent. A.D.) [Villa (1995)].

This could be a sign that the Roman transportation system, developed in ancient times, was based on a common-sense logic but also possessed all features of bio-logic, at least as related to reasoning at the level of amorphous spatially distributed living creatures.

The results are encouraging and inspire us to undertake further investigations. For example, during laboratory experiments we noticed striking similarities between sprouting, scouting and space searching executed by *P. polycephalum* and evolution of transport networks in the Tuscany region of the Italian peninsula (Fig. 10.13). The pre-Roman road networks in this region show signs of rather extensive exploration of the space (Fig. 10.13c). The same strategy is explored by slime mould during its initial stage of substrate colonisation (Fig. 10.13a). With time — due to implicit competition between roads — only a few roads are selected by the masses of travellers, such as e.g. Etrurian paths in the Tuscany region (Fig. 10.13d). The same selection of few transport links occurs in protoplasmic networks of *P. polycephalum* (Fig. 10.13b), as protoplasmic tubes with lesser flux are eradicated. We believe that our experiments will lead to further extensive experiments on bioinspired urban developments boosted by paleotopography, particle-based modes of large-scale collectives and nature-inspired computing.

Chapter 11

Malaysian expressways: is there a logic behind them?

Andrew Adamatzky, Zuwairie Ibrahim, Amar Faiz Zainal Abidin and Badaruddin Muhammad

Transportation in Malaysia had started to develop during British colonial rule and the country's transport network is now diverse and developed. The main modes of transport in peninsular Malaysia include buses, trains, cars and aeroplanes. Malaysia is served by an excellent transport system. In this country, transportation is always available even in remote areas. Travelling by road in peninsular Malaysia is popular as it has a well-developed network of roads. The road network covers 98,721 km, of which 80,280 km is paved, and 1,821 km is expressways [Transport in Malaysia (2011)]. Meanwhile, the North–South Expressway extending over 800 km (497 mi) between the Thai border and Singapore is the longest highway in the country.

The unique feature of the transportation system in Malaysia is the Halal logistic hub that provided the transportation. Currently, there are rapidly growing Halal industries (food, cosmetics, pharmaceuticals, clothing and so on) that demand logistic systems that can ensure the integrity of products right to the point of consumption. In Halal logistics, it is important to segregate Halal goods from non-Halal goods to avoid cross contamination. Segregation is needed from non-Halal products such as pork, dogs, alcoholic drinks, carrion or Halal animals that are not slaughtered according to the Shariah law and blood. Over the years, Halal parks have been set up in several states as part of the government's plan to encourage investments in Halal businesses.

In 2008, an area of the Port Klang Free Zone was dedicated to Halal logistics and was named the Halal Flagship Zone. Launched in 2003, the Selangor Halal Hub (SHH) is equipped with facilities that support a comprehensive range

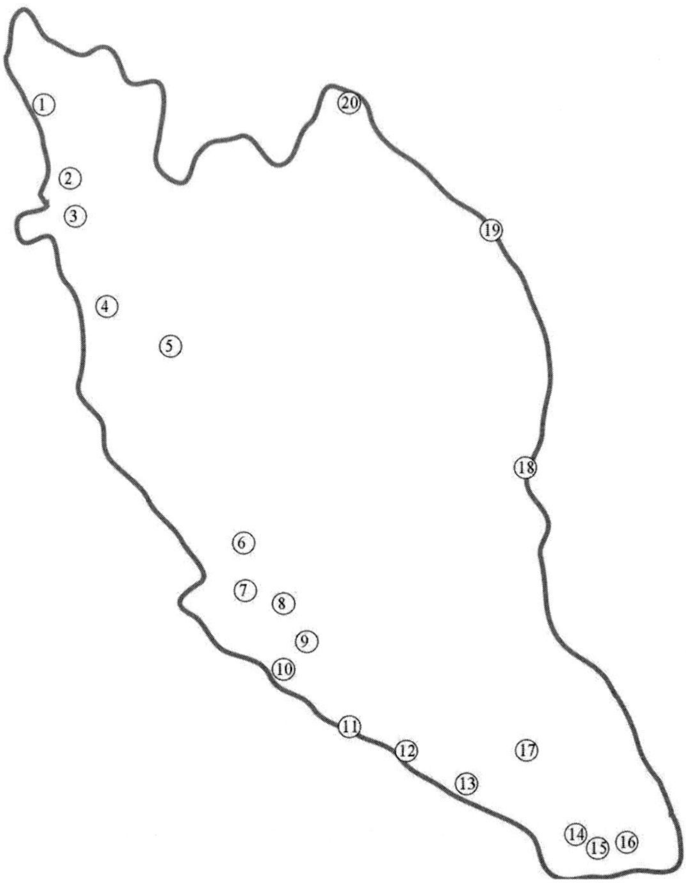

Fig. 11.1 Outline map of Malaysia with major urban areas **U** shown by encircled numbers.

of logistic services from storage and transportation to export certification. One of the most attractive features of SHH is the fast-track approval for foreign and local investors located within the zone. In east Malaysia, the Tanjung Manis Halal Park in Sarawak focuses on upstream and downstream Halal food manufacturing activities.

We selected 20 most populated major urban areas listed below (see configuration of the areas in Fig. 11.1):

(1) Alor Star,
(2) Sungai Petani,
(3) Kulim area, including Kulim, Bukit Mertajam and Butterworth,

Fig. 11.2 Experimental setup. Urban areas, represented by oat flakes, are colonised by slime mould *P. polycephalum*.

(4) Taiping,
(5) Ipoh,
(6) Rawang,
(7) Kuala Lumpur area, including Subang Jaya, Kuala Lumpur, Klang, Ampang, Shah Alam, Petaling Jaya, Cheras, Selayang and Batu Caves,
(8) Kajang area, including Kajang and Semenyih,
(9) Seremban,
(10) Port Dickson,
(11) Bandar Melaka,
(12) Muar area, including Muar and Ayer Itam,

(13) Batu Pahat,
(14) Skudai area, including Skudai and Senai,
(15) Johor Bahru area, including Johor Bahru and Ulu Tiram,
(16) Pasir Gudang,
(17) Kluang,
(18) Kuantan,
(19) Kuala Terengganu,
(20) Kota Bahru.

At the beginning of each experiment an oat flake colonised by plasmodium was placed in the Kuala Lumpur area. We undertook 30 experiments (Fig. 11.2).

11.1 The coastal routes

In all trials, we observed four scenarios of slime mould colonising urban areas **U** (snapshots of experiments are shown in Figs. 11.5, 11.4 and 11.6):

- SENES: south-east then north-east and then south (Fig. 11.3a), observed in 13 experiments,
- SENW: south-east then north and north-west (Fig. 11.3b), observed in 11 experiments,
- SE-T-NW: south-east and at the same time north-west (Fig. 11.3c), observed in five experiments,
- SEWN: south-east and at the same time west and then north/north-west (Fig. 11.3d), observed in one experiment.

The SENES scenario is executed by slime mould in over 43% of experiments. The slime mould propagates from Kuala Lumpur to the Rawang and Kajang areas (Fig. 11.4a). It does not propagate further north from Rawang but heads south and colonises Seremban, Port Dickson, Bandar Melaka, Muar, Batu Pahat, Skudai, Kluang, Johor and Pasir Gudang. The plasmodium becomes 'blocked' in Pasir Gudang and does not make any feasible attempts to move towards the close unoccupied area of Kuantan (Fig. 11.4b). When 'blocked' in Pasir Gudang the plasmodium reactivates its active zone in Rawang and builds a transport link connecting Rawang with Ipoh. It then propagates further south and colonises the Taiping and Kuliam areas (Fig. 11.4c). At this point of spanning **U** the slime mould branches: usually the northern branch travels towards the Sungai Petani and Alor Star areas and the eastern branch propagates towards Kota Bahru. Transport links between Kota Bahru and Kuala Terengannu and Kuanta complete colonisation of **U** (Fig. 11.4d).

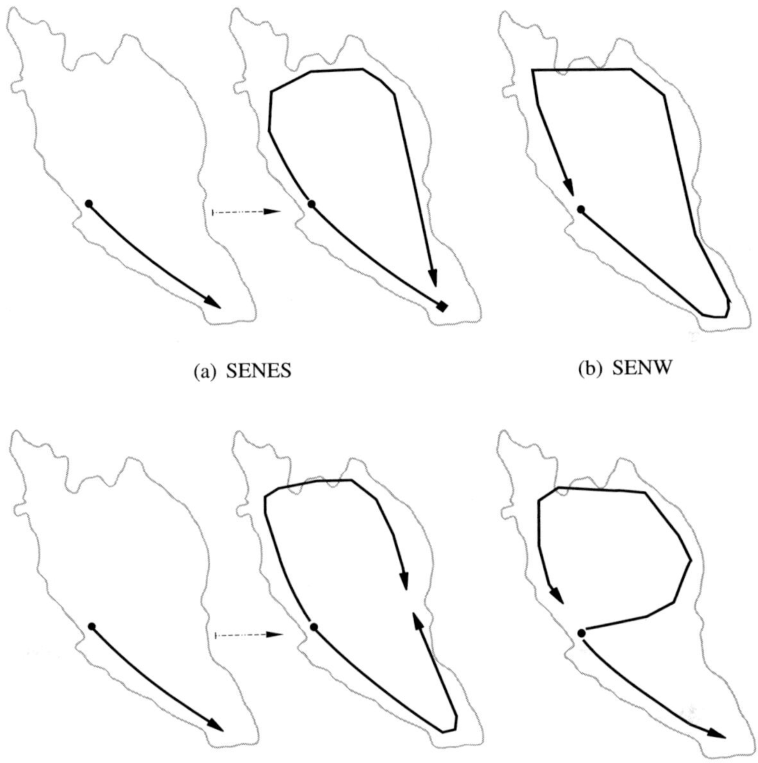
(a) SENES (b) SENW

(c) SE-T-NW (d) SEWN

Fig. 11.3 Scenario of colonisation of **U** by plasmodium of *P. polycephalum*.

The SENW scenario occurs in 36% of experiments. The initial stage of the colonisation (Fig. 11.5a) is similar to the SENES scenario: urban areas from Kuala Lumpur to Pasir Gudang are colonised. From Kluang and/or Pasir Gudang the plasmodium propagates towards Kuantan and then colonises Kuala Terengannu and Kota Bahru (Fig. 11.5b). Then the plasmodium propagates towards the Alor Star and Kulim areas and colonises the remaining urban areas (Fig. 11.5c).

Both clockwise and anticlockwise colonisations of **U** are performed by plasmodium in scenario SE-T-NW, in 17% of experiments. Starting in Kuala Lumpur, slime mould colonised **U** simultaneously along the following routes (Fig. 11.6):

- Kuala Lumpur–Kajang–Seremban–Port Dickson–Bandar Melaka–Muar–Batu Prahat–Skudai and Kluang–Johor Bahru–Pasir Gudang–Kuantan–Kuala Terengganu,

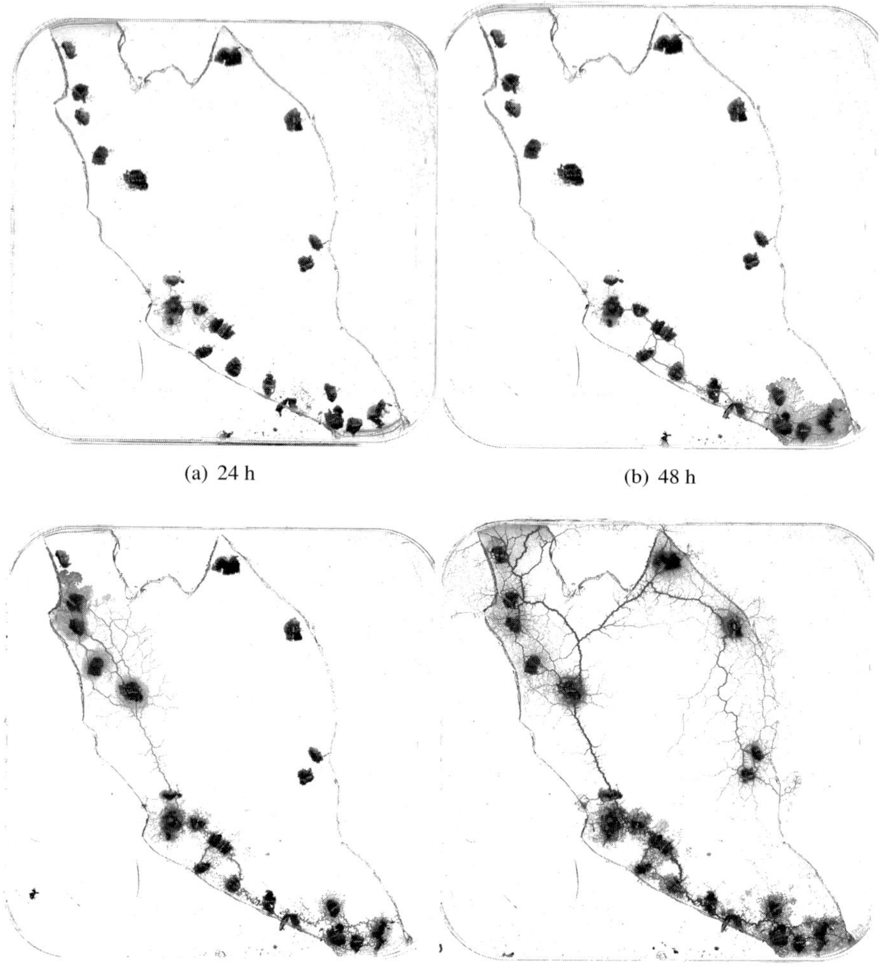

(a) 24 h (b) 48 h

(c) 72 h (d) 96 h

Fig. 11.4 Snapshots of slime mould colonising urban areas **U** by SENES scenario. The images are recorded at 24-h intervals after inoculation of plasmodium in Kuala Lumpur.

- Kuala Lumpur–Ipoh–Taiping–Kulim–Sungai Petani and Alor Star–Kota Bahru.

Examples of Physarum graphs for various values of θ are shown in Fig. 11.7.

A Physarum graph $\mathbf{P}(\frac{1}{30})$ consists of a planar southern part and a non-planar northern part (Fig. 11.8a). The non-planar component is located north of Ipoh in

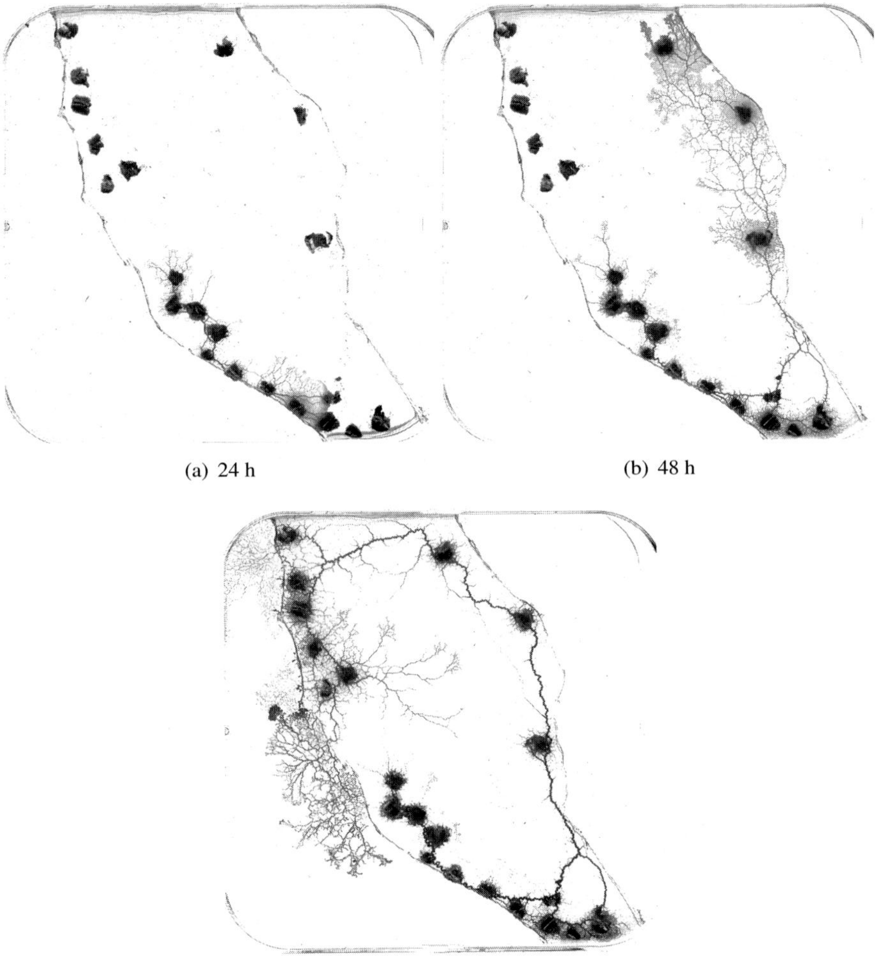

(a) 24 h (b) 48 h

(c) 72 h

Fig. 11.5 Snapshots of slime mould colonising urban areas **U** by SENW scenario. The images are recorded in 24-h intervals after inoculation of plasmodium in Kuala Lumpur.

the east and north of Kuantan in the west. The graph becomes planar when θ increases to $\frac{4}{30}$ (Fig. 11.8b). When we remove edges appearing in less than half of experiments, i.e. those with weights less than $\frac{15}{30}$, the Physarum graph becomes almost linear and the only cycle in the south is formed by the Batu Prahat, Skudai and Kluang urban areas (Fig. 11.8b).

Finding 72. *The transport network built by plasmodium of* P. polycephalum *on*

(a) 24 h

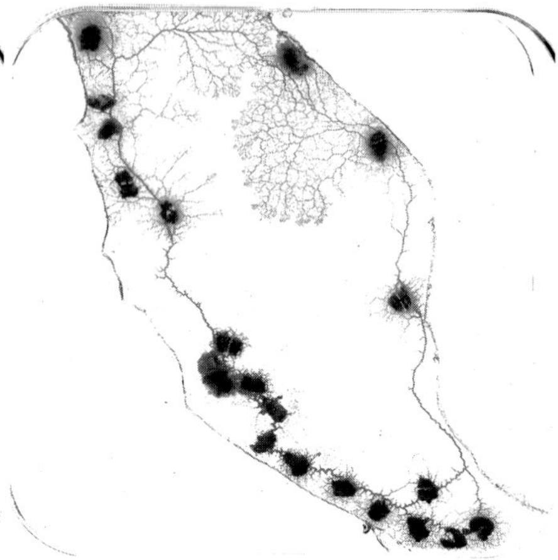

(b) 48 h

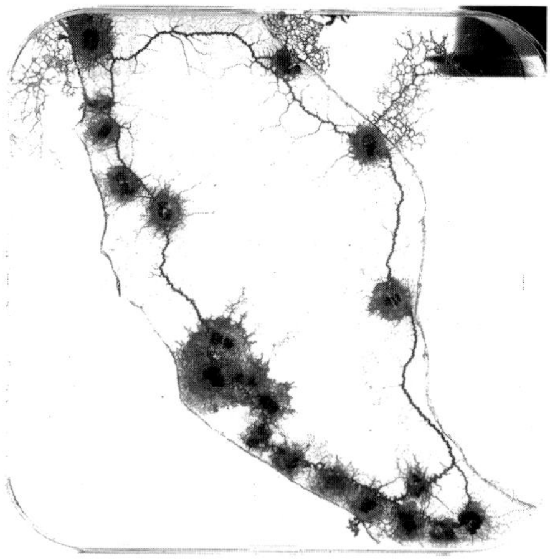

(c) 72 h

Fig. 11.6 Snapshots of slime mould colonising urban areas **U** by SE-T-NW scenario. The images are recorded in 24-h intervals after inoculation of plasmodium in Kuala Lumpur.

20 major urban areas of Malaysia consists of a linear chain spanning urban areas along the coast and a single cycle.

The Physarum graph becomes disconnected when $\theta = \frac{16}{30}$ (Fig. 11.8d). One component is a three-node chain, spanning Kuantan to Kuala Terengganu to Kota Bahru. The second component is a chain from Alor Star to Batu Prahat, a cycle Batu Prahat, Skudai and Kluang and a segment Johor Bahru to Pasir Gudang. The

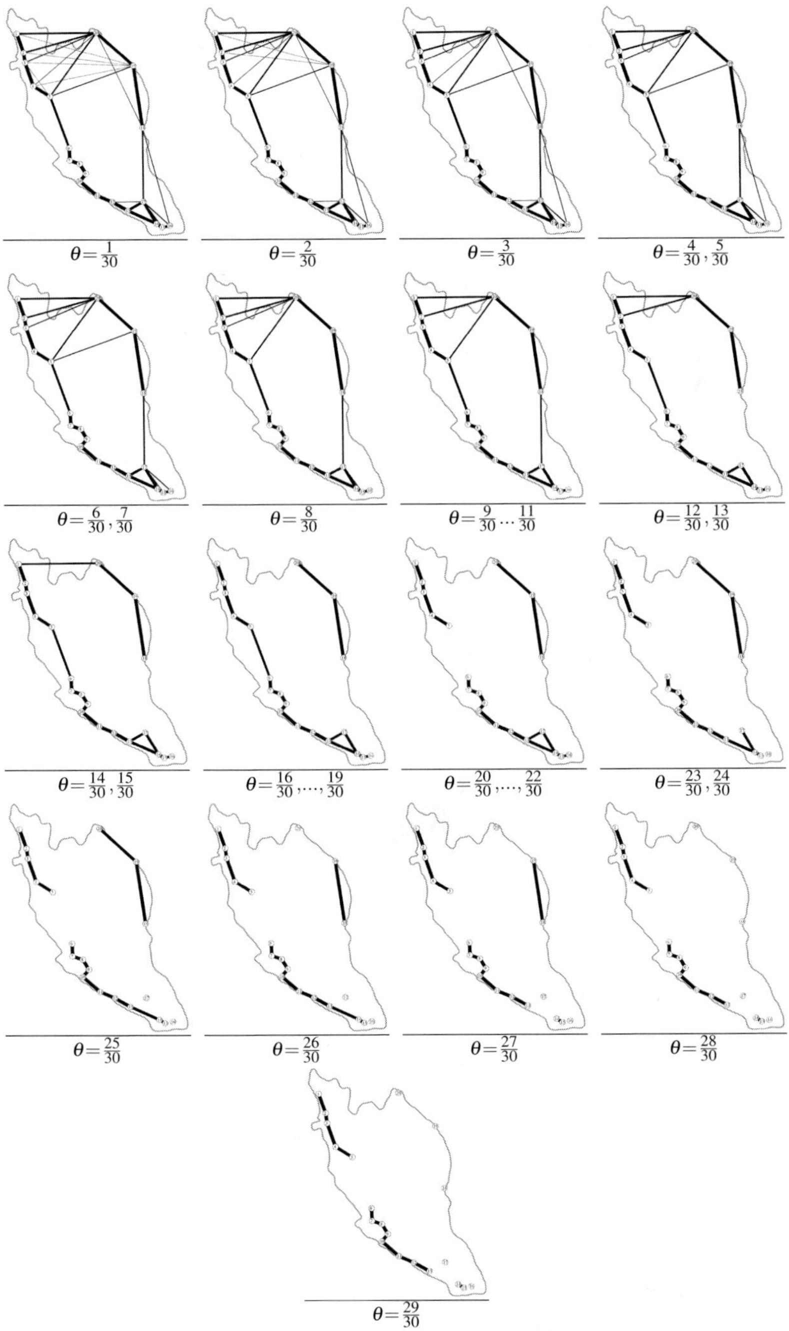

Fig. 11.7 Physarum graphs $\mathbf{P}(\theta)$ for $\theta = \frac{1}{30}, \ldots, \frac{29}{30}$.

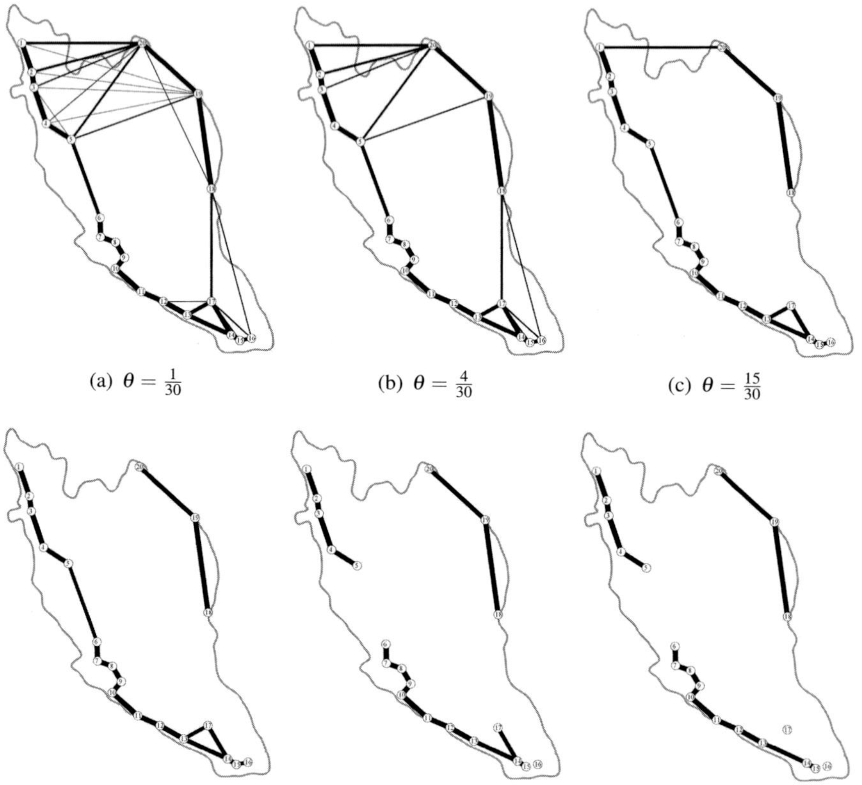

(a) $\theta = \frac{1}{30}$ (b) $\theta = \frac{4}{30}$ (c) $\theta = \frac{15}{30}$

(d) $\theta = \frac{16}{30}$ (e) $\theta = \frac{23}{30}$ (f) $\theta = \frac{25}{30}$

Fig. 11.8 Physarum graphs $\mathbf{P}(\theta)$ for selected values of θ. Weights of edges are reflected in the width of their line representations.

reason for the formation of two disconnected components is not because slime mould builds disconnected protoplasmic networks but because in almost every experiment plasmodium chooses different ways to connect eastern and western urban areas in the north of Malaysia; thus, no single edge there is represented in over half of experiments.

The graph becomes planar, yet disconnected but without isolated nodes, for $\theta = \frac{23}{30}$ (Fig. 11.8e). The urban area Kluang becomes an isolated vertex for $\theta = \frac{23}{30}$ (Fig. 11.8f). We call an edge stable when it is represented by slime mould's protoplasmic tubes in over 80% of laboratory experiments.

Finding 73. *A stable transport network of Malaysia, represented by plasmodium of* P. polycephalum, *consists of three disconnected chains:*

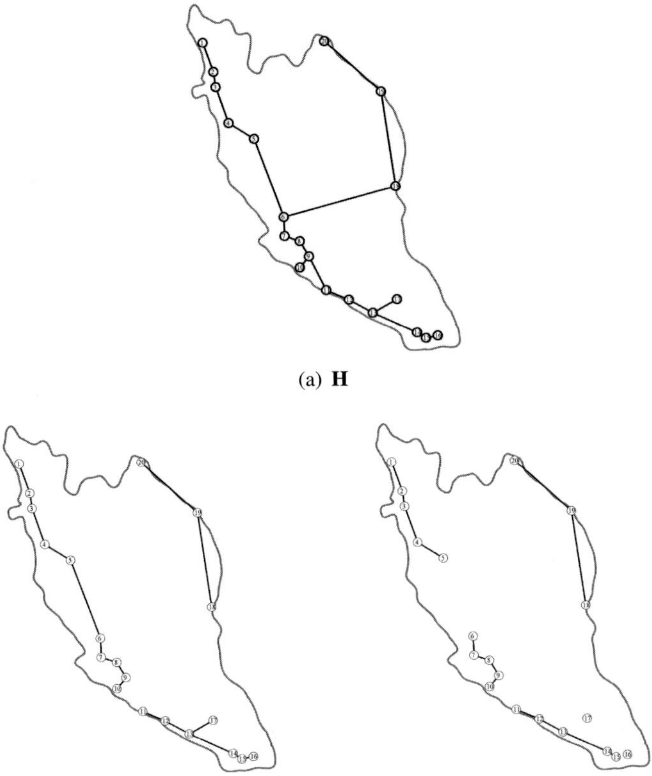

(a) **H**

(b) $\mathbf{H} \cap \mathbf{P}(\theta)$, $\theta = \frac{1}{30}, \ldots, \frac{15}{30}$

(c) $\mathbf{H} \cap \mathbf{P}(\theta)$, $\theta = \frac{23}{30}, \ldots, \frac{25}{30}$

Fig. 11.9 (a) Graph **H** of Malaysian expressway network. (b, c) Intersections of the expressway graph **H** and Physarum graph $\mathbf{P}(\theta)$.

- *Rawang, Kuala Lumpur, Kajang, Seremban, Port Dickson, Bandar Melaka, Muar, Batu Prahat, Skudai, Johor Bahru and Pasir Gudang,*
- *Alor Star, Sungai Petani, Kulim, Taiping and Ipoh,*
- *Kota Bahru, Kuala Terengganu and Kuantan.*

11.2 Strong chains and isolated cities

The graph **H** of the Malaysian expressway network is shown in Fig. 11.9.

Few cities of **U** have no outgoing or incoming expressways. Therefore, in these few cases we were forced to use federal roads. The following edges of the graph **H** correspond to federal roads:

- Batu Pahat–Kluang,
- Kuantan–Kuala Terengganu,
- Kuala Terengganu–Kota Bahru.

Finding 74. *Slime mould* P. polycephalum *approximates almost all edges of the expressway graph. Man-made transport links Rawang–Port Dickson and Seremban–Bandar Melaka are never represented by protoplasmic tubes.*

The above finding comes from the straightforward comparison of the Physarum graph (Fig. 11.8) and the expressway graph **H** (Fig. 11.9). Essentially *P. polycephalum* approximates 17 of 19 edges of the expressway graph **H**. Such a good degree of approximation may be due to the very particular arrangement of major urban areas along coastlines of Malaysia. This configuration of the sources of nutrients causes slime mould to link the sources by a single chain with few if any branchings.

Finding 75. *Chains of man-made expressways Alor Star–Port Dickson and Bandar Melaka–Pasir Gudang are approximated by slime mould of* P. polycephalum *in over half of laboratory experiments.*

Finding 76. *Chains of the man-made federal road Kuantan–Kota Bahru and expressways Alor Start–Ipoh, Bandar Melaka–Skudai and Rawang–Port Dickson are approximated by slime mould* P. polycephalum *in over 80% of experiments.*

11.3 Trees rooted in Rawan and Kuala Lumpur are minimal

Finding 77. *The minimum spanning tree rooted in Kuala Lumpur and the relative neighbourhood graph would be subgraphs of the Physarum graph* $\mathbf{P}(\frac{4}{30})$ *if the Physarum graph had the edge Kajang–Kuantan.*

Finding 78. *The Gabriel graph would be a subgraph of the Physarum graph* $\mathbf{P}(\frac{4}{30})$ *if the Physarum graph had two edges: Kajang–Kuantan and Sermban–Kuantan.*

This can be demonstrated by direct comparisons of intersections of proximity graphs and the Physarum graph $\mathbf{P}(\frac{4}{30})$ (Fig. 11.10d–f) with the original proximity graphs (Fig. 11.10a–c) and taking into account Toussaint hierarchy. With regard to the spanning tree, an even stronger claim can be made: $\mathbf{MST}(\mathrm{KL}) \subset \mathbf{P}(\frac{15}{30})$. That is, a minimum spanning tree rooted in Kuala Lumpur is almost a subgraph of the robust slime mould transport network except for the edge Kajang–Kuantan.

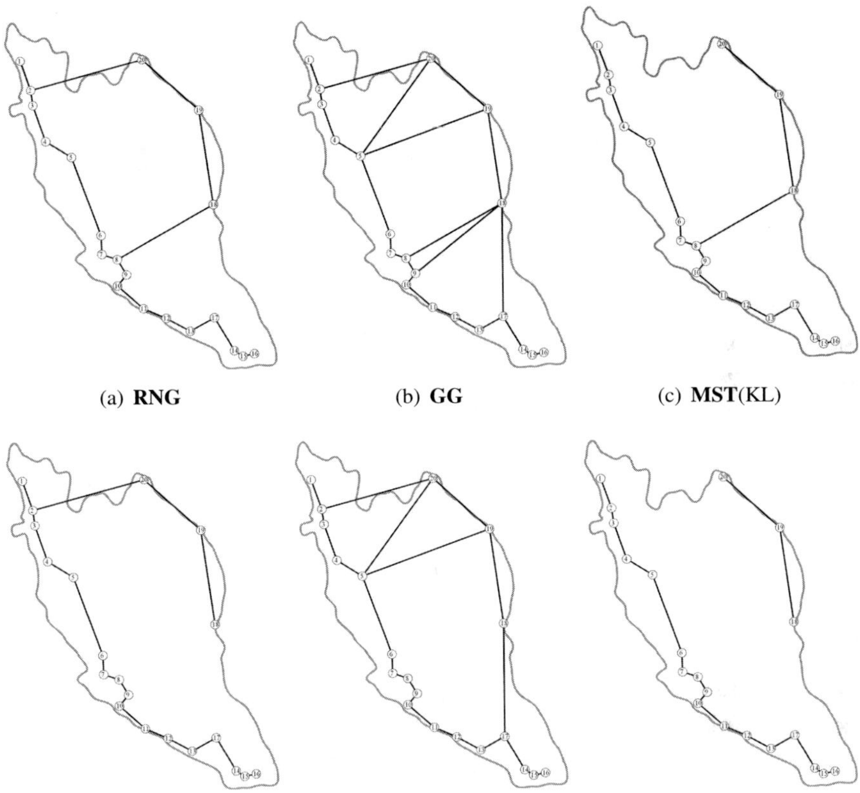

(a) **RNG** (b) **GG** (c) **MST**(KL)

(d) **RNG**$\cap$**P**($\frac{4}{30}$) (e) **GG**$\cap$**P**($\frac{4}{30}$) (f) **MST**(KL)$\cap$**P**($\frac{4}{30}$)

Fig. 11.10 Proximity graphs constructed on sites of **U** and their intersections with planar Physarum graph **P**($\frac{4}{30}$): (a) relative neighbourhood graph, (b) Gabriel graph, (c) minimum spanning tree rooted in Kuala Lumpur. Intersections of **P**($\frac{4}{30}$) with (d) relative neighbourhood graph, (e) Gabriel graph, (f) minimum spanning tree rooted in Kuala Lumpur (KL).

It looks like the transport link from Kajang to Kuantan is essential from the planar proximity graph's point of view but it is not important for slime mould based transport networks.

Is the tree rooted in Kuala Lumpur really a minimum spanning tree? To get an answer, we built spanning trees rooted in all sites of **U** (Fig. 11.11). We found that there are six morphologies of trees with lengths (sum of edge lengths) varying from normalised 1, trees rooted in Rawan and Kuala Lumpur (Fig. 11.11a), to 1.13, trees rooted in Alor Star and Sungai Petani (Fig. 11.11f). Amongst all possible trees, only the tree rooted in Kulim or Taiping (Fig. 11.11b) does not

Fig. 11.11 Spanning trees rooted in sites of **U** and their relative lengths l: (a) Rawang and Kuala Lumpur, $l = 1.0$, (b) Kulim and Taiping, $l = 1.01$, (c) Ipoh, $l = 1.02$, (d) Kuala Terengannu and Kota Bahru, $l = 1.04$, (e) Kajang, Seremban, Port Dickson, Bandar Melaka, Muar, Batu, Skudai, Johor Bahru, Pasir, Kluang and Kuantan, $l = 1.10$, (f) Alor Star and Sungai Petani, $l = 1.13$.

contain the edge Kajang–Kuantan. This tree is not strictly minimal but it is just 1% longer than the minimum spanning tree rooted in Rawan and Kuala Lumpur. This means that the edge Kajang–Kuantan does not really contribute to minimality of the spanning trees constructed on **U**.

Finding 79. *The minimum spanning tree rooted in Kulim or Taipin is a subgraph of the robust Physarum graph* $\mathbf{P}(\frac{15}{30})$.

Thus, we can conclude that the slime mould based transport network on **U** includes a minimum network.

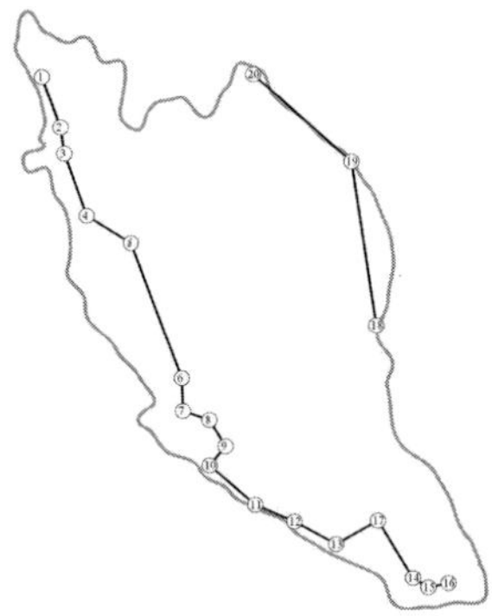

Fig. 11.12 β-skeleton $\mathbf{B}(\beta)$ constructed on $\mathbf{U}$, $2.7 \leq \beta \leq 6.4$.

We use β-skeletons in a hope to uncover an intrinsic relation between the θ parameter of the Physarum graph $\mathbf{P}(\theta)$ and the β parameter of β-skeletons. The parameter θ governs reliability of an edge by removing edges whose frequency in laboratory experiments is less than θ. The parameter β reflects the size and shape of a lune determining the proximity neighbourhood of nodes [Kirkpatrick and Radke (1985)]. In layman terms, the larger β the more narrow and longer is the lune between two neighbouring points. Increases in θ and β lead to a decrease in a graph's connectivity. By comparing Physarum graphs for θ up to 30, and β-skeletons for β up to 20, we did not find any convincing correlations. However, the following result was obtained.

Finding 80. $\mathbf{B}(\beta) \cup$ *(Batu Prahat, Skudai)* $= \mathbf{P}(\frac{26}{30})$, $2.7 \leq \beta \leq 6.4$.

Compare Figs. 11.12 and 11.8d.

Finding 81. *If the man-made transport links Rawang–Kuantan, Seremban–Bandar Melaka and Batu Pahat–Skudai were non-existent and the man-made routes Kajang–Kuantan and Kluang–Skudai built the Malaysian expressways, the graph would be the minimum spanning tree grown from Kuala Lumpur.*

The above results arise from direct comparison of **H** (Fig. 11.9a) and spanning trees (Fig. 11.11).

Finding 82. *Let* $\mathbf{L}$ *be intersections of proximity graphs with the expressway graph* $\mathbf{L} = \mathbf{RNG} \cap \mathbf{H} = \mathbf{GG} \cap \mathbf{H} = \mathbf{MST}(KL) \cap \mathbf{H}$*; then* $\mathbf{P}(\frac{25}{30}) \in \mathbf{L}$.

The finding announces the amazing fact that intersections of the expressway graph with proximity graphs are the same. Moreover, this 'intersection graph' is

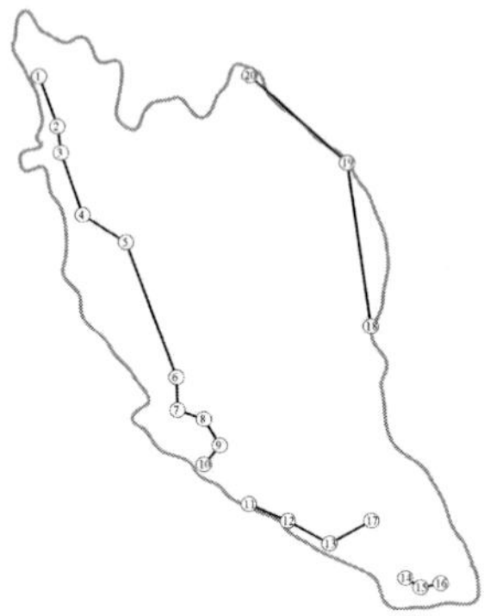

Fig. 11.13 Proximity graphs constructed on sites of **U** and their intersections with planar expressway graph **H**: $\mathbf{RNG} \cap \mathbf{H} = \mathbf{GG} \cap \mathbf{H} = \mathbf{MST}(\mathrm{KL}) \cap \mathbf{H}$.

a subgraph of the strong Physarum graph. In other words, the intersection of the expressway graph with proximity graphs is represented by protoplasmic tubes in over 83% of laboratory experiments.

11.4 Contamination in Kuantan

To study the reaction of Physarum-grown transport networks on major disasters, we placed crystals of sodium chloride in the approximate position of Kuantan (Fig. 11.14a). The salt diffuses in the substrate outwards from its original application site, the epicentre of disaster. It can therefore imitate radioactive and/or chemical pollution and subsequent disturbance spreading along Malaysian transport networks.

Patterns of responses recorded in laboratory experiments are uniform and repeatable. Transport links in domains affected by contamination are destroyed and abandoned. These links are visible as whitish protoplasmic tubes (Fig. 11.14b). Parts of the transport network positioned far away from the epicentre of contamination become hypertrophied (Fig. 11.14c) due to increase of exploratory activity in the urban areas not affected by contamination (Fig. 11.14d). Based on the experiments with slime, we can propose that the following scenario would develop.

Finding 83. *If the epicentre of contamination is located in Kuantan and the speed of propagating contamination is about 100 miles per 24 h, transport functionality will be significantly diminished in X14, X3 and X63 roads; a substantial increase of traffic would be observed between Kota Bahru and the Alor Star, Sungai Petani and Kulim areas; a significant increase in migration and economic activity would*

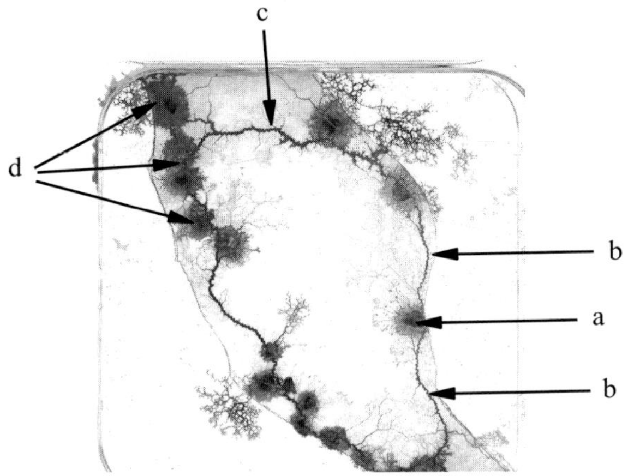

Fig. 11.14 Reconfiguration in response to contamination: (a) epicentre of contamination, (b) abandoned transport links, (c) hypertrophied transport links, (d) increase of exploratory activity. Image is recorded 24 h after contamination.

also be observed in Kedah and north Perak and Kelantan states.

11.5 Summary

In laboratory experiments we found that protoplasmic networks constructed by slime mould *P. polycephalum* mainly consist of linear chains spanning urban areas along the coastlines. We demonstrated that the slime mould approximates almost all edges of the expressway graph. Only two of 19 transport links are never represented by protoplasmic tubes: Rawang–Port Dickson and Seremban–Bandar Melaka. With regard to particular transport routes, we found that chains of man-made expressways Alor Star–Port Dickson and Bandar Melaka–Pasir Gudang are imitated by the slime mould in over half of laboratory experiments. Chains of the man-made federal road Kuantan–Kota Bahru and expressways Alor Start–Ipoh, Bandar Melaka–Skudai and Rawang–Port Dickson are approximated by the slime mould *P. polycephalum* in 80% of experimental trials. In experiments with slime mould, we found that the structure of the Malaysian expressway network is biological from *P. polycephalum*'s point of view. Is it optimal and common-sense logical? Indeed. If the man-made transport links Rawang–Kuantan, Seremban–Bandar Melaka and Batu Pahat–Skudai were non-existent and the man-made routes Kajang–Kuantan and Kluang–Skudai built the Malaysian expressways, the graph would be the minimum spanning tree grown from Kuala Lumpur. We also experimentally found that if a very-large-scale contamination occurs with epicentre in Kuantan, then a substantial increase of traffic will be observed along Kota Bahru to Alor Start and Sungai Petani to Kulim routes, as well as a substantial increase in migration and economic development in Kedah and north Perak and Kelantan states.

Chapter 12

Physarum narcotráficum: Mexican highways and slime mould

Andrew Adamatzky, Genaro J. Martínez, Sergio V. Chapa-Vergara, René Asomoza-Palacio and Christopher R. Stephens

Mexico is the most populated Spanish-speaking country with diverse landscapes, sudden north–south changes in land size, highest level of public transport in Latin America and high concentration of pollutants in the central part of the country. These features are not unique but enhanced and, what is most important, they provide a very good test bed for studies of nature-inspired approaches to road planning.

Mexico ranks seventh in highway length with 6,335 km (3,935 miles), after (1) United States with 91,541 km (56,859 mi), (2) China with 24,474 km (15,202 mi), (3) Germany with 11,400 km (7,081 mi), (4) France with 10,300 km (6,398 mi), (5) Spain with 9,063 km (5,629 mi) and (6) Italy with 8,957 km (5,563 mi)[1], Mexico comprises 31 states and one Federal District (DF or Mexico City) which constitutes the centre of government; for this reason, each city in Mexico has a fairly direct route via main highways to Mexico City. With a total population of 113,724,226 (July 2011 estimation) on a territory of 1,964,375 km^2 (758,450 mi^2)[2], Mexico has a salient and strategic geographical position because it is a natural land transit between North America (mainly the United States and Canada) and Mexico itself and Central and South America, i.e. all Latin American countries. This contributes towards essential components of economical structure for the United States in transporting prime materials, natural resources, imports and exports and labour migration. For this reason, years ago extensive highways were

[1] `http://www.publicpurpose.com/hwy-worldmotorway.htm`

[2] National Institute of Statistics and Geography (INEGI Spanish abbreviation) `http://inegi.org.mx/`

built from Central America across Mexico to the United States, and from coast to coast.

We selected 19 highly populous urban areas, based on a selection of relevant features, including economic impact, sea- and airports and tourist attractions:

(1) Tijuana,
(2) Nogales,
(3) Ciudad Juárez,
(4) Hermosillo,
(5) Chihuahua,
(6) Nuevo Laredo,
(7) Monterrey,
(8) Mazatlán,
(9) Ciudad Victoria,
(10) San Luis Potosí,
(11) Guadalajara,
(12) Léon and Guanajuato,
(13) Morelia,
(14) Edo. México, DF, Puebla,
(15) Xalapa and Veracruz,
(16) Chilpancingo and Acapulco,
(17) Oaxaca and Huatulco,
(18) Tuxtla Gutiérrez,
(19) Merida and Cancún.

Further, we refer to the urban regions as **U**. These regions of **U** are projected onto gel and oat flakes, of size and shape approximately matching the size and shape of the regions, are placed in the positions of the regions (Fig. 12.1b). At the beginning of each experiment a piece of plasmodium, usually already attached to an oat flake, was placed, or inoculated, in the region corresponding to Edo. México, DF, Puebla (region 14 in Fig. 12.1a).

12.1 Mexico City to Monterrey in 12 h

Snapshots of plasmodium foraging patterns during a typical experiment are shown in Fig. 12.2. Initially a piece of plasmodium was placed onto an oat flake representing Mexico City. The plasmodium grows and propagates in a south-eastern direction occupying the Oaxaca–Huatulco region and then travels towards the Tuxtla Gutiérrez region (Fig. 12.2a). The plasmodium reaches the Tuxtla

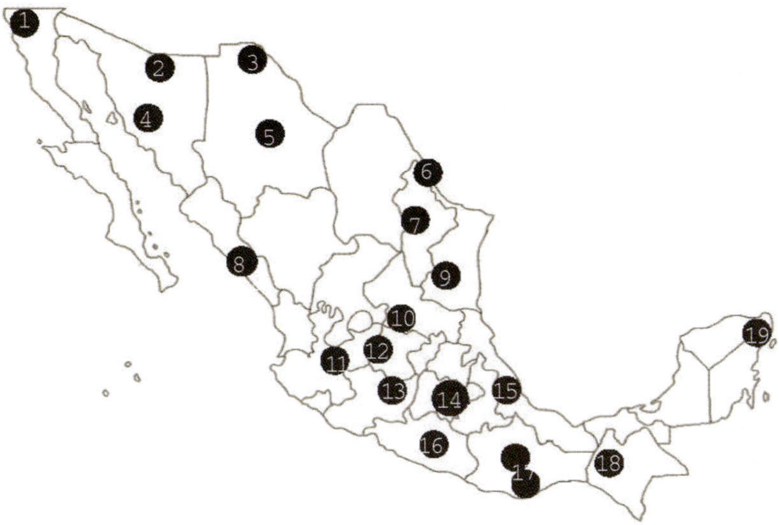

(a)

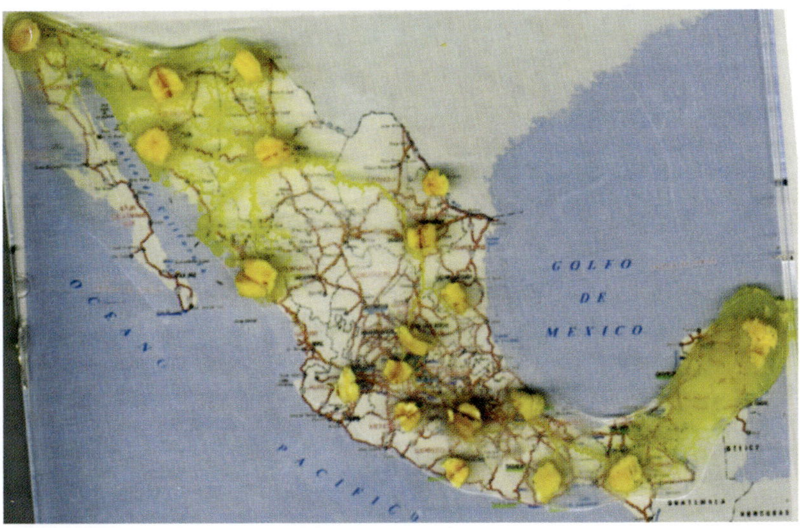

(b)

Fig. 12.1 Experimental basics: (a) contour map of Mexico with 19 sources of nutrients indicated, (b) snapshot of a typical setup: urban areas are represented by oat flakes, plasmodium is inoculated in Mexico City, the plasmodium protoplasmic transport network grows to reach the oat flakes. Note that configurations of protoplasmic tubes are different in these two experiments.

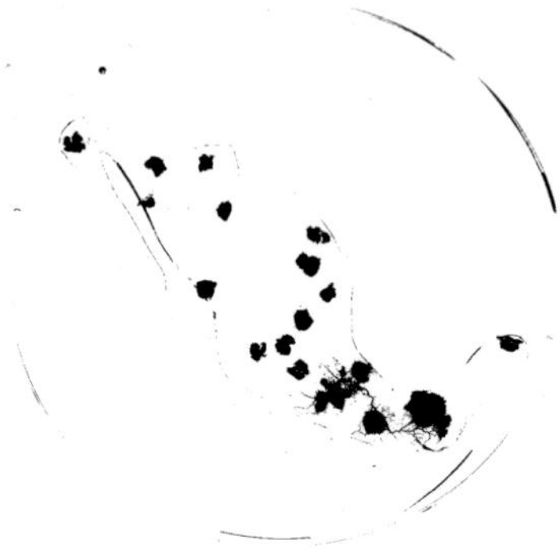

(a) $t = 12$ h

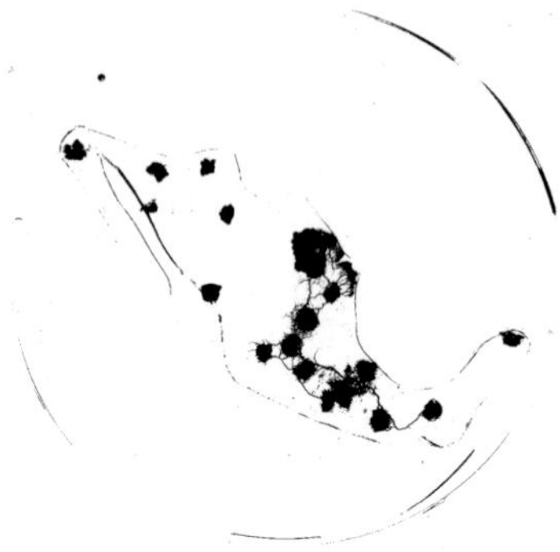

(b) $t = 24$ h

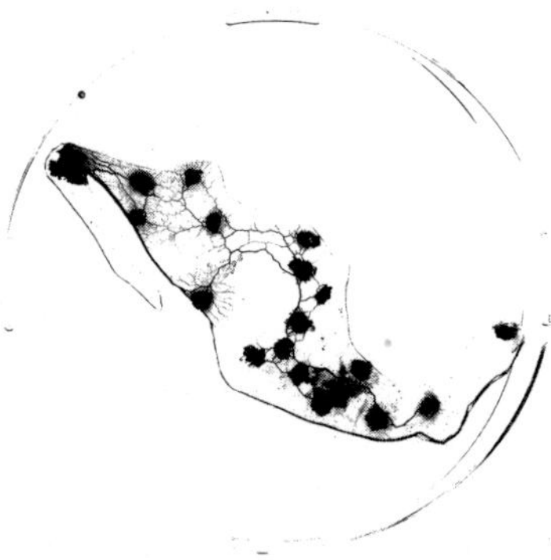

(c) $t = 38$ h

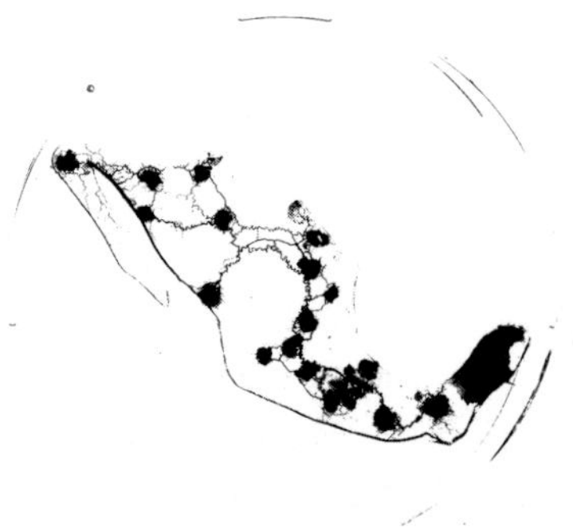

(d) $t = 52$ h

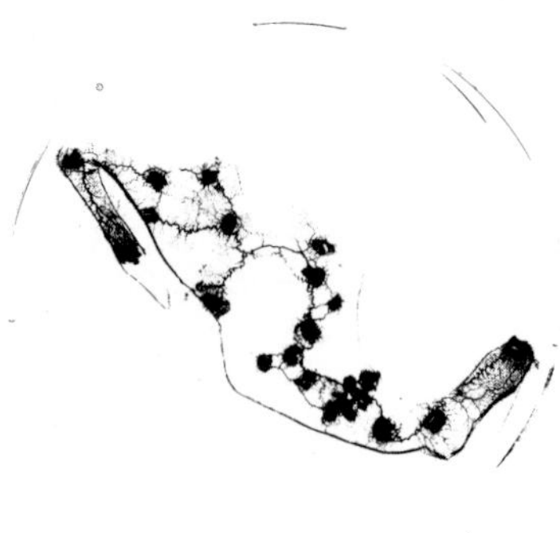

(e) $t = 61$ h

Fig. 12.2 Typical plasmodium development. Time elapsed from inoculation is shown in the subfigure captions.

Gutiérrez region usually in 12 h after inoculation (real-time distance is 11 h). At the same time the plasmodium colonises the Xalapa–Veracruz and Chilpancingo–Acapulco regions and moves north-north-west. In a normal situation (vigorous plasmodium, fresh agar gel, air not contaminated by bacteria), south-eastern and north-north-western developments occur simultaneously. In 12–14 h after inoculation the plasmodium can reach as far as the Monterrey and Nuevo Laredo regions (Fig. 12.2b). The plasmodium colonises almost the whole of Mexico in 38 h, spanning major urban regions from Tijuana in the north-west to Tuxtla Gutiérrez in the south-east (Fig. 12.2c). The colonisation is truly completed in 52 h, when an oat flake representing Merida–Cancún is covered by plasmodium mass (Fig. 12.2d). Despite spanning all cities marked by oat flakes, the plasmodium continues exploring the territory outlined by the agar gel and propagates along the Baja California peninsula (Fig. 12.2e). There are no cities represented by food sources in that region; therefore, the plasmodium retreats from the peninsula.

In some situations the plasmodium does not span all cities, represented by oat flakes. An example is shown in Fig. 12.2. In a few hours after being inoculated in Mexico City the plasmodium propagates in all directions. In 12 h the plasmodium reaches the Ciudad Victoria region in the north and Xalapa–Veracruz in the south (Fig. 12.2a). On the 24th hour of development the plasmodium expands its occupation until the Nuevo Laredo region in the north and Tuxtla Gutiérrez in the south (Fig. 12.2b). The plasmodium continues then to the Chihuahua and Merida–Cancún regions (Fig. 12.2c) and abandons the oat flake representing the Nuevo Laredo region. Even when the whole territory of Mexico becomes colonised by plasmodium the Nuevo Laredo region remains free (Fig. 12.2d).

The plasmodium may not stop its foraging activity even when all sources of nutrients are occupied and the whole agar plate is explored. As illustrated in Fig. 12.2, a vigorous plasmodium can spread over surrounding Petri dishes, trying to settle on bare plastic. Plasmodium parts residing on a non-agar substrate suffer from the lack of humidity and therefore cease to be sustained. Abandoned protoplasmic tubes are clearly visible in the southern half of the Petri dish in Fig. 12.2d.

As illustrated in Figs. 12.2, 12.2 and 12.2, plasmodium rarely develops exactly the same foraging pattern twice. Even during a single experiment, the plasmodium sometimes may change the topology of its protoplasmic networks, abandoning and recolonising sources of food. Thus, we could never consider a stationary configuration of a protoplasmic network but a probabilistic graph, representing all possible configurations of protoplasmic networks occurring in experiments for a given configuration of sources of nutrients.

Physarum graphs extracted from 26 laboratory experiments are shown in Fig. 12.2. The graphs become planar when we remove edges with weights below

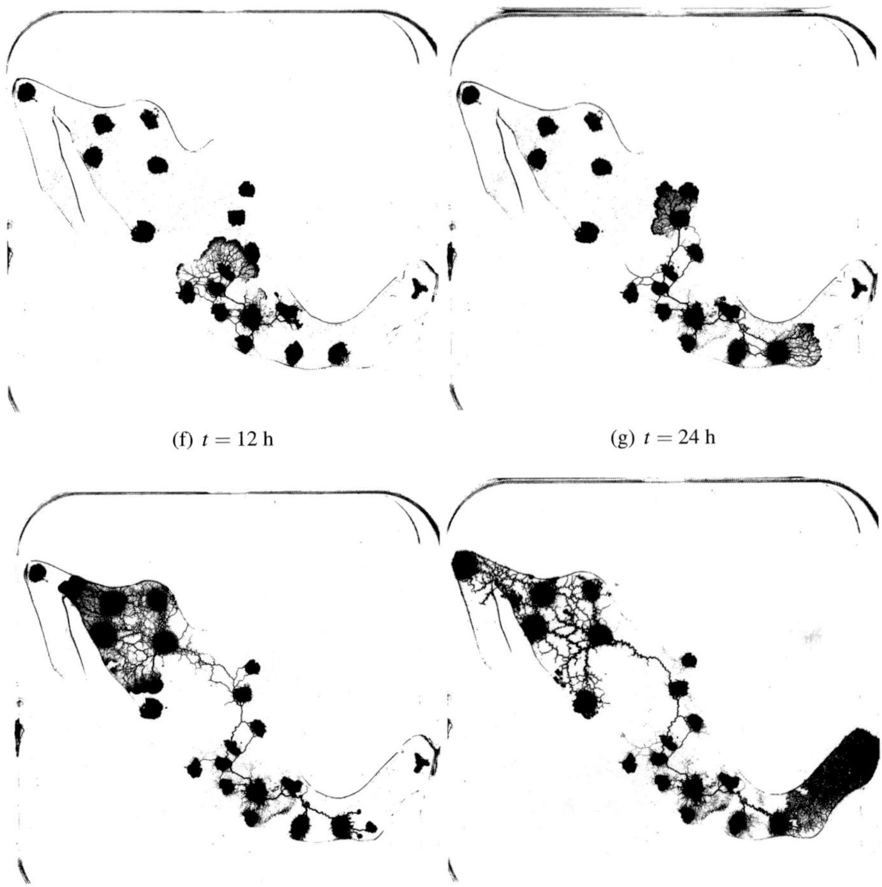

(f) $t = 12$ h (g) $t = 24$ h

(h) $t = 35$ h (i) $t = 47$ h

Fig. 12.2 (Continued) Plasmodium does not always span all cities (sources of food). Time elapsed from inoculation is shown in the subfigure captions.

0.19 (Fig. 12.2b). The graph is acyclic, or a tree, when only edges with probability weights exceeding 0.58 are taken into consideration (Fig. 12.2d). Thus, edges of the spanning tree are represented by protoplasmic tubes in over half of the experimental trials.

Finding 84. *A spanning tree rooted in Mexico City is a stable core structure of a Physarum foraging and nutrient transport network built on a configuration of urban regions* **U**.

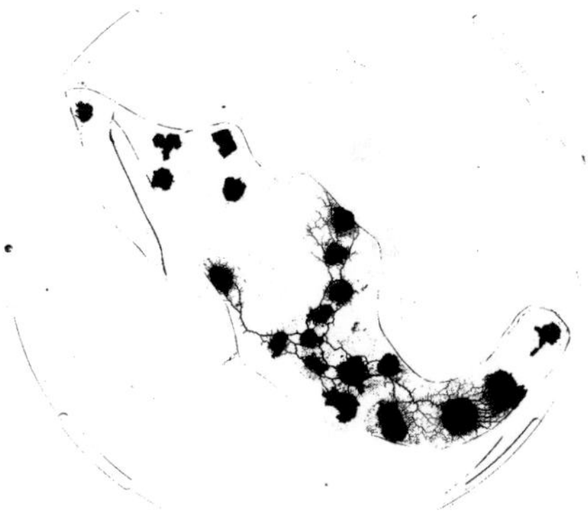

(j) $t = 24$ h

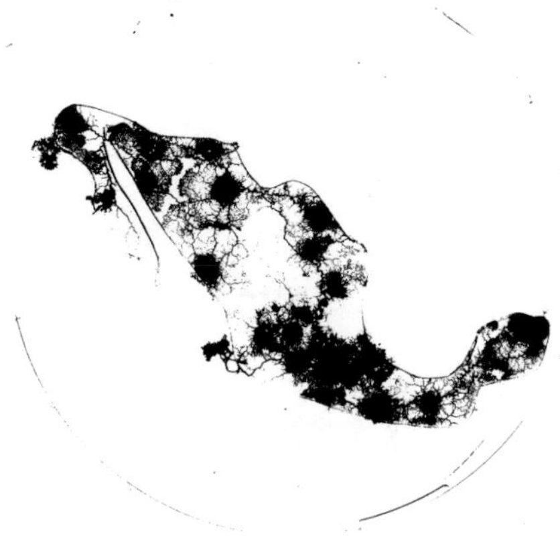

(k) $t = 36$ h

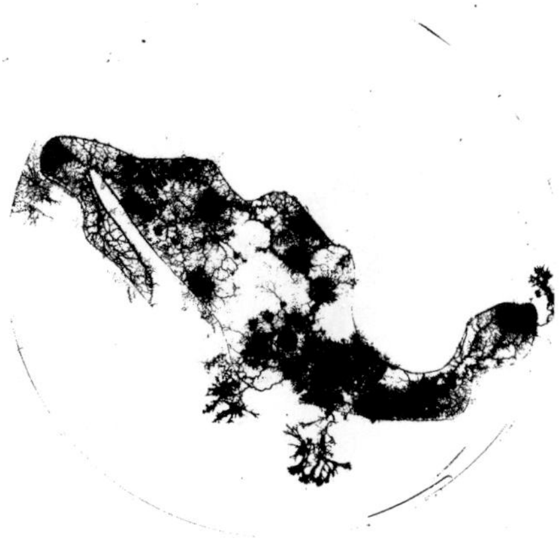

(l) $t = 48$ h

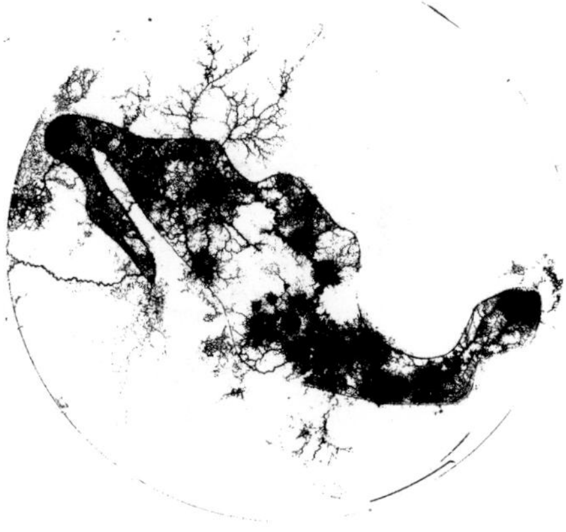

(m) $t = 70$ h

Fig. 12.2 (Continued) Plasmodium spreads beyond the 'dedicated' experimental domain. Time elapsed from inoculation is shown in the subfigure captions.

If we increase the value of θ to 0.61, the Physarum transport graph becomes disconnected (Fig. 12.3a). The nodes corresponding to the Tijuana and Mazatlán regions become isolated. There is no direct man-made highway between such cities, mainly with main big highways such as M15 or M2, and this isolation, of course, represents geographical limitations.

The graph is split into two components when $\theta = 0.65$ (Fig. 12.3b). The larger component corresponds to the highway that dominates the north-west of Mexico,

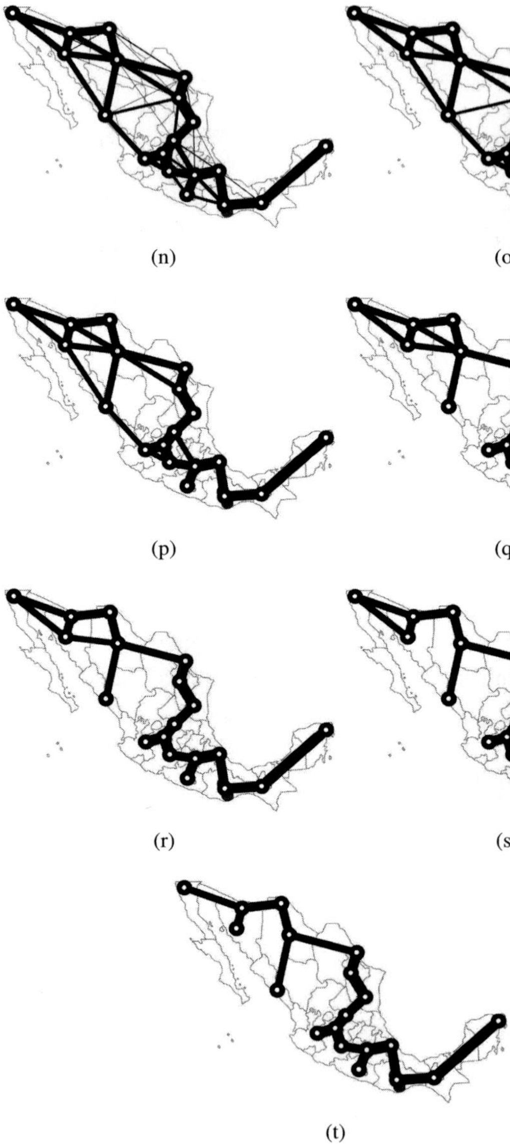

Fig. 12.2 (Continued) Configurations of Physarum graph $\mathbf{P}(\theta)$ for (a) $\theta = 0$, (b) $\theta = 0.19$, (c) $\theta = 0.38$, (d) $\theta = 0.33$, (e) $\theta = 0.5$, (f) $\theta = 0.54$ and (g) $\theta = 0.58$. Thickness of each edge is proportional to the edge's weight.

(a)

(b)

Fig. 12.3 Configurations of Physarum graph $\mathbf{P}(\theta)$ for (a) $\theta = 0.61$ and (b) $\theta = 0.65$. Thickness of each edge is proportional to the edge's weight.

that is, M2 crossing from Tijuana to Ciudad Juárez. As is the case with the real M2, the corresponding structure in the gel does not reach the 'east coast'.

The other component displays a strong connection between the regions corresponding to Mexico City and the south-east of Mexico. This reflects the existence of real highway links connecting the north-eastern and south-eastern regions of Mexico with Mexico City, namely the motorways M85 from Nuevo Laredo to Mexico City and M180 that runs from Xalapa–Veracruz to Cancún.

The table in Fig. 12.4 provides information from 'Guía Roji' publications[3] about the major highways connecting the cities that correspond to the regions in **U**.

Finding 85. *Transport links to the Tijuana and Mazatlán urban regions are weakly represented by slime mould.*

These are implications of Finding 1 and the table in Fig. 12.4 connectivity. One needs to travel along at least two or three highways to reach Mazatlán from Tijuana.

We now consider the question of how well Physarum graphs are seen to approximate the Mexican highway network. A sketch of the highway network and **H**, a graph derived from it, are shown in Fig. 12.5.

Finding 86. **H** *contains only three edges that are not also contained in* $\mathbf{P}(0.58)$.

The intersection of **P** and **H** of Physarum and highway graphs is shown in Fig. 12.5c. The three edges of **H** that are not present in $\mathbf{P}(0.58)$ are transport

[3]Source: Guía Roji 'Por las Carreteras de México 2011', 17th edition. Web site: `http://www.guiaroji.com.mx`

Areas One way	1	2	3	4	5	6	7	8	9	10	11	12	13	14	15	16	17	18	19
1	1	0	M2	0	0	0	0	0	0	0	0	0	0	0	0	0	0	0	0
2	0	1	0	M15	0	0	0	M15	0	0	M15	0	M15	M15	0	0	0	0	0
3	M2	0	1	0	M45	0	0	0	0	0	0	M45	0	0	0	0	0	0	0
4	0	M15	0	1	M16	0	0	M15	0	0	M15	0	M15	M15	0	0	0	0	0
5	0	0	M45	M16	1	0	0	0	0	M49	0	M45	0	0	0	0	0	0	0
6	0	0	0	0	0	1	M85	0	M85	0	0	0	0	M85	0	0	0	0	0
7	0	0	0	0	0	M85	1	M40	M85	0	0	0	0	M85	0	0	0	0	0
8	0	M15	0	M15	0	0	M40	1	0	0	M15	0	M15	M15	0	0	0	0	0
9	0	0	0	0	0	M85	M85	0	1	0	0	0	0	M85	0	0	0	0	0
10	0	0	0	0	M49	0	0	0	0	1	M80	0	0	M57	0	0	0	0	0
11	0	M15	0	M15	0	0	0	M15	0	M80	1	0	M15	M15	0	0	0	0	0
12	0	0	M45	0	M45	0	0	0	0	0	0	1	0	0	0	0	0	0	0
13	0	M15	0	M15	0	0	0	M15	0	0	M15	0	1	M15	0	0	0	0	0
14	0	M15	0	M15	0	M85	M85	M15	M85	M57	M15	0	M15	1	M150	M95	M190	M190	0
15	0	0	0	0	0	0	0	0	0	0	0	0	0	M150	1	0	0	0	M180
16	0	0	0	0	0	0	0	0	0	0	0	0	0	M95	0	1	M200	0	0
17	0	0	0	0	0	0	0	0	0	0	0	0	0	M190	0	M200	1	M190	0
18	0	0	0	0	0	0	0	0	0	0	0	0	0	M190	0	0	M190	1	0
19	0	0	0	0	0	0	0	0	0	0	0	0	0	0	M180	0	0	0	1

Fig. 12.4 Main highway connections between urban areas of **U**.

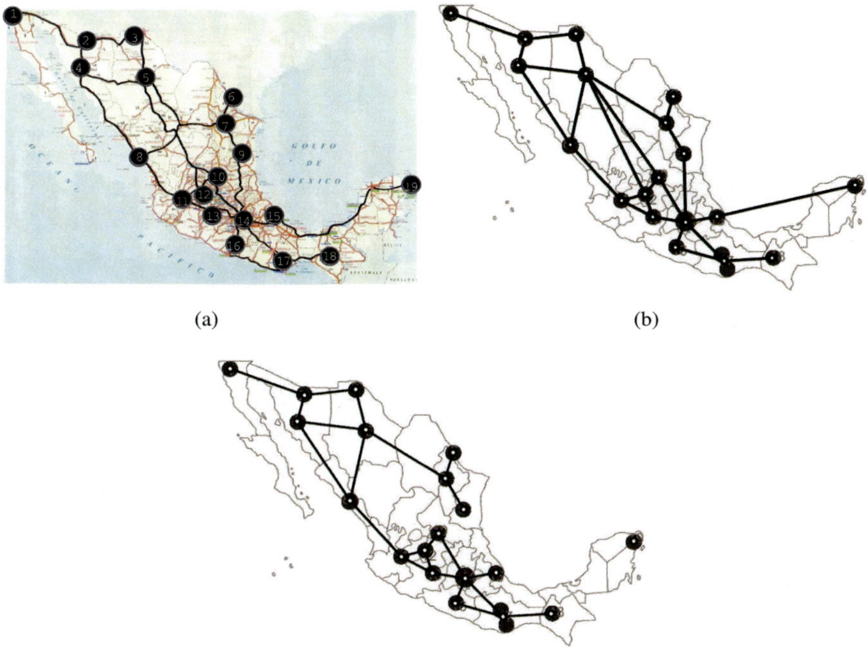

(a) (b)

(c)

Fig. 12.5 Highway network is highlighted in (a) and graph **H** of man-made motorway network is shown in (b). Intersection $\mathbf{P} \bigcap \mathbf{H}$ of Physarum **P** and highway **H** graphs is shown in (c).

links connecting the Chihuahua and Nuevo Laredo regions, Xalapa–Veracruz and Oaxaca–Huatulco regions and Tuxtla Gutiérrez and Merida–Cancún regions. Such differences could possibly be related to the fact that in laboratory experiments we did not represent features of the geographical landscape in the agar gel structure but just used a uniform flat gel plate cut in the shape of Mexico.

Indeed, we have considered natural restrictions. Missing connections between Chihuahua and Nuevo Laredo are due to the Chihuahua desert, while connections between Xalapa–Veracruz and Oaxaca–Huatulco are restricted by geographical features such as the Juárez mountain range. The missing connection between Tuxtla Gutiérrez and Merida–Cancún may be explained by a large border over the Gulf of Mexico going up to motorway M180 and crossing the Chiapas mountain range.

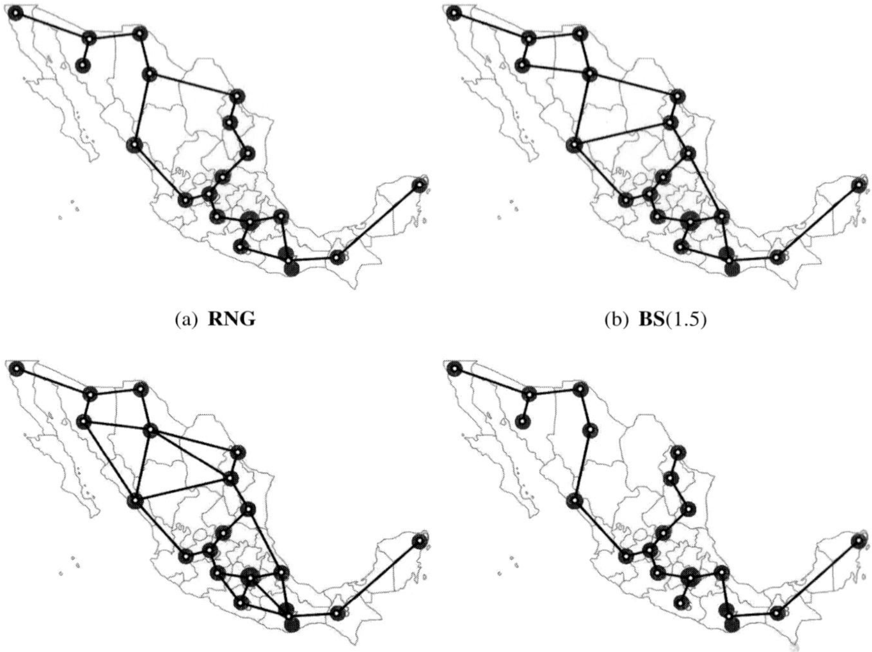

(a) **RNG** (b) **BS**(1.5)

(c) **GG** (d) **ST**

Fig. 12.6 Proximity graphs constructed on regions **U**: (a) relative neighbourhood graph **RNG**, (b) β-skeleton with control parameter 1.5, (c) Gabriel graph **GG**, (d) minimum spanning tree **ST**.

12.2 Spanning trees, Physarum and conquistadors

Let us compare the Physarum graph **P** and the highway graph **H** to the relative neighbourhood graph **RNG**, β-skeleton **BS**(1.5) (β-skeleton **BS** was calculated for $\beta = 1.5$ to make an 'intermediate' graph between **RNG** and **GG**), Gabriel graph **GG** and minimum spanning tree **MST** (Fig. 12.7). From now on, we mean **P**(0.19) when writing **P**.

Finding 87. **RNG** $\subset$ **P**.

This is a direct outcome of calculating intersections of the graphs (Fig. 12.7). The relative neighbourhood graph is considered to be generally optimal or near optimal in the context of road networks [Watanabe (2005); Watanabe (2008)] in terms of minimising travel distances and total length of the road network. It is interesting that the relative neighbourhood graph is a subgraph of the Physarum

(a) **P**⋂**RNG**

(b) **P**⋂**BS**(1.5)

(c) **P**⋂**GG**

(d) **P**⋂**MST**

(e) **H**⋂**RNG**

(f) **H**⋂**BS**(1.5)

(g) **H**⋂**GG**

(h) **H**⋂**MST**

Fig. 12.7 Intersections of Physarum graph **P** (a–d) and highway graph **H** (e–h) with proximity relative neighbourhood graph **RNG**, β-skeleton **BS**(1.5), Gabriel graph **GG** and minimum spanning tree **MST**.

graph, suggesting a similarity between the formation process of the plasmodium transport network and the human design process in the case of the road network.

In all cases in Fig. 12.7a–d, the pathway between Mexico City and the USA is preserved in the three following ways. Mexico City via Ciudad Juárez (M45), Mexico City via Nogales by motorway M15 and Mexico City via Nuevo Laredo by motorway M85. These are historically significant routes since Mexican colonisation, representing ancient ways to transport materials and natural resources to Spanish monarchs and later British colonies (now the USA). The transport routes were initially built to connect Xalapa–Veracruz to Mexico City (in the Aztec age, when the Spanish tried to reach Tenochtitlan); in modern times they are represented by the motorway M15.

In this context, it is not surprising that Ciudad Juárez occupies a strategic position in reaching the USA directly from central Mexico and beyond. This is because motorway M45 is connected with M190 and runs across all the Mexican Republic south-to-north from Central America. It is also important to highlight that Ciudad Juárez has undergone an explosive growth in the last few years, partly due to its dramatic life at the intersection of main drug transport arteries.

South–east intersections represent connectivity with highways M180 and M190. Both are the main large highways connecting principal cities in their regions with Mexico City.

Finding 88. MST $\subset$ P.

The result is expected because a spanning tree is a subgraph of a relative neighbourhood graph [Toussaint (1980)]; however, we intentionally highlight it to show that — in terms of minimal-length travel — a graph built by plasmodium of *P*hysarum polycephalum offers an optimal solution for transportation of nutrients. Experiment-wise, we found that the spanning tree **MST** is a subgraph of the Physarum graph $\mathbf{P}(\theta)$ for $\theta \leq 0.58$; compare Figs. 12.2 and 12.7d.

As we demonstrated previously by cutting edges of $\mathbf{P}(\theta)$ with $\theta \leq 0.58$, we transform a cyclic graph to a spanning tree (Fig. 12.2g). By comparing Figs. 12.2g and 12.6d, we find that if Physarum built a transport link connecting the Mazatlán and Guadalajara regions instead of connecting the Chihuahua and Nuevo Laredo regions, then the Physarum graph would approximate an ideal minimum spanning tree. The actual travel distance makes the Physarum spanning tree $\mathbf{P}(0.58)$ just a bit longer than the ideal spanning tree **MST**, constructed by the conventional algorithm (Fig. 12.6d).

In contrast to the Physarum graph, the highway graph **H** poorly matches proximity graphs (Fig. 12.7e–h).

Finding 89. *Let* $\mathbf{G} \in \{\mathbf{RNG}, \mathbf{BS}(1.5), \mathbf{GG}, \mathbf{MST}\}$*; then* $\mathbf{G} \bigcap \mathbf{H}$ *is a disconnected graph.*

The Merida–Cancún region stays isolated in all four intersections (Fig. 12.7e–h) of the highway graph **H** with proximity graphs. A connected cluster of regions Nuevo Laredo, Monterrey and Ciudad Victoria is disconnected from the rest of the other vertices of **U** in $\mathbf{H} \bigcap \mathbf{RNG}$, $\mathbf{H} \bigcap \mathbf{BS}(1.5)$ and $\mathbf{H} \bigcap \mathbf{GG}$. Of course, this cluster represents the most important economic activity of the region. But in this case the connectivity result is poor with respect to the intersection with **P**. A similar situation takes place with the Merida–Cancún connection in the intersection of the minimum spanning tree **MST** with the highway graph **H**.

Let a be a transport link connecting the San Luis Potosí region with Mexico City–Puebla State; then we see that links of the motorway network **H** which are also edges of the Gabriel graph **GG** correspond to the common links of the motorway network **H** and the Physarum graph $\mathbf{P}(0.19)$. The San Luis Potosí connection represents a special case linked directly to the Mexico City region with motorway M57.

Finding 90. $\mathbf{H} \bigcap \mathbf{GG} \bigcup \{a\} = \mathbf{H} \bigcap \mathbf{P}(0.19)$.

We have lost a link between Merida–Cancún and central Mexico and also a centre-to-north link from San Luis Potosí to Chihuahua. This means that motorways M85 and M190 are partly misrepresented. Having said that, the Pacific-long highway connecting Nogales to Mexico City by M15 is preserved. Thus, at least, we have preserved one of the largest and most important connections between Mexico City and the north of Mexico. In this way, $\mathbf{P}(0.58)$ is the best approximation, as we saw previously.

12.3 Summary

To approximate, or rather reconstruct, the development of the transport network in Mexico, we cut a Mexico-shaped plate of agar, represented 19 major urban regions by oat flakes and placed a plasmodium of *P. polycephalum* in the place of Mexico City. The plasmodium developed into a fully fledged plasmodium spanning all, or almost all, oat flakes on the agar plate. We compared graphs of protoplasmic networks developed by Physarum with the man-made federal highway network and with basic types of proximity graphs.

We found that the Physarum-made graph is a subgraph of the highway network apart from a few edges. This means that, in principle, slime mould based approximation of man-made transport networks works even in very simple experimental

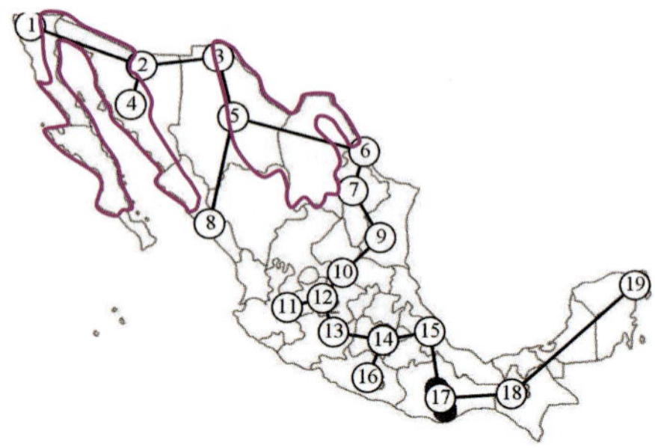

Fig. 12.8 Sonoran and Chihuahua deserts are shown together with Physarum graph $\mathbf{P}(\frac{15}{19})$.

setups. The slight mismatch between Physarum and highway graphs may be because we did not take the geographical profile of Mexico into account completely; no mountains or rivers were mapped onto the plasmodium growth substrate. Also, inoculation of plasmodium in Mexico City was backed up only by the fact that Mexico City is the most populated region of the country. Having said that, we must admit that some features of the Mexican geography, especially deserts, are taken into account by the slime mould without being physically present in the growth substrate.

For example, edges of Physarum graphs in Fig. 12.2d–g 'detect' presence of the Sonoran and Chihuahua deserts (Fig. 12.8); see also Fig. 12.7d–f and h. It would be difficult to build a transport route via Baja California or the mountains of Sierra De Juárez, and thus the slime mould does not develop such links (see e.g. Fig. 12.2a).

The slime mould reconstructs principal routes of transport between Central and South America and Mexico. The principal routes imitated by *P. polycephalum* include the routes from Chiapas to Ciudad Juárez:

- Tuxtla Gutiérrez → Oaxaca and Huatulco → Xalapa and Veracruz → Mexico and Puebla → Morelia → Léon and Guanajuato → San Luis Potosí → Monterrey → Chihuahua → Ciudad Juárez,
- Tuxtla Gutiérrez → Oaxaca and Huatulco → Xalapa and Veracruz → Mexico and Puebla → Morelia → Léon and Guanajuato → San Luis Potosí → Ciudad Victoria → Monterrey → Chihuahua → Ciudad Juárez,
- Tuxtla Gutiérrez → Oaxaca and Huatulco → Xalapa and Veracruz → Mexico and Puebla → Morelia → Léon and Guanajuato → San Luis Potosí → Monterrey → Nuevo Laredo → Chihuahua → Ciudad Juárez (via Nuevo Laredo),

- Tuxtla Gutiérrez $\rightarrow$ Oaxaca and Huatulco $\rightarrow$ Xalapa and Veracruz $\rightarrow$ Mexico and Puebla $\rightarrow$ Morelia $\rightarrow$ Léon and Guanajuato $\rightarrow$ San Luis Potosí $\rightarrow$ Ciudad Victoria $\rightarrow$ Monterrey $\rightarrow$ Nuevo Laredo $\rightarrow$ Chihuahua $\rightarrow$ Ciudad Juárez (via Nuevo Laredo)

and from Chiapas to Nuevo Laredo:

- Tuxtla Gutiérrez $\rightarrow$ Oaxaca and Huatulco $\rightarrow$ Xalapa and Veracruz $\rightarrow$ Mexico and Puebla $\rightarrow$ Morelia $\rightarrow$ Léon and Guanajuato $\rightarrow$ San Luis Potosí $\rightarrow$ Monterrey $\rightarrow$ Nuevo Laredo,
- Tuxtla Gutiérrez $\rightarrow$ Oaxaca and Huatulco $\rightarrow$ Xalapa and Veracruz $\rightarrow$ Mexico and Puebla $\rightarrow$ Morelia $\rightarrow$ Léon and Guanajuato $\rightarrow$ San Luis Potosí $\rightarrow$ Monterrey $\rightarrow$ Ciudad Victoria $\rightarrow$ Nuevo Laredo.

Both routes from Chiapas to Ciudad Juárez or Nuevo Laredo are main arteries for *narcotráfico*.

Chapter 13

Physarum in The Netherlands: responding to the flood

Andrew Adamatzky, Michael Lees and Peter M. A. Sloot

In this chapter we hope to move towards a more general understanding of slime mould's capability to compute road networks by investigating the roads in The Netherlands. The Netherlands presents an excellent case study for the two following reasons.

First, The Netherlands is amongst the top countries with the highest population density because only 12% of the territory is allocated to built-up areas [Alpokin (2012)]. The country has the highest-density motorway network in Europe. Moreover, the demand on the system is at levels which are reaching current limits, with a total length of 132,397 km and usage of 140×10^9 people per km per year[1]. Such high occupancy may pose a need for urgent expansion of the transport networks and a better understanding of the limitations to that growth. As analysed in [Alpokin (2012)], the Dutch national spatial strategy puts a stress on developing transport networks amongst the major cities. Six national 'networks of cities' might lead to increased congestion and increased travel times. Sprawling of urban areas along major transport arteries is a common feature, which also could be imitated by slime mould due to its active branching behaviour.

Second, The Netherlands is also at risk of significant flooding[2] because over 40% of the country's land lies below sea level. The largest flooding disaster in 1953 cost almost two thousand human lives [Goemans and Visser (1987)], and only by chance were larger cities such as Amsterdam, Rotterdam and The Hague not flooded at that time [Wesselink (2007)]. Despite a wide range of technologically advanced systems, The Netherlands are not immune from mass-scale floods.

[1] `www.autosnelwegen.nl`

[2] `http://urbanflood.eu/default.aspx`

The chance-flood defence systems are small

> "However, on a human time scale it means a chance of at least 1% of the population having to face flooding once in its lifetime — a tiny percentage, yes, but not negligible. More importantly, if things do go wrong, in many places they are apt to go badly wrong. Millions of people live several meters below sea level. In the worst scenario, parts of these areas will be below several meters of water within a few hours, and many casualties will result. Economic, cultural, and societal damage will be unimaginable." [Wesselink (2007)].

This is why it is important to imitate a worst-case scenario when a substantial area of The Netherlands is flooded and mass migration of those who manage to stay alive starts. We imitate this mass migration of slime mould.

All experiments consider the 21 most populous urban areas in The Netherlands (Fig. 13.6a):

(1) Leeuwarden,
(2) Groningen,
(3) Den Helder,
(4) Lelystad,
(5) Zwolle,
(6) Haarlem,
(7) Amsterdam,
(8) Utrecht,
(9) Amersfoort,
(10) Apeldoorn,
(11) Enschede,
(12) Den Haag,
(13) Rotterdam,
(14) Dordrecht,
(15) Nijmegen,
(16) 's-Hertogenbosch,
(17) Breda,
(18) Tilburg,
(19) Middelburg,
(20) Eindhoven,
(21) Maastricht.

At the beginning of each experiment a piece of plasmodium, usually already attached to an oat flake, was placed in Amsterdam (region 7 in Fig. 13.6a). A total of 62 experiments were conducted.

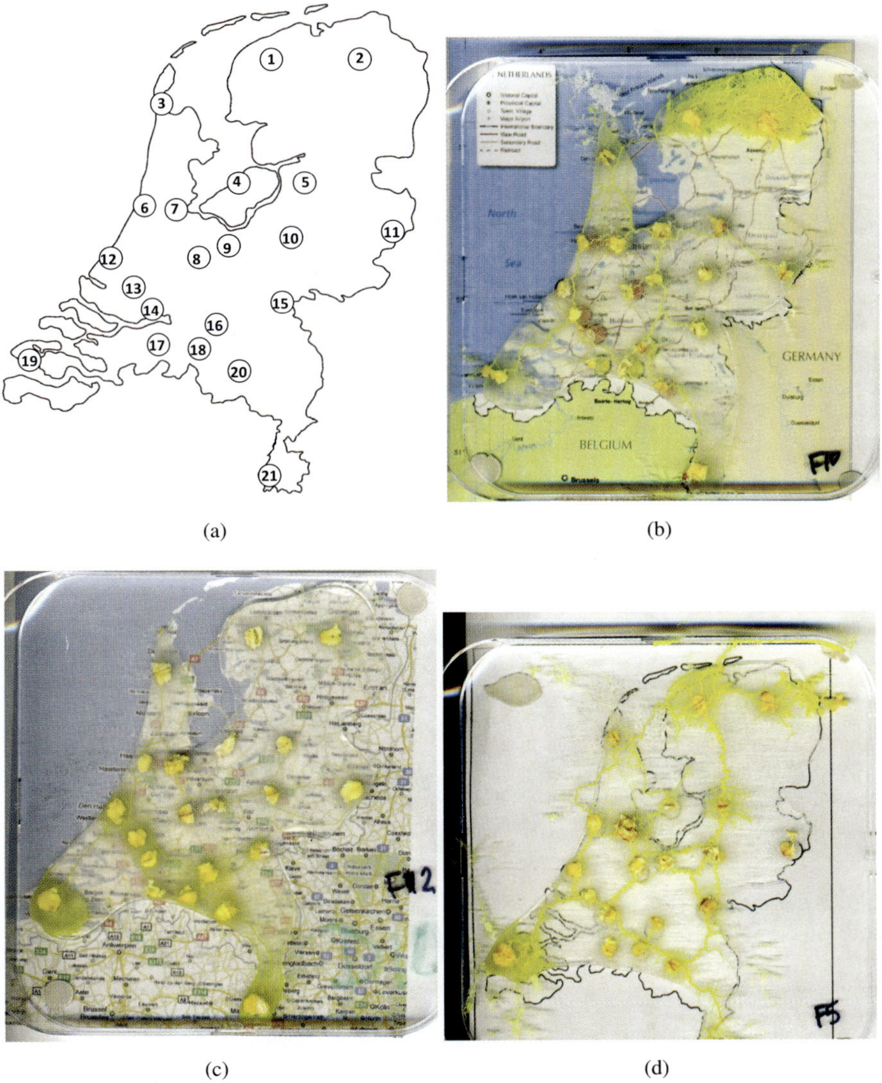

Fig. 13.1 Experimental basics: (a) outline map of The Netherlands with 21 sources of nutrients indicated, (b–d) snapshots of typical setups: urban areas are represented by oat flakes, plasmodium is inoculated in Amsterdam, the plasmodium spans the oat flakes by protoplasmic transport network.

13.1 Amersfoort–Lelystad–Leeuwarden–Groningen

In a laboratory experiment, illustrated in Fig. 13.2, the following chain of events

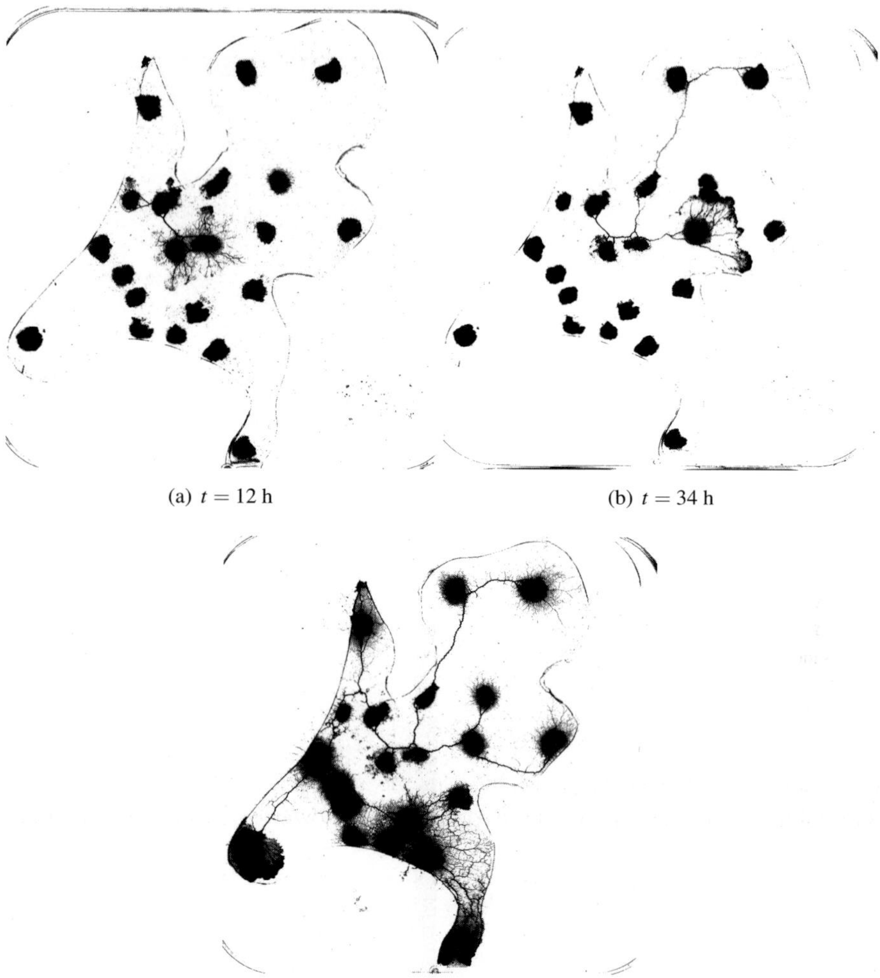

(a) $t = 12$ h (b) $t = 34$ h

(c) $t = 57$ h

Fig. 13.2 Illustrative example of plasmodium development on configuration of cities represented by oat flakes. Time elapsed from inoculation is shown in the subfigure captions.

unfolds (dynamics of colonisation is schematically represented in Fig. 13.3). An oat flake colonised by plasmodium was placed on top of the oat flake representing Amsterdam. In 12 h the plasmodium follows gradients of chemoattractants, links Amsterdam with Haarlem and propagates towards Utrecht and Amersfoort, spreading in all directions except the north-west (Fig. 13.2a). After 34 h the plasmodium colonises Leeuwarden and Groningen. It develops clearly vis-

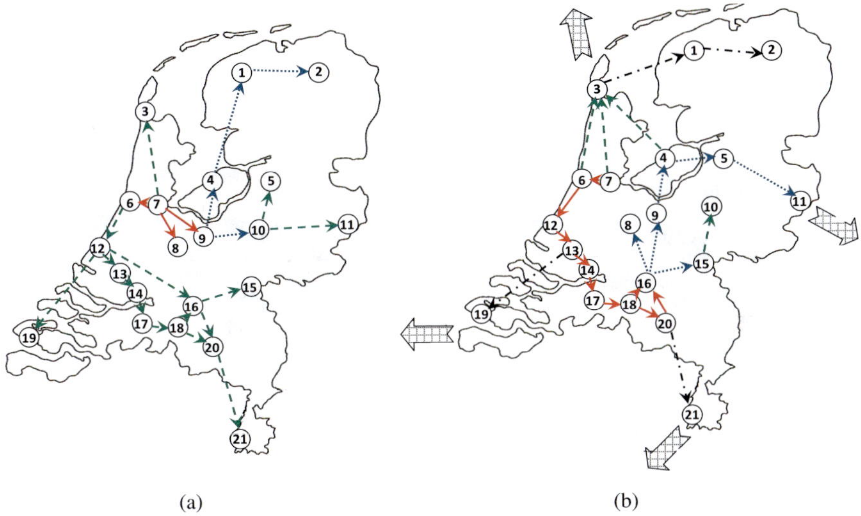

Fig. 13.3 Diagram of colonisation dynamics derived from experiments of Fig. 13.2 (a) and Fig. 13.4 (b): links developed in 12 h after inoculation are shown by red solid arrows, in 34 h by blue dotted arrows, in 57 h by green dashed lines and in 80 h by dash-dotted lines. Large mesh-patterned arrows indicate migration of plasmodium outside the country.

ible protoplasmic tubes, which represent a transport link Amersfoort–Lelystad–Leeuwarden–Groningen (Fig. 13.2b). In the same time interval the plasmodium colonises Apeldoorn and starts propagations towards Zwolle and Enschede (Fig. 13.2b).

After a total of 57 h the plasmodium connects Apeldoorn with Enschede and Zwolle by protoplasmic tubes and colonises the south-western part of the country. Namely, the plasmodium links Haarlem and The Hague and builds a route from The Hague to Middelburg and a link Hague–Rotterdam–Dordrecht–Breda–Tilburg–'s-Hertogenbosch (Fig. 13.2c and Fig. 13.3a, green dashed lines). At the same time the plasmodium forms a protoplasmic tube directly connecting Amsterdam, Den Helder and The Hague with 's-Hertogenbosch, and develops the links 's-Hertogenbosch–Nijmegen and Tilburg–'s-Hertogenbosch–Eindhoven–Maastricht (Fig. 13.2c).

We observe that the dynamics of colonisation is non-uniform (Fig. 13.3a). The plasmodium does not spread or diffuse in all directions simultaneously but rather colonises the north-north-western part of the country first and only then explores the south-south-west. This may be due to the fact that centres of activity (biochemical oscillators) form during propagation, and the contractive waves evoked

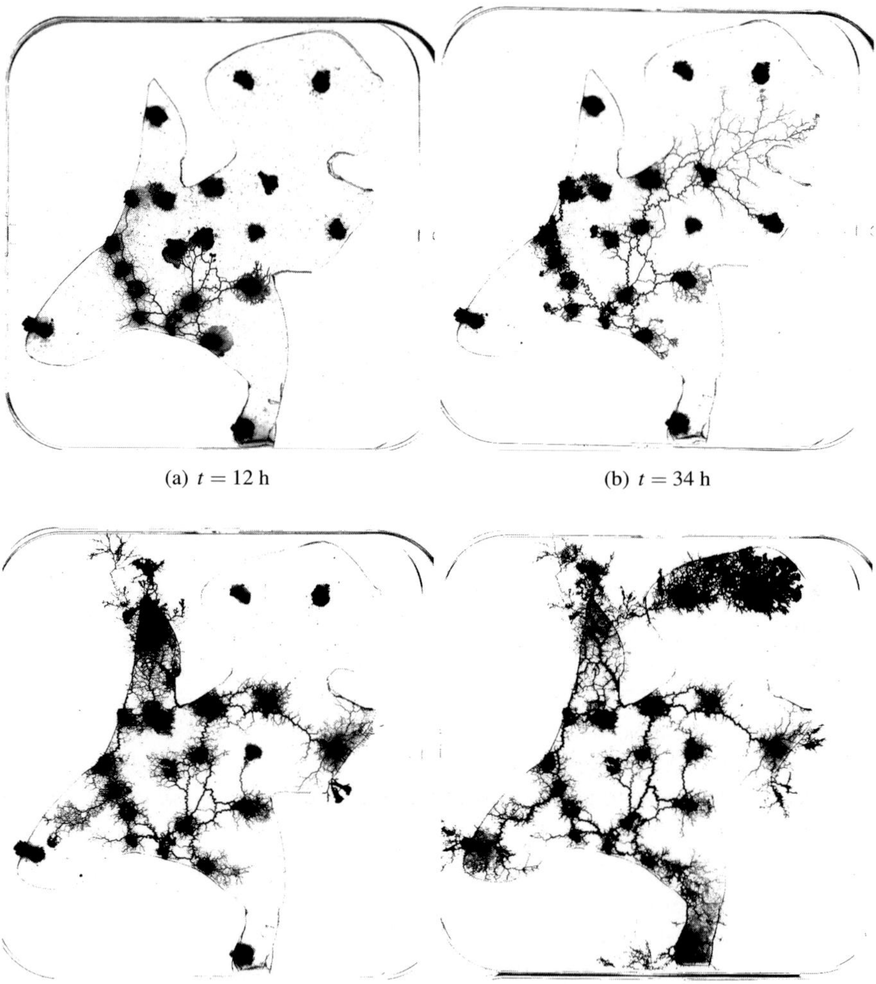

(a) $t = 12$ h (b) $t = 34$ h

(c) $t = 57$ h (d) $t = 80$ h

Fig. 13.4 Plasmodium spreads beyond 'dedicated' experimental domain. Time elapsed from inoculation is shown in the subfigure captions.

by the oscillators force the protoplasm to move towards the oscillators. Therefore, if an oscillator is formed in the northern part of the plasmodium, then propagation in all other directions would be suppressed.

The plasmodium of *P. polycephalum* rarely repeats itself in experimental trials. The overall or average pattern may be the same but a myriad of variations are possible in the course of plasmodium's spatial development. Outperforming (spreading out of the dedicated area) and underperforming (not colonising the whole area) are typical examples of the varieties in plasmodium behaviour. These two examples are illustrated and discussed below.

In a substantial number of laboratory experiments, the plasmodium did not stop its foraging activity even when all sources of nutrients were occupied and the whole agar plate was explored. As shown in Fig. 13.4, a vigorous plasmodium can spread over surrounding Petri dishes, trying to settle on bare plastic.

In these experiments plasmodium starts its colonisation in Amsterdam as before. Its colonisation is more aggressive in this case and it colonises Haarlem, Den Haag, Rotterdam, Dordrecht, Breda, Tilburg, 's-Hertogenbosch and Eindhoven within the first 12 h. A pronounced protoplasm transport link is established connecting these cities in a chain (first 12 h from the moment of inoculation, Fig. 13.4a). In 34 h after inoculation the plasmodium sprawls from 's-Hertogenbosch to Utrecht, Amersfoort and Apeldoorn, and then builds a transport link Amersfoort–Lelystad–Zwolle–Enschede (Fig. 13.4b).

Protoplasmic tubes connecting Haarlem, Amsterdam and Lelystad with Den Helder are grown simultaneously after 57 h of the experiment. By the same time the plasmodium also connects Nijmegen with Apeldoorn (Fig. 13.4c). Protoplasmic transport links Den Helder–Leeuwarden–Groningen, Rotterdam–Middelburg and Eindhoven–Maastricht are developed by the 80th hour of plasmodium's foraging activity (Fig. 13.4d). A schematic illustration of the colonisation dynamics is shown in Fig. 13.3b.

Plasmodium starts to show overperformance after 57 h of the experiment. It sprawls from Den Helder north-westward and from Enschede south-eastward onto bare plastic of the experimental container (Fig. 13.4c). The plasmodium does not propagate on the plastic long enough and retracts in a few hours (this can be seen in Fig. 13.4d). Another sprawling takes place by the 80th hour of experimentation, when the plasmodium propagates westward of Middelburg and south-eastward of Maastricht (Fig. 13.4d). See also diagrams of sprawling outside the country in Fig. 13.3b. Also notice how the plasmodium dynamically changes its foraging strategy (Fig. 13.4). It first attempts to colonise cities in the north-eastern part of the country but then abandons the attempt and moves to the north-east later via the IJsselmeer lake.

In some experiments the plasmodium never manages to span all cities, and fails to colonise some oat flakes. An example is shown in Fig. 13.4; in 65 h after inoculation the plasmodium colonises the majority of The Netherlands and establishes a network of protoplasmic tubes over most of the cities represented by oat flakes (Fig. 13.4c). Later it goes into a hibernation stage and forms a

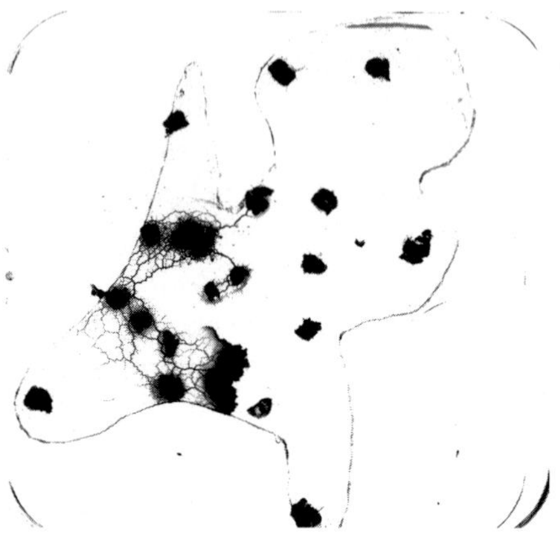

(e) $t = 22$ h

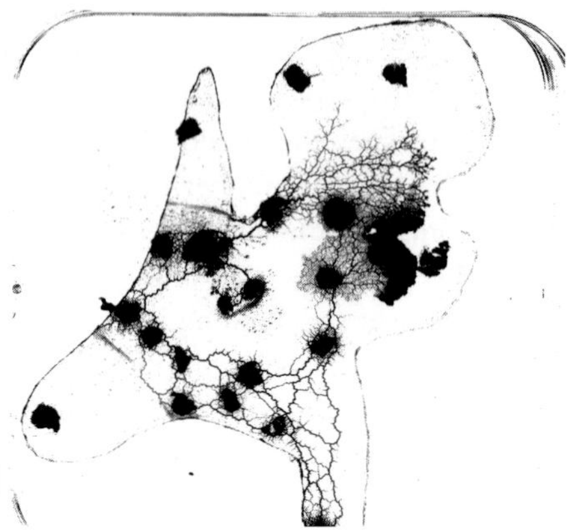

(f) $t = 43$ h

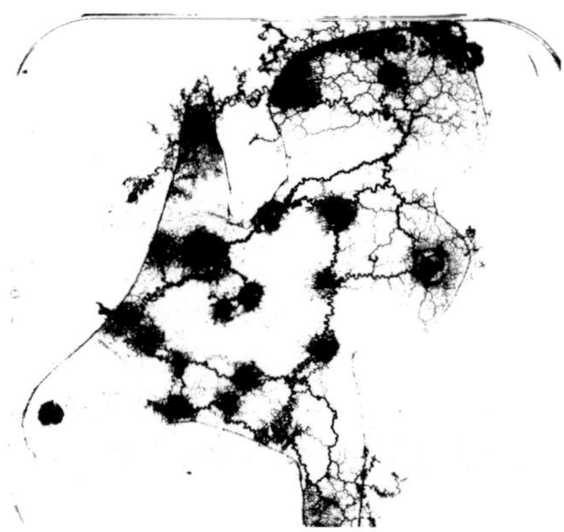

(g) $t = 65$ h

Fig. 13.4 (Continued) Plasmodium does not always span all cities (sources of food). Time elapsed from inoculation is shown in the subfigure captions.

sclerotium. However, at no moment of its development does the plasmodium even approach Middelburg. Such situations are rather atypical and did not happen often in our experiments.

Physarum graphs extracted from 62 laboratory experiments are shown in Figs. 13.4 and 13.5. The graph becomes planar only for $\theta = \frac{9}{62}$ (Fig. 13.4p), i.e. when edges occurred in over 15% of the experiments. We can therefore infer that the Physarum graph is planar. However, with acquiring planarity the graph becomes disconnected: one node, Den Helder city, becomes isolated.

The Physarum graphs become acyclic for $\theta = \frac{26}{62}$ (Fig. 13.5h), i.e. when their

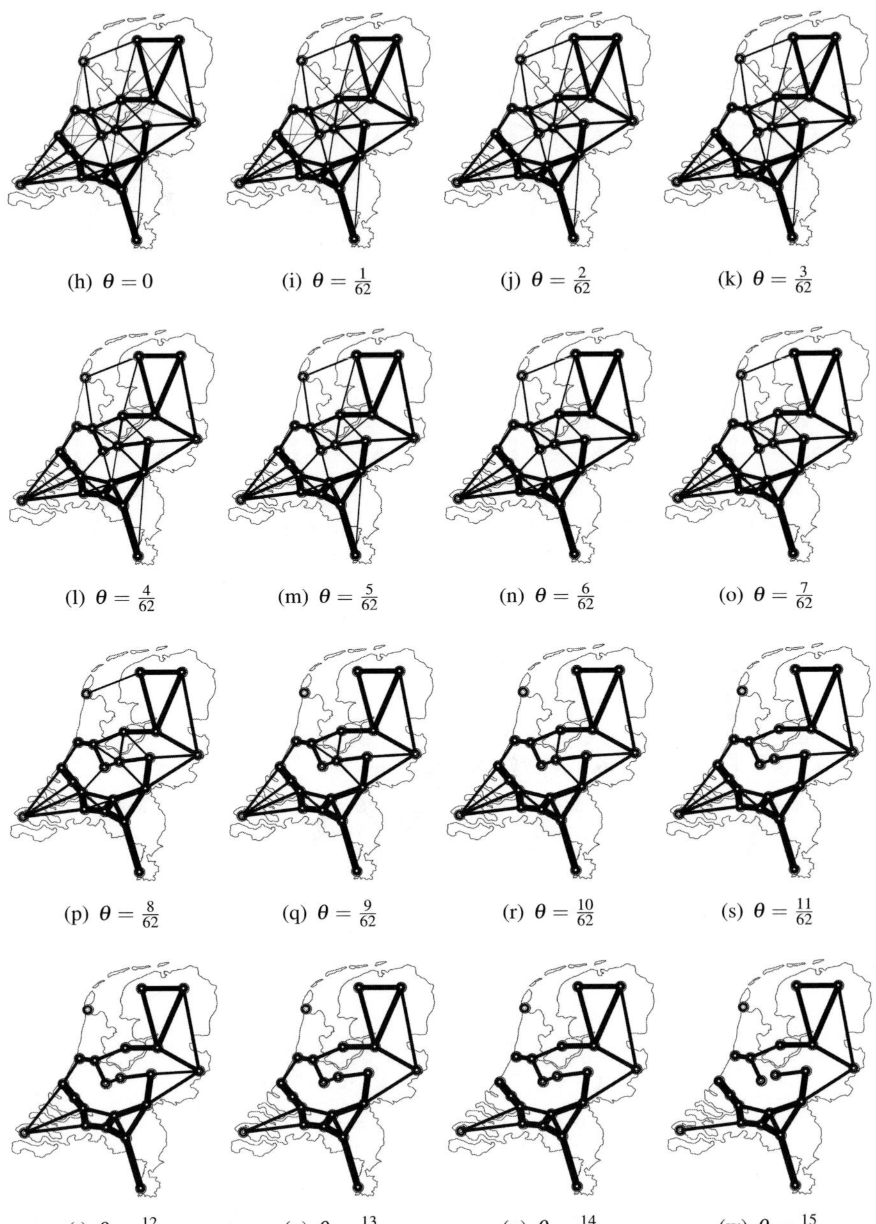

(h) $\theta = 0$ (i) $\theta = \frac{1}{62}$ (j) $\theta = \frac{2}{62}$ (k) $\theta = \frac{3}{62}$

(l) $\theta = \frac{4}{62}$ (m) $\theta = \frac{5}{62}$ (n) $\theta = \frac{6}{62}$ (o) $\theta = \frac{7}{62}$

(p) $\theta = \frac{8}{62}$ (q) $\theta = \frac{9}{62}$ (r) $\theta = \frac{10}{62}$ (s) $\theta = \frac{11}{62}$

(t) $\theta = \frac{12}{62}$ (u) $\theta = \frac{13}{62}$ (v) $\theta = \frac{14}{62}$ (w) $\theta = \frac{15}{62}$

Fig. 13.4 (Continued) Configurations of Physarum graph $\mathbf{P}(\theta)$ for $\theta = 0, \frac{1}{62}, \ldots, \frac{15}{62}$. Thickness of an edge is proportional to the edge's weight.

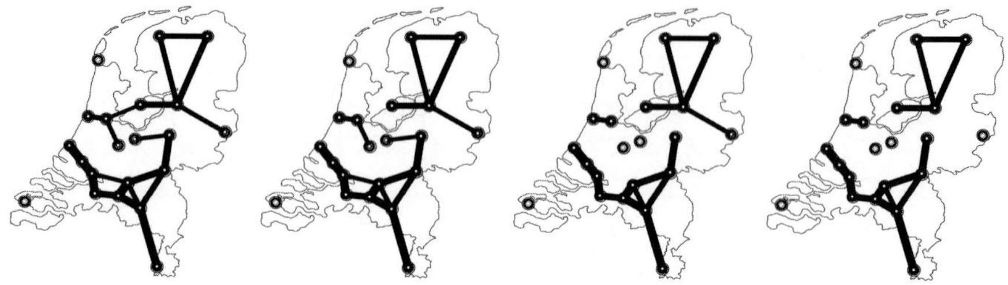

(a) $\theta = \frac{16}{62}$ (b) $\theta = \frac{17}{62}$ (c) $\theta = \frac{19}{62}$ (d) $\theta = \frac{21}{62}$

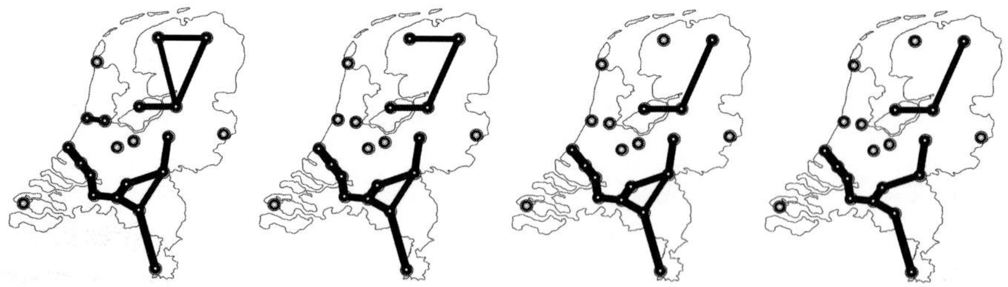

(e) $\theta = \frac{22}{62}$ (f) $\theta = \frac{23}{62}$ (g) $\theta = \frac{25}{62}$ (h) $\theta = \frac{26}{62}$

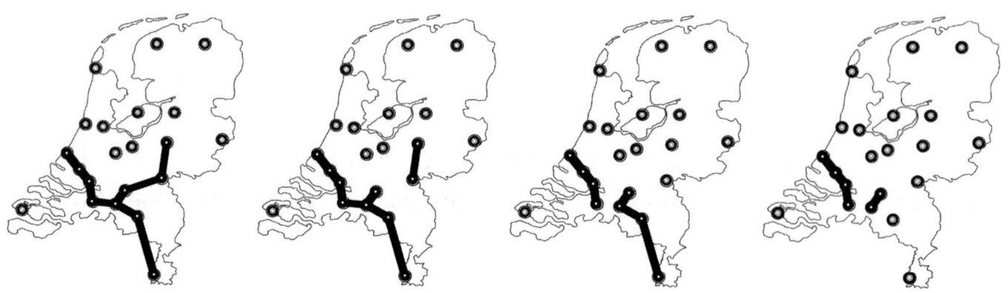

(i) $\theta = \frac{27}{62}$ (j) $\theta = \frac{30}{62}$ (k) $\theta = \frac{31}{62}$ (l) $\theta = \frac{34}{62}$

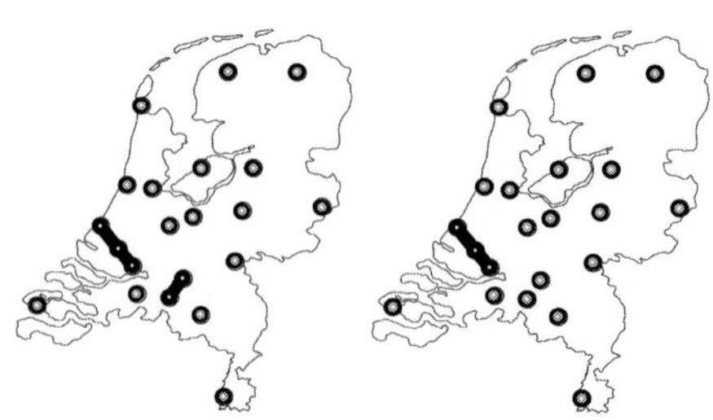

(m) $\theta = \frac{35}{62}$ (n) $\theta = \frac{37}{62}$

Fig. 13.5 Configurations of Physarum graph $\mathbf{P}(\theta)$ for $\theta = \frac{16}{62}, \frac{17}{62}, \ldots, \frac{37}{62}$. Thickness of an edge is proportional to the edge's weight.

edges appear as protoplasmic tubes in over 41% of the experiments. When the graph becomes acyclic it is split into a set of isolated nodes: Den Helder, Leeuwarden, Haarlem, Amsterdam, Utrecht, Amersfoort, Enschede, Middelburg and two additional components. One component is a chain of three cities: Lelystad, Zwolle and Groningen. The second component is a tree rooted in Tilburg. The tree has three linear branches:

- Tilburg–Breda–Dordrecht–Rotterdam–Den Haag,
- Tilburg–'s-Hertogenbosch–Nijmegen–Apeldoorn,
- Tilburg–Eindhoven–Maastricht.

This tree is a characteristic feature of the Physarum graph and it appears in over 60% of experiments. The tree is 'destroyed' when $\theta \geq \frac{30}{62}$ (Fig. 13.5j); then only chains remain, which give away isolated nodes with further increase of θ. Some chains are more stable than others. Thus, the chain Breda–Dordrecht–Rotterdam–Den Haag appears in over 54% of experiments (Fig. 13.5l), while the chain Dordrecht–Rotterdam–Den Haag appears in almost 60% of experiments.

The experiments have now provided a reasonably consistent set of connections between the various urban centres in The Netherlands. The next question is to assess how well these Physarum graphs approximate The Netherlands' motorway network. A graph **H** of Dutch motorways is given in Fig. 13.6a.

The intersections $\mathbf{P}(\theta) \bigcap \mathbf{H}$ of Physarum and motorway graphs are shown in Fig. 13.6b–d for $\theta = 0$, $\frac{8}{16}$ and $\frac{15}{16}$. A relaxed probabilistic Physarum graph $\mathbf{P}(0)$, where an edge appears in the graph if it is recorded in at least one experiment, matches the motorway graph **H** almost perfectly. Just three edges of **H** are not present in $\mathbf{P}(0) \bigcap \mathbf{H}$:

- Amsterdam–Der Helder,
- Zwolle–Apeldoorn,
- Rotterdam–Dordrecht (Fig. 13.6b).

$\theta = \frac{8}{16}$ is the highest value for which the Physarum graph $\mathbf{P}(\theta)$ remains connected (Fig. 13.4). The graph $\mathbf{P}(\frac{8}{62}) \bigcap \mathbf{H}$ loses a few more edges (presented in $\mathbf{P}(\frac{0}{62}) \bigcap \mathbf{H}$):

- Den Helder–Lelystad,
- Den Helder–Haarlem,
- Leeuwarden–Lelystad,
- Utrecht–Dordrecht,
- Utrecht–Nijmegen,
- Breda–Enschede,
- Utrecht–Enschede (Fig. 13.6c).

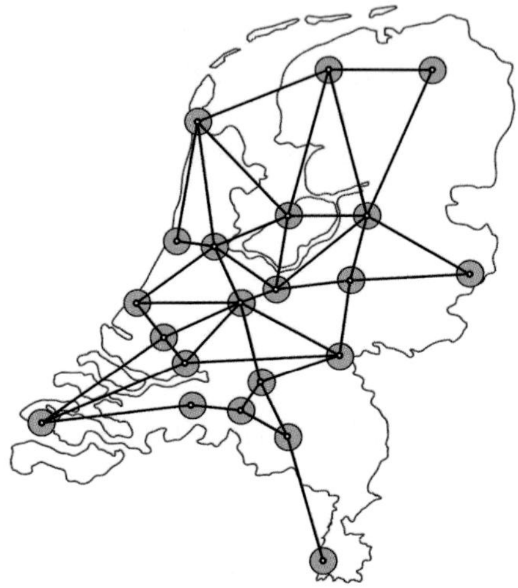

(a) $\mathbf{H}$

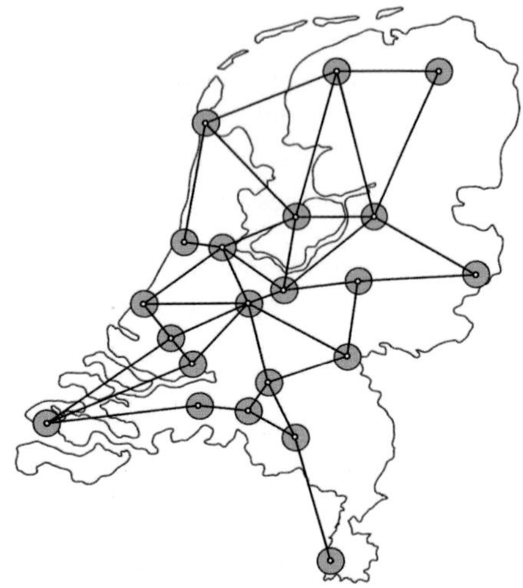

(b) $\mathbf{P}(0)\bigcap\mathbf{H}$

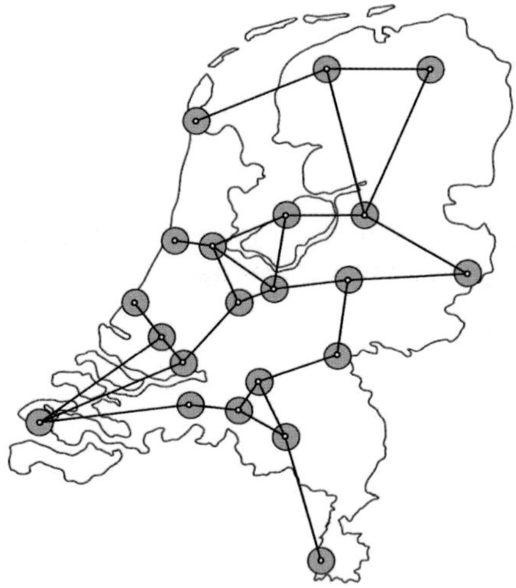

(c) $\mathbf{P}(\frac{8}{62})\bigcap\mathbf{H}$

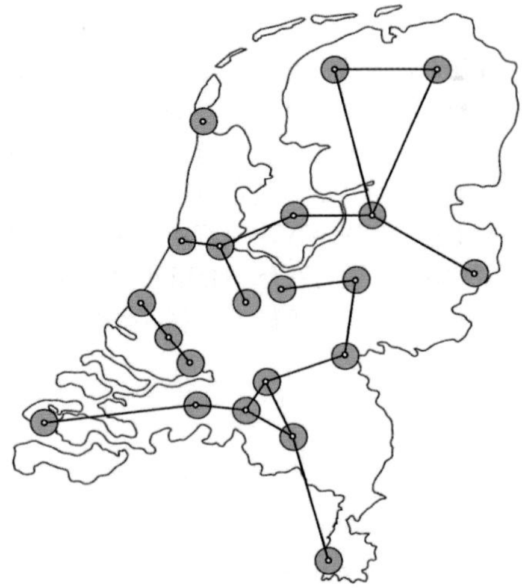

(d) $\mathbf{P}(\frac{15}{62})\bigcap\mathbf{H}$

Fig. 13.6 Graph $\mathbf{H}$ of man-made motorway network is shown in (a). Intersections $\mathbf{P}(\theta)\bigcap\mathbf{H}$, $\theta = 1$, $\frac{8}{16}$ and $\frac{15}{16}$ of Physarum $\mathbf{P}$ and motorway $\mathbf{H}$ graphs are shown in (b–d).

As soon as θ reaches the value $\frac{15}{16}$, the graph $\mathbf{P}(\theta)\bigcap\mathbf{H}$ becomes separated on the isolated node Den Helder, and there are components:

- the chain Enschede–Den Haag–Rotterdam,
- the cycle Nijmegen–Breda–Middelburg with branches Breda–'s-Hertogenbosch–Tilburg, Middelburg–Eindhoven and Nijmegen–Dordrecht–Apeldoorn–Amersfoort,
- the cycle Leeuwarden–Groningen–Zwolle with branches Zwolle–Enschede and Zwolle–Lelystad–Amsterdam–{Haarlem, Utrecht} (Fig. 13.6d).

13.2 Redundancy of the transport network

We constructed a relative neighbourhood graph [Toussaint (1980)] **RNG** (Fig. 13.7a), a Gabriel graph [Gabriel and Sokal (1969); Matula and Sokal (1984)] **GG** (Fig. 13.7c), a β-skeleton (Fig. 13.7b) and a minimum spanning tree **MST** (Fig. 13.7d) (we rooted **MST** in Amsterdam) over nodes corresponding to centres of urban areas. We then calculated intersections of these graphs with the Physarum graph $\mathbf{P}(0)$ and the motorway graph **H**, see Fig. 13.8.

The following is a list of edges of the proximity graphs that are not present in Physarum or motorway graphs:

- $\mathbf{P}\bigcap\mathbf{MST} = \mathbf{MST}$ – {(Den Helder, Haarlem), (Zwolle, Apeldoorn)},
- $\mathbf{P}\bigcap\mathbf{RNG} = \mathbf{RNG}$ – {(Den Helder, Haarlem), (Zwolle, Apeldoorn)},
- $\mathbf{P}\bigcap\mathbf{BS}(1.5) = \mathbf{BS}(1.5)$ – {(Den Helder, Haarlem), (Zwolle, Apeldoorn), (Utrecht, Den Haag)},
- $\mathbf{P}\bigcap\mathbf{GG} = \mathbf{GG}$ – {(Den Helder, Haarlem), (Zwolle, Apeldoorn), (Utrecht, Den Haag), (Den Helder, Amsterdam), (Den Helder, Lelystad)},
- $\mathbf{H}\bigcap\mathbf{MST} = \mathbf{MST}$ – {(Haarlem, Enschede), (Rotterdam, 's-Hertogenbosch) },
- $\mathbf{H}\bigcap\mathbf{RNG} = \mathbf{RNG}$ – {(Haarlem, Enschede), (Rotterdam, 's-Hertogenbosch) },
- $\mathbf{H}\bigcap\mathbf{BS}(1.5) = \mathbf{BS}(1.5)$ – {(Haarlem, Enschede), (Rotterdam, 's-Hertogenbosch)},
- $\mathbf{H}\bigcap\mathbf{GG} = \mathbf{GG}$ – {(Haarlem, Enschede), (Rotterdam, 's-Hertogenbosch), (Enschede, Tilburg), (Rotterdam, Nijmegen), (Amersfoort, Dordrecht), (Dordrecht, Tilburg)}.

The motorway graph closely matches the spanning tree, relative neighbourhood graph and β-skeleton. Only two edges Haarlem– Enschede and Rotterdam–'s-Hertogenbosch, presented in **RNG**, **MST** and $\mathbf{BS}(1.5)$, do not exist in **H**. The fact that the relative neighbourhood graph is *almost* a subgraph of the motorway graph indicates the intrinsically logical organisation of the transport networks in

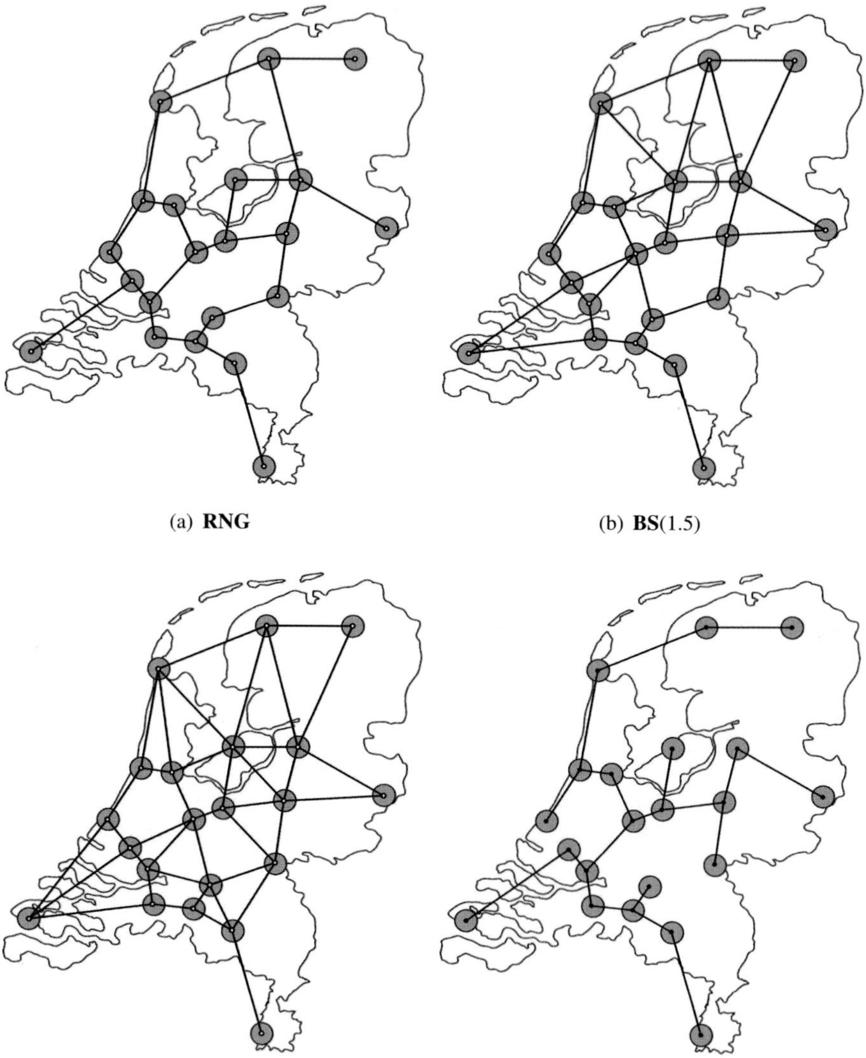

(a) **RNG** (b) **BS**(1.5)

(c) **GG** (d) **MST**

Fig. 13.7 Proximity graphs constructed on regions **U**: (a) relative neighbourhood graph **RNG**, (b) β-skeleton with control parameter 1.5, **BS**(1.5) (c) Gabriel graph **GG**, (d) minimum spanning tree **MST**.

The Netherlands.

It can be said that the transport networks in The Netherlands are redundant, from slime mould's point of view, because there is a substantial number of edges

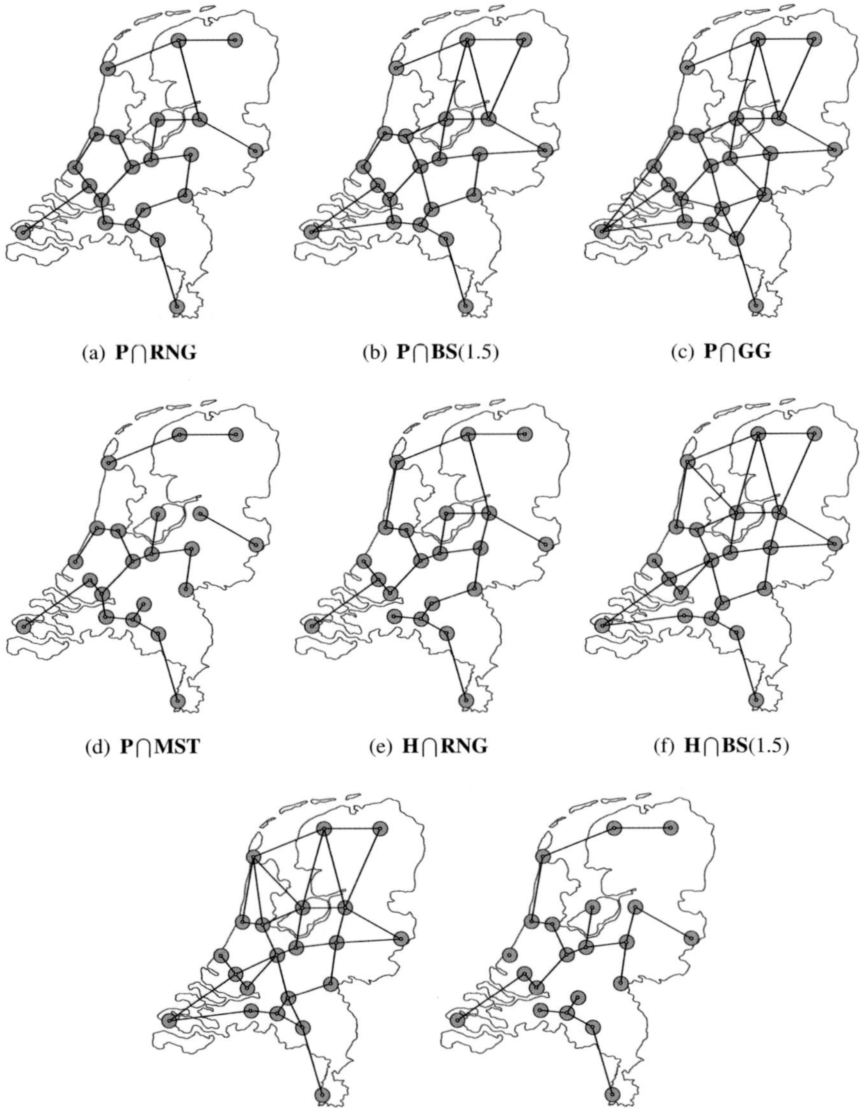

(a) **P**$\bigcap$**RNG** (b) **P**$\bigcap$**BS**(1.5) (c) **P**$\bigcap$**GG**

(d) **P**$\bigcap$**MST** (e) **H**$\bigcap$**RNG** (f) **H**$\bigcap$**BS**(1.5)

(g) **H**$\bigcap$**GG** (h) **H**$\bigcap$**MST**

Fig. 13.8 Intersections of Physarum graph **P**(0) (a–d) and motorway graph **H** (e–h) with proximity relative neighbourhood graph **RNG**, β-skeleton **BS**(1.5), Gabriel graph **GG** and minimum spanning tree **MST**.

of **H** not present in **RNG**.

The same can be said about the Physarum graph, because only edges Den Helder–Haarlem and Zwolle–Apeldoorn of **RNG** are not represented by protoplasmic tubes. Under-representations of β-skeleton and **GG** are much more substantial: three and five edges, respectively.

13.3 Flooding: *Nederlanders* migrate to *Deutschland*

In this section, we describe the experiments which investigate the capability of the plasmodium to adapt its network during induced flooding of the Petri dish. We imitate the flooding due to the rising of the North Sea level by climate change or by incoming freak storms from the west. In essence, we investigate how the plasmodium would calculate and adapt the transport network, if The Netherlands were to suffer similar flooding of its cities and roads. Experiments on flooding were conducted in 12×12 cm^2 Petri dishes. A Petri dish was raised by 1–2 cm in its south-eastern corner and partly filled with liquid (distilled water, either pure or coloured with one drop of food colouring) (Fig. 13.9a). The area flooded roughly corresponds to the territories of The Netherlands with 1/10000–1/4000 chances of the protection from flooding failure [Wesselink (2007); Huisman (1998)].

In most cases the flooded area included Middelburg in the south-west and Groningen in the north-east. In the central part of the country the flooding often reached Zwolle. The exact flooded area varied between experiments due to slight variations in the thickness of agar gel substrate and minor differences in inclinations of Petri dishes; the flooding scheme shown in (Fig. 13.9) is rather indicative.

Initially plasmodium reacts to flooding with increased activity. During the first few hours of flooding the plasmodium typically increases its branching at the boundary of the flooded area (Fig. 13.10a). Often there are indications of indiscriminate increase of foraging, namely panic foraging. For example, in Fig. 13.10b we can see active sprawling of plasmodium in the areas around Apeldoorn, Dordrecht and Enschede: no protoplasmic tubes are formed but rather uniform sheets of plasmodium propagate in these areas. Eventually the activity ceases and flooded transport links become abandoned (Fig. 13.10c).

In some cases no 'panic' branching occurs, but non-flooded protoplasmic tubes become thicker due to increased transport of nutrients and relocation of masses of protoplasm from areas affected by flooding (Fig. 13.11a). Often only protoplasmic tubes located along flood lines are hypertrophied; for example, there is an increased thickness of tubes along the route 's-Hertogenbosch–Apeldoorn–Zwolle–Groningen in Fig. 13.11b.

(a)

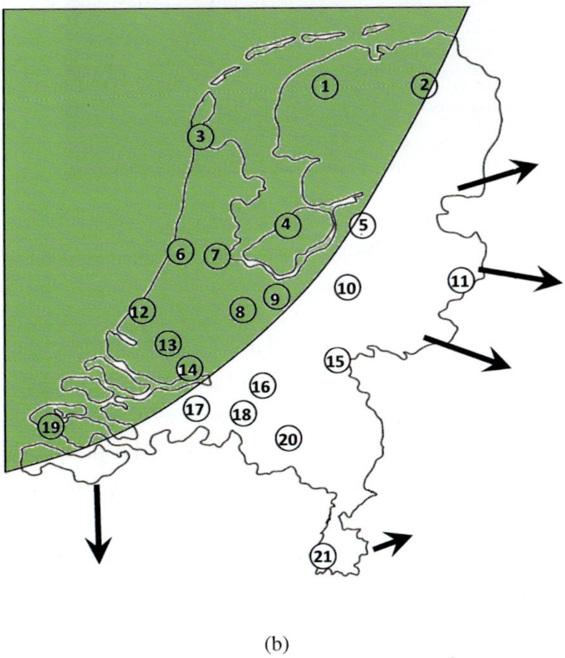

(b)

Fig. 13.9 Flooding setup: (a) array of Petri dishes during imitated flooding, (b) flooding scheme; flooded part is shaded (green), directions of outward migration are shown by arrows.

(a) $t = 0$ h (b) $t = 30$ h

(c) $t = 46$ h

Fig. 13.10 Illustration of plasmodium behaviour during experiment on flooding. Photographs are taken at different angles.

With time, water is absorbed by the gel and is sucked under the gel plate due to capillary forces, which in turn causes the overall humidity to increase. This is associated with a reduced concentration of nutrients and an increased concentration of metabolites ejected in the agar plate; these force the plasmodium to abandon the agar plate and migrate beyond. Often the plasmodium attempts complete evacuation by crawling onto the water surface (for example see Fig. 13.12).

Outward migration, calculated in nine experiments, is shown in Fig. 13.9b. We conclude that if a flooding of The Netherlands were to happen the major impact of migration (60%) will be felt by West Germany, with just a minor impact on northern France (20%) and a very slight impact on Belgium (10%). If The Netherlands were flooded, West Germany would accept the largest impact of mass migration.

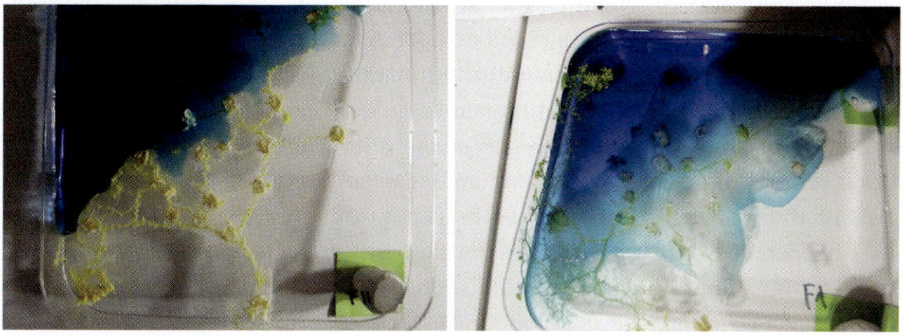

(a) (b)

Fig. 13.11 Compensation of transport networks (a) and increase in borderline transport (b).

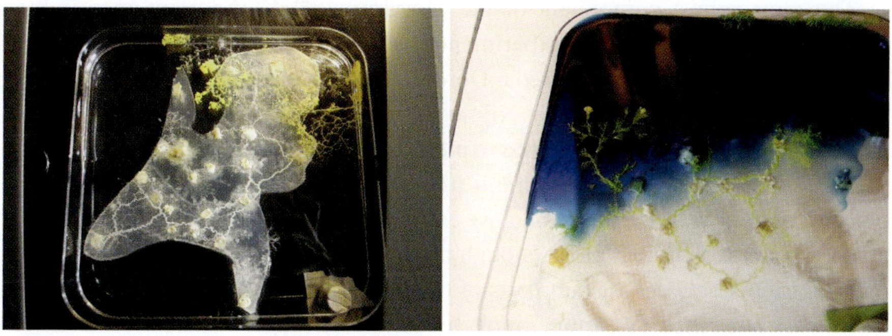

(a) (b)

Fig. 13.12 Examples of evacuation.

13.4 Summary

In laboratory experiments with plasmodium of *Physarum polycephalum* we discovered that the Physarum protoplasmic network forms a subnetwork of the man-made Netherlands' motorway networks, i.e. every transport link represented by Physarum can also be found as a segment of the motorway network. However, the converse does not hold. The motorway network is not a subgraph of the Physarum network; there are edges of the motorway graph not represented by edges of the Physarum graph. Transport links Amsterdam to Der Helder, Zwolle to Apeldoorn and Rotterdam to Dordrecht are never represented in the Physarum graph. This slight redundancy of the man-made motorway network, compared to the Physarum network, could be due to several reasons. For example, transport networks in many countries are based on economical and political factors, which do not necessarily possess a feature of optimality. Also, some motorway links were designed to reduce congestion and to reduce commuting for people living outside major urban areas. Congestion *per se* is not an issue for *P. polycephalum*, whose protoplasmic tubes are elastic and adaptable to almost all acceptable degrees of throughput. Moreover, the branching of tubes and formation of new tubes happen only when new sources of nutrients are detected (via chemoattractants) by the slime mould or existing sources of nutrients are depleted and the mould switches to its exploration mode.

Also, in many experiments Den Helder city remains disconnected from other cities. This is either a mishap of the Physarum approach, e.g. plasmodium prefers not to enter the narrow peninsula of North Holland, or an indication of a somewhat inefficient location of Den Helder.

The most robust component of the Physarum graph, the component which is present in the majority of experiments, is a tree with three linear branches. The first branch is Tilburg–Breda–Dordrecht–Rotterdam–Den Haag, the second is Tilburg–'s-Hertogenbosch–Nijmegen–Apeldoorn and the third is Tilburg–Eindhoven–Maastricht. A possible explanation would be that relative positions and distances between the cities in this tree are optimal for plasmodium's physiological functioning. Cities are not close enough to be contaminated by products of plasmodium's activity but close enough not to put significant strain on the pumping of nutrients between distant parts of the plasmodium's body.

If we prune the Physarum graph by removing edges which occur in less than a quarter of experiments and look at the intersection (i.e. the set of edges present in both graphs) of this graph with a graph of motorways, we find that the intersection consists of three disconnected components. The first component is a chain Enschede–Den Haag–Rotterdam. The second component is a cycle Nijmegen–Breda–Middelburg with branches Breda–'s-Hertogenbosch–Tilburg, Middelburg–Eindhoven and Nijmegen–Dordrecht–Apeldoorn–Amersfoort. Finally, the third component is a cycle Leeuwarden–Groningen–Zwolle with branches Zwolle–Enschede and Zwolle–Lelystad–Amsterdam–Haarlem–Utrecht.

With respect to the proximity graph, the key finding is that relative neighbourhood graphs (which are commonly recognised as a best approximation of urban streets and transport networks) are almost (apart from two edges) subgraphs of the Physarum graph and motorway graph. Only two edges Haarlem–Enschede and Rotterdam–'s-Hertogenbosch, present in the relative neighbourhood graph, minimum spanning tree and β-skeleton, do not exist in the motorway graph $\mathbf{H}$. The Physarum graph $\mathbf{P}(0)$ is not a subgraph of the minimum spanning tree. This

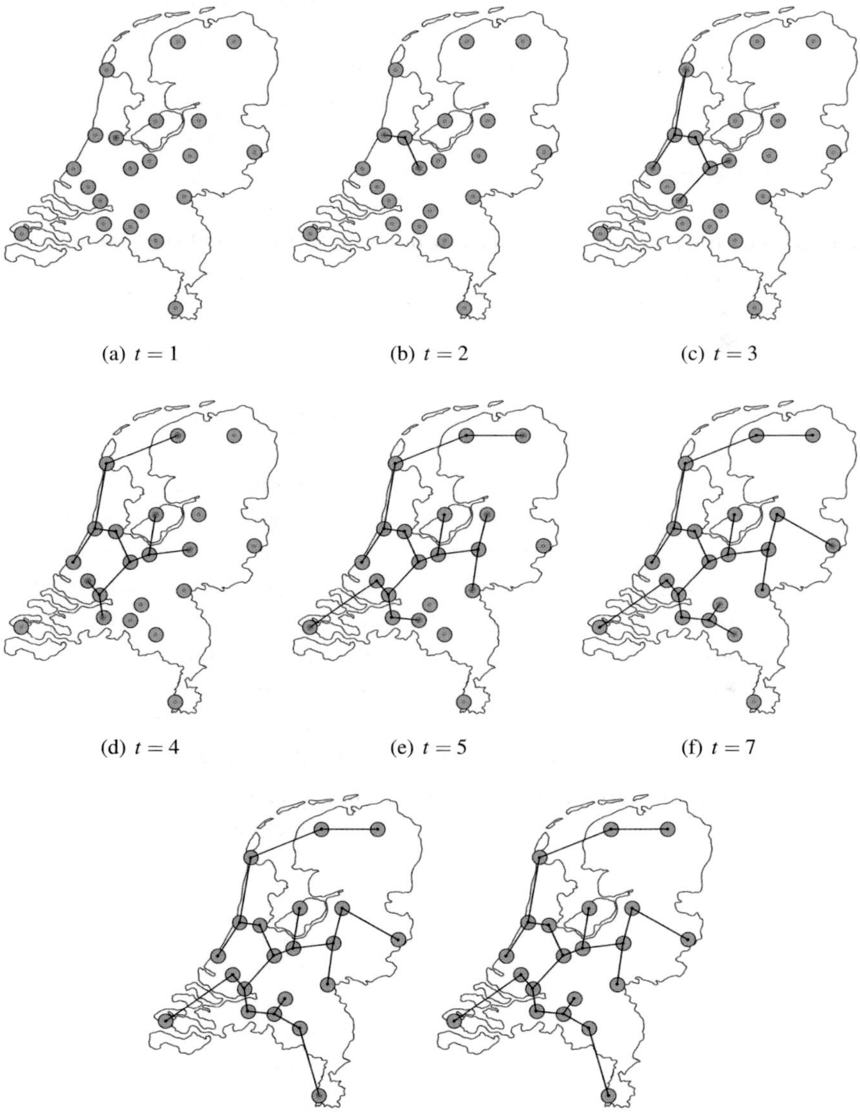

(a) $t = 1$ (b) $t = 2$ (c) $t = 3$

(d) $t = 4$ (e) $t = 5$ (f) $t = 7$

(g) $t = 8$ (h) $t = 9$

Fig. 13.13 Snapshots of a growing spanning tree of **U** rooted in Amsterdam. Vertices active at time step t' are shown by black disc at snapshot $t = t'$.

means that despite being a good approximation of the motorway network the slime mould, in the particular case of The Netherlands, does not include a minimum transport network, represented by a spanning tree.

By physically imitating flooding of some parts of The Netherlands, we predicted that if a real flooding were to occur, the following events will take place: a substantial increase in traffic on the parts of motorway networks close to the boundary between flooded and non-flooded areas, propagation of the traffic congestion to all non-flooded parts of the country, complete paralysis and abandonment of the transport network and migration of population from The Netherlands to Germany, France and Belgium.

The plasmodium is incredibly similar, especially in wave-like behaviour, to subexcitable nonlinear media [Adamatzky (2007)c] and is mainly guided by gradients of chemoattractants [Adamatzky (2009)a]. Based on these two facts, we could assume that the plasmodium must spread from Amsterdam omnidirectionally and then start forming branches between cities, such as e.g. illustrated in Fig. 13.13. Such phenomena do not occur; rather the plasmodium behaves more like a 'single-headed' creature; it chooses one direction of movement, explores it, then chooses another direction and explores it again, see Fig. 13.3. Such behaviour of the plasmodium may explain the differences between ideal planar proximity graphs and protoplasmic networks constructed by the plasmodium.

Chapter 14

Rebuilding Iberian motorways with slime mould

Andrew Adamatzky and Ramon Alonso-Sanz

The Iberian peninsula is the continental part of the states of Spain and Portugal. Both Spain and Portugal are members of the European Union (EU) that adopted the common EU currency (euro) and signed the Schengen agreement, which abolishes passport controls among most of the EU state members. Both states belong to the NATO military alliance. Thus, both countries committed themselves to join relevant supranational entities, which, beyond pure geographical reasons, support a common treatment of the *Iberian peninsula*, albeit such a being not having any real political entity. Let us stress here that when referring to Spain and Portugal in the text, we should properly refer to continental Spain and Portugal; thus, importantly, not including the Canary, Balearic, Azores and Madeira archipelagos.

We consider the 23 most populous urban areas in the Iberian peninsula (Fig. 14.1a):

(1) A Coruna,
(2) Gijon–Oviedo,
(3) Santander,
(4) Bilbao,
(5) San Sebastian,
(6) Vigo,
(7) Valladolid,
(8) Zaragoza,
(9) Porto,
(10) Tarragona,
(11) Barcelona,

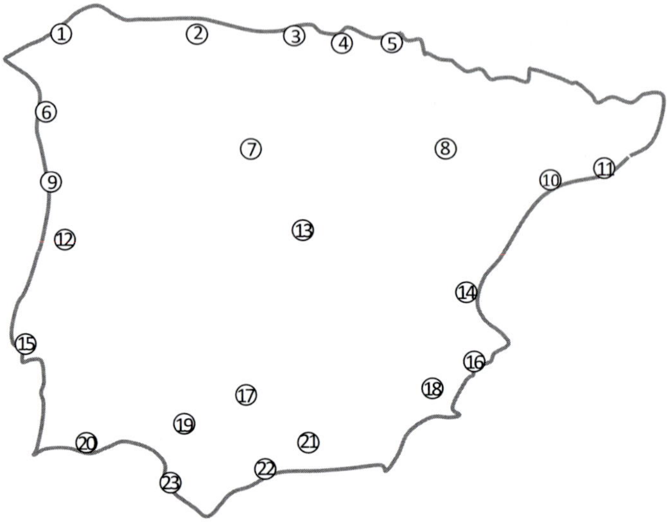

(a)

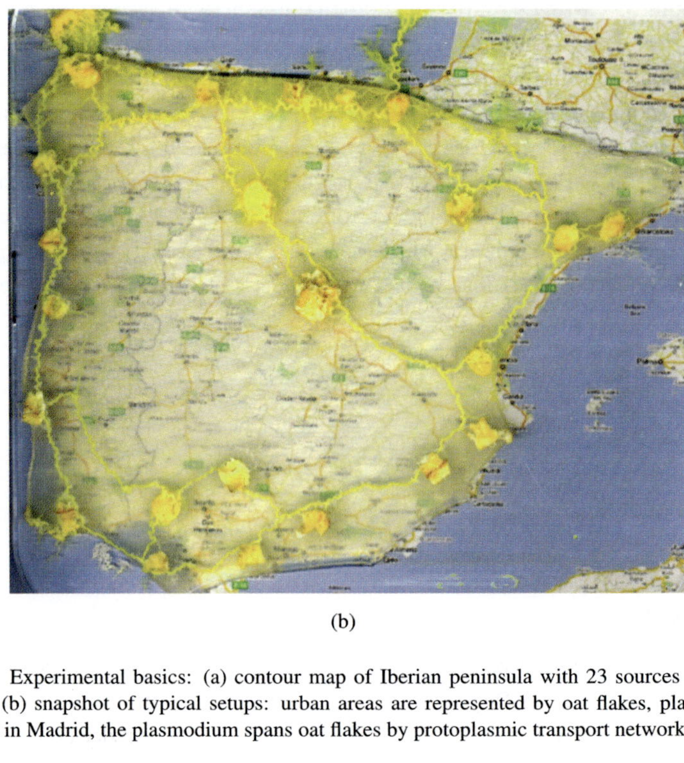

(b)

Fig. 14.1 Experimental basics: (a) contour map of Iberian peninsula with 23 sources of nutrients indicated, (b) snapshot of typical setups: urban areas are represented by oat flakes, plasmodium is inoculated in Madrid, the plasmodium spans oat flakes by protoplasmic transport network.

(12) Pombal,
(13) Madrid (capital of Spain),
(14) Valencia,
(15) Lisbon (capital of Portugal),
(16) Alicante,
(17) Cordoba,
(18) Murcia,
(19) Sevilla,
(20) Faro,
(21) Granada,
(22) Malaga–Marbella,
(23) Cadiz.

At the beginning of each experiment a piece of plasmodium, usually already attached to an oat flake, was placed in Madrid (region 13 in Fig. 14.1a).

14.1 Segregating Portugal and Spain

A typical experiment is illustrated in Fig. 14.2. Plasmodium is inoculated on an oat flake representing Madrid. The plasmodium initially produces a short-range scouting pattern, by propagating omnidirectionally (Fig. 14.2a). As soon as the plasmodium starts feeling attractants, emitted by oat flakes representing Valencia, the plasmodium forms a typical wave-like pattern propagating towards Valencia (Fig. 14.2a). This travelling pattern looks like and, as we demonstrated experimentally and in computational experiments in [Adamatzky et al. (2009); Adamatzky (2009)a], behaves like a wave fragment in a subexcitable nonlinear medium. Such plasmodium wave fragment may travel for a long distance conserving its size and shape, without expanding or collapsing. After reaching and colonising Valencia the plasmodium colonises Alicante, Murcia and Granada and then follows clockwise along Iberia's boundaries. In 30–35 h from the beginning of the experiment the plasmodium colonises almost all cities (Fig. 14.2b) and reaches Zaragoza. In the interval 34–58 h, the plasmodium colonises Tarragona and Barcelona, and also builds a link between Madrid and Valladolid (Fig. 14.2c).

Finding 91. *Plasmodium chooses clockwise and anticlockwise directions of propagation equiprobably.*

Plasmodium foraging patterns in the Iberian peninsula are surprisingly (compare with our results on Physarum approximation of motorways in the United Kingdom [Adamatzky and Jones (2009)] and Mexico [Adamatzky et al. (2011)b])

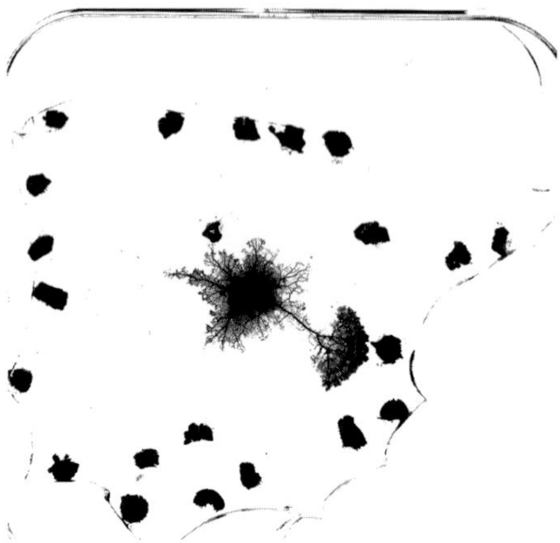

(a) $t = 10$ h

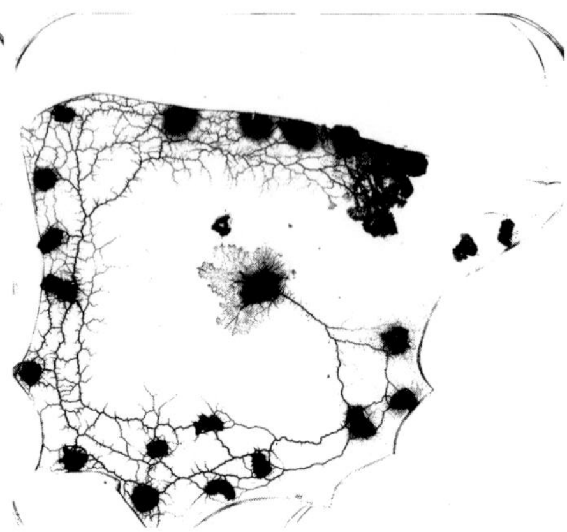

(b) $t = 34$ h

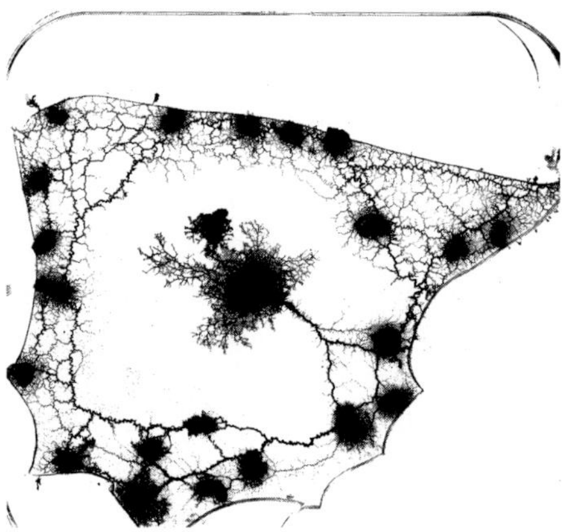

(c) $t = 58$ h

Fig. 14.2 Typical plasmodium development. Time elapsed from inoculation is shown in the subfigure captions.

consistent in all experiments. The plasmodium reaches the cities lying along the shore and then propagates along the shore. In 16 of 30 experiments, the plasmodium colonises the Iberian peninsula moving clockwise, in 14 anticlockwise. We personally feel more comfortable with the clockwise direction, so we selected images only with clockwise propagation.

Finding 92. *When colonising urban regions of the Iberian peninsula, plasmodium of* P. polycephalum *does* **not** *grow as a minimum spanning tree rooted in Madrid.*

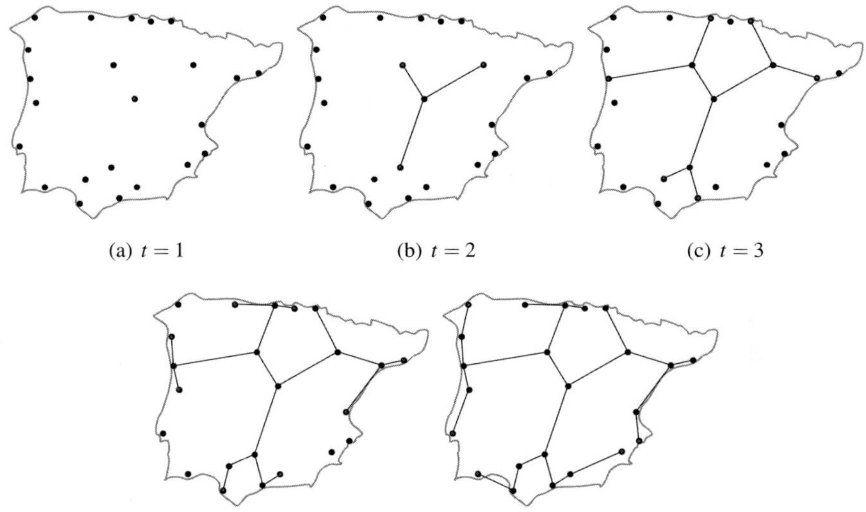

(a) $t = 1$ (b) $t = 2$ (c) $t = 3$

(d) $t = 4$ (e) $t = 5$

Fig. 14.3 Snapshots of a spanning tree of **U** rooted in Madrid. Vertices active at each time step are shown by black disc at snapshot.

The foraging activity of Physarum in the experimental setup of the Iberian peninsula surprised us. We expected that — due to the favourable location of Madrid — the plasmodium would first sprout from Madrid to the closest shore cities, e.g. will *simultaneously* build links from Madrid to Bilbao/San Sebastian in the north, Zaragoza and Valencia in the east, Cordoba in the south and Pombal in the west. This ideal expected scenario of plasmodium development is shown in Fig. 14.3. This did not happen. The plasmodium did not follow the classical growth of a spanning tree but rather chose one shore region close to Madrid and then grew along the shore.

In some experiments the plasmodium does not span all cities. Three examples are shown in Fig. 14.4; in these experiments the plasmodium had more than enough time (up to 4–5 days) to discover and colonise all oat flakes if the plasmodium 'wanted to'.

In the experiment shown in Fig. 14.4a, plasmodium develops two branches rooted in Madrid. The first branch grows from Madrid to Valladolid and then sprouts to Santander and Bilbao. Plasmodium propagates from Santander to Gijon–Oviedo and stops then; it does not propagate further to A Coruna. From Bilbao the plasmodium grows to San Sebastian and then to Zaragoza. At Zaragoza the plasmodium branches into protoplasmic tubes following to Tarragona and

(a) $t = 82$ h

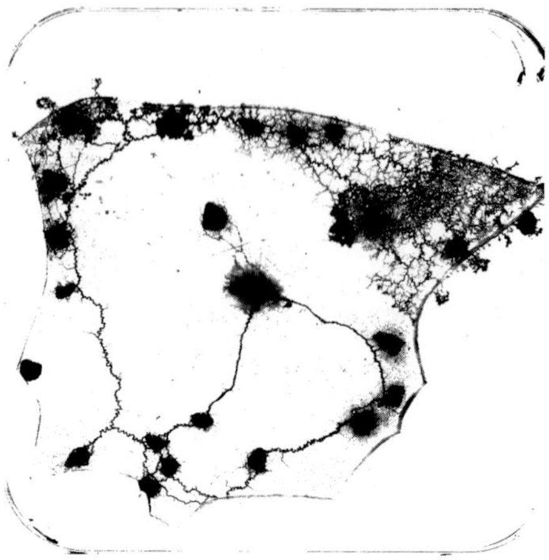

(b) $t = 82$ h

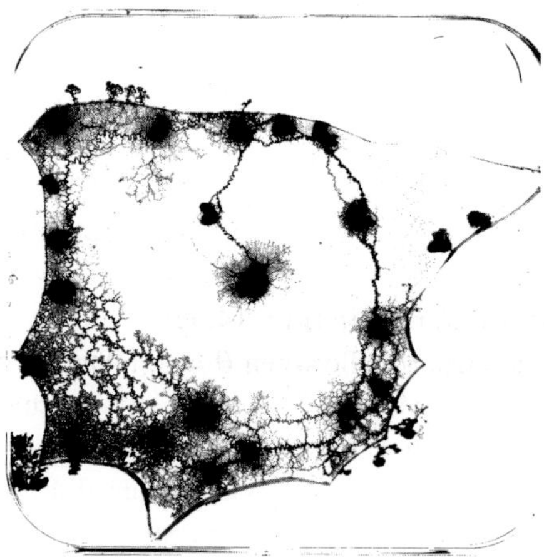

(c) $t = 82$ h

Fig. 14.4 Plasmodium does not always span all cities. Snapshots from three experiments. Time elapsed from beginning of experiments (inoculation of plasmodium in Madrid) is shown in subfigure captions.

Barcelona, and a tube leading to Valencia and Alicante (Fig. 14.4a, eastern part of plasmodium network). The second branch grows from Madrid to Cordoba, then a protoplasmic tube follows to Sevilla and Granada. Part of the plasmodium residing in Sevilla colonises Faro and Cadiz, while the plasmodium which colonised Granada propagates to the Malaga–Marbella region. Even after 58 h of foraging activity the plasmodium 'refuses' to colonise Murcia in the east and several urban regions in the west: A Coruna, Vigo, Porto, Pombal and Lisbon. The example (Fig. 14.4a) where so many cities remain uncolonised is rather unusual. For example, in the experiment of Fig. 14.4b the plasmodium did not colonise Lisbon, and in the experiment of Fig. 14.4c the plasmodium missed Tarragona and Barcelona.

Finding 93. *Plasmodium rarely ventures outside Iberia.*

In our previous experiments with approximating transport links in the United Kingdom [Adamatzky and Jones (2009)], Mexico [Adamatzky et al. (2011)b] and The Netherlands [Adamatzky et al. (2012)], we found that often plasmodium got carried away with its foraging performance and propagated outside the country-shaped agar plate. We never observed such a thing in experiments with the Iberian peninsula: the plasmodium never ventured outside the Iberia-shaped agar plate. One possible explanation could be because the shape of Iberia is more close to a convex shape than the shapes of the United Kingdom [Adamatzky and Jones (2009)], Mexico [Adamatzky et al. (2011)b] or The Netherlands [Adamatzky et al. (2012)].

Graphs $\mathbf{P}(\theta)$ extracted from 30 laboratory experiments are shown in Fig. 14.5. An unconstrained graph $\mathbf{P}(0)$ is shown in Fig. 14.5a; it has an outer 'shell' of heavyweight edges and a mostly lightweight interior. The graph becomes planar when we remove edges with weights below $\frac{5}{30}$ (Fig. 14.5b). The graph becomes disconnected, with the isolated edge Valladolid–Madrid, when only edges weighing at least $\frac{16}{30}$ are allowed to remain (Fig. 14.5c).

The graph $\mathbf{P}(\theta)$ becomes acyclic when $\theta = \frac{22}{30}$. The acyclic graph $\mathbf{P}(\frac{22}{30})$ consists of three isolated nodes: Madrid, Valladolid and Barcelona, one isolated segment: Zaragoza–Tarragona and an almost linear chain of nodes starting in San Sebastian and finishing in Valencia (Fig. 14.5c). The acyclic graph $\mathbf{P}(\frac{22}{30})$ is a minimal structure, or a core, of the Physarum graph.

Let us check how well Physarum graphs approximate motorway networks.

Assuming that edges which appear in at least in two of 30 experiments are taken into account, we compare the motorway graph $\mathbf{H}$ with the Physarum graph $\mathbf{P}(\frac{1}{30})$ in Fig. 14.6b. We see that $\mathbf{P}(\frac{1}{30})$ almost matches the motorway graph $\mathbf{H}$ apart from a few edges. The edges of $\mathbf{H}$ not represented in $\mathbf{P}(\frac{1}{30})$ are mainly transport links from Valladolid and Madrid to peripheral cities of Portugal and Spain: Valladolid–A Coruna, Valladolid–Porto, Madrid–Lisbon, Madrid–Sevilla, Malaga–Marbella–Murcia, Madrid–Bilbao, Madrid–San Sebastian (Fig. 14.6a and b).

When we increase the minimal allowed weight of edges to $\frac{2}{30}$, the intersection of the motorway and Physarum graphs becomes disconnected (Fig. 14.6c).

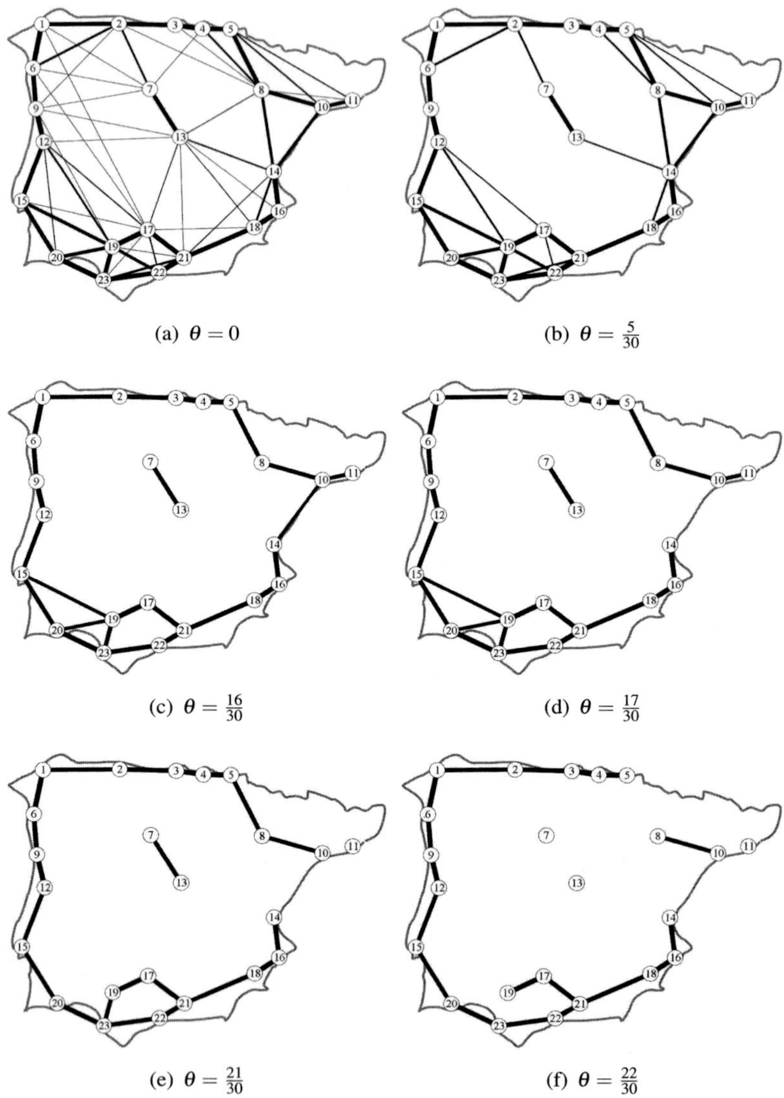

Fig. 14.5 Configurations of Physarum graph $\mathbf{P}(\theta)$ for various values of θ. Thickness of each edge is proportional to the edge's weight.

$\mathbf{P}(\frac{2}{30}) \bigcap \mathbf{H}$ consists of a western component of the transport network: starting in A Coruna and ending in the Malaga–Marbella area and Cadiz, and an eastern component, which includes the rest of the cities. The western part belongs to Portugal,

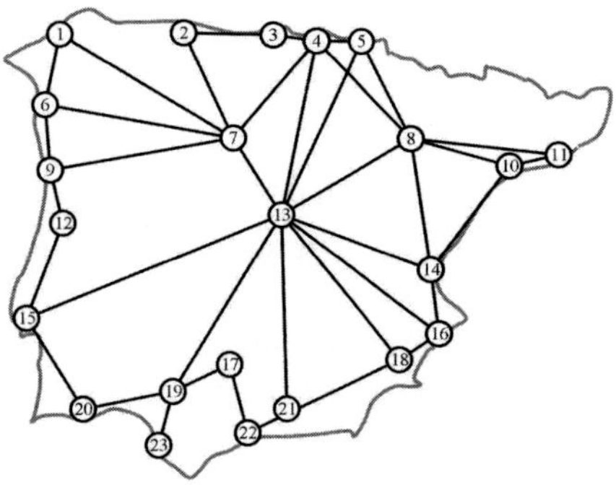

(a) **H**

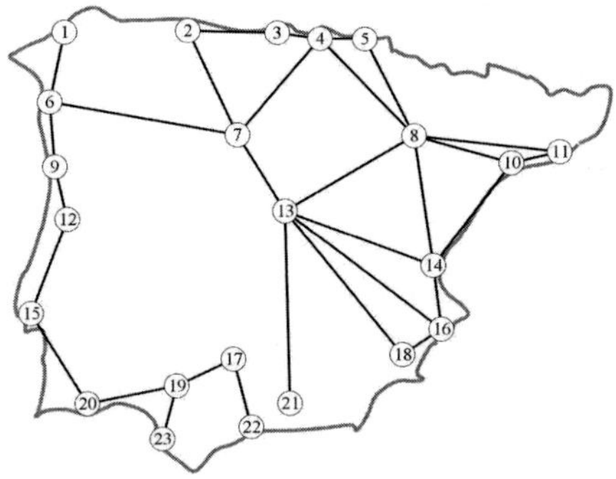

(b) $\mathbf{P}(\frac{1}{30}) \bigcap \mathbf{H}$

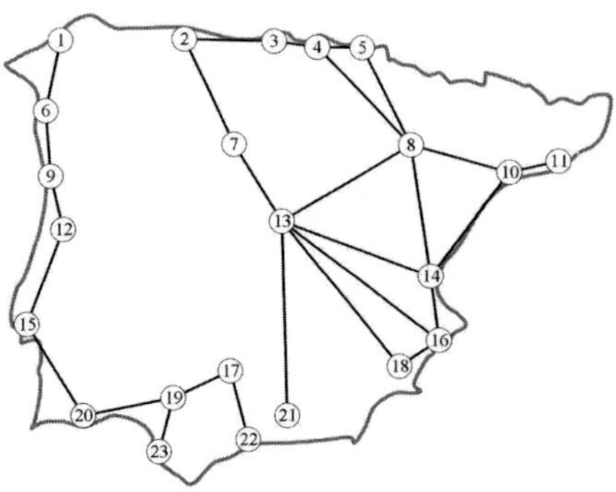

(c) $\mathbf{P}(\frac{2}{30}) \bigcap \mathbf{H}$

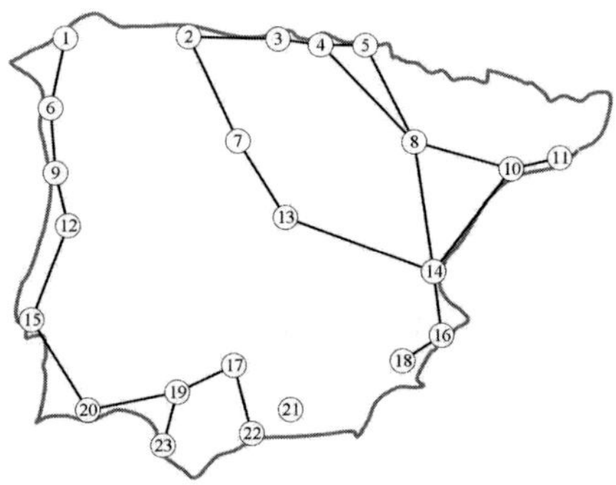

(d) $\mathbf{P}(\frac{5}{30}) \bigcap \mathbf{H}$

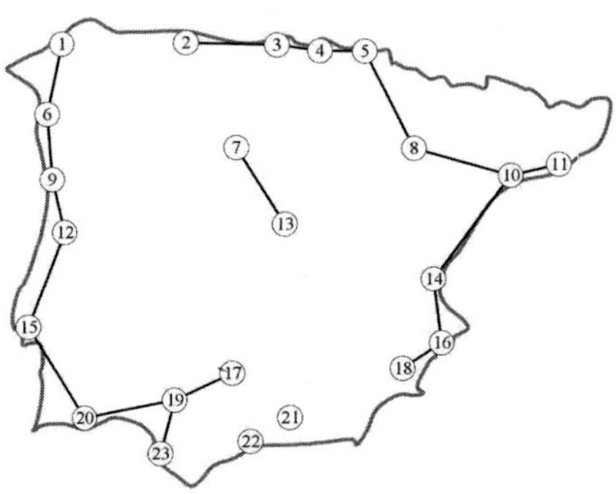

(e) $\mathbf{P}(\frac{16}{30}) \bigcap \mathbf{H}$

Fig. 14.6 Graph **H** of man-made motorway network is shown in (a). Intersection of Physarum graph and motorway graph is shown in (b).

apart from a few cities, and the eastern part is entirely Spanish.

Finding 94. *Plasmodium of* Physarum polycephalum *segregates transport networks in Spain and Portugal.*

This is because a restriction on over-$\frac{2}{30}$ weight for the Physarum graph's edges means that these edges appear in over 6% of experiments, i.e. just above the possible level of noise. Further increase of θ leads to isolation of Granada and further simplification of the western part of the network (Fig. 14.6d). For $\theta = \frac{5}{30}$, the western network, A Coruna–Vigo–Porto–Pombal–Lisbon–Faro–Cadiz–Sevilla, remains undisturbed. The eastern network is transformed to a main cycle: Madrid–Valladolid–Gijon–Oviedo–Santander–Bilbao–San Sebastian–Zaragoza–Tarragona–Valencia–Madrid. The cycle includes two embedded subcycles: Bilbao–San Sebastian–Zaragoza–Bilbao and Zaragoza–Tarragona–Valencia–Zaragoza. Two small chains attached to the major cycle are Tarragona–Barcelona and Valencia–Alicante–Murcia (Fig. 14.6d).

As soon as θ rises to $\frac{16}{30}$, the intersection of the motorway and Physarum graphs splits into two isolated nodes: Malaga–Marbella and Granada, a two-node segment: Madrid–Valladolid and two trees. One tree is rooted in Sevilla and consists of three chains: Sevilla–Cordoba, Sevilla–Cadiz and Sevilla–Faro–Lisbon–Pombal–Porto–Vigo–A Coruna. The second tree is rooted in Tarragona and consists of three chains: Tarragona–Barcelona, Tarragona–Valencia–Alicante–Murcia and Tarragona–Zaragoza–San Sebastian–Bilbao–Santander–Gijon–Oviedo (Fig. 14.6e).

Finding 95. *Plasmodium of* P. polycephalum *does not consider Madrid (13) and Valladolid (7) as a part of the connected transport network in the Iberian peninsula.*

Why is it so? Plasmodium is never concerned with political games; it only tries to optimise its foraging activity. Keeping Madrid and Valladolid as a part of the integrated network puts unnecessary strains on plasmodium functioning. Thus, we can speculate that a star-shaped motorway network rooted in Madrid is rather an artificial, or man-made, feature than an intrinsically biological phenomenon.

The motorway and Physarum graphs have the same labelling; therefore, we can compare them straightforwardly. Let $d_i^{\mathbf{P}}(\theta)$ and $d_i^{\mathbf{H}}$ be degrees of node i in graphs **P** and **H**. A degree mismatch between $\mathbf{P}(\theta)$ and **H** is a sum of absolute values of node mismatches. The correlation between θ and degree mismatch is illustrated in Fig. 14.7a. Lowest mismatches are observed for $\theta = \frac{8}{30}, \frac{9}{30}, \frac{12}{30}, \frac{13}{30}$ and $\frac{16}{30}$. The lowest degree mismatch in the Physarum graph $\mathbf{P}(\frac{16}{30})$ bears particular importance because when the threshold θ exceeds $\frac{16}{30}$ the graph becomes

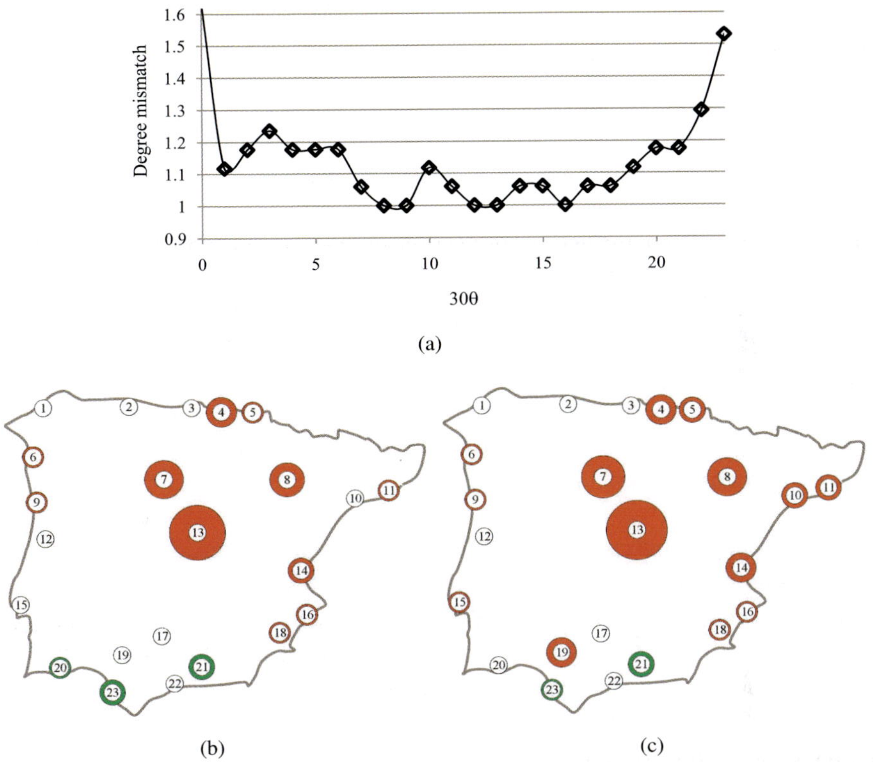

Fig. 14.7 Mismatches between node degrees of Physarum and motorway graphs: (a) degree mismatch vs threshold θ. Values of mismatches are normalised by dividing them by minimal mismatch for the studied range of θ, (b, c) spatial representation of degree mismatches for $\mathbf{P}(\theta)$ and $\mathbf{H}$ for $\theta = \frac{16}{30}$ (b) and $\theta = \frac{22}{30}$ (c). A node i has green (light grey) band if $d_i^{\mathbf{P}} > d_i^{\mathbf{H}}$ and red (dark grey) band if $d_i^{\mathbf{P}} < d_i^{\mathbf{H}}$. Width of the band at i is proportional to $|d_i^{\mathbf{P}} - d_i^{\mathbf{H}}|$.

disconnected yet fitting well the motorway graph in terms of node degrees.

How does the spatial distribution of degree mismatches change during the transition from $\mathbf{P}(\frac{16}{30})$ to $\mathbf{P}(\frac{22}{30})$? The degree mismatch in Madrid, Valladolid and Zaragoza slightly increases (Fig. 14.7b and c). Tarragona and Sevilla start showing some level of degree mismatch while Faro reverts to a fully matching state.

Finding 96. *From the man-made motorway perspective, Physarum underdevelops transport links in the north-east of Iberia and overdevelops transport links in the south of Iberia.*

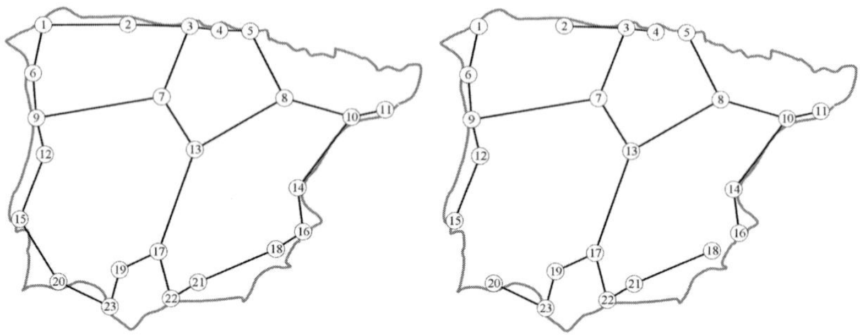

(a) **RNG** (b) **MST**

Fig. 14.8 Proximity graphs constructed on regions **U**: (a) relative neighbourhood graph **RNG**, (b) minimum spanning tree **MST**.

14.2 Physarum shows higher optimality

A protoplasmic network constructed in any particular experiment is planar; a Physarum graph **P** may be non-planar; however, it becomes planar when we put constraints on a minimal weight of edges, $\mathbf{P}(\frac{5}{30})$.

Strictly speaking, a spanning tree rooted in Madrid (Fig. 14.8d) is not a minimum spanning tree. The minimal length tree is rooted in Alicante. However, differences in lengths are negligible: the tree rooted in Madrid is just over 9% longer than the spanning tree rooted in Alicante (Fig. 14.9). Therefore, further on we will address the spanning tree rooted in Madrid as the minimum spanning tree.

Finding 97. $\mathbf{RNG}(\mathbf{U}) - \mathbf{ST}(\mathbf{U})$ = *A Coruna– Gijon–Oviedo, Bilbao–San Sebastian, Alicante–Murcia, Lisbon–Faro.*

This finding shows that acyclic minimal spanning of urban areas **U** is just three links shorter than cyclic spanning. Amongst the edges of $\mathbf{RNG}(\mathbf{U}) - \mathbf{ST}(\mathbf{U})$, only the link A Coruna–Gijon–Oviedo is not present in the Physarum graph for the highest value of $\theta = \frac{16}{30}$. We say that a planar graph on a set **U** is *optimal* if it matches well the minimum spanning tree constructed on **U** or at least a relative neighbourhood graph of **U**.

Finding 98. *The man-made motorway network in the Iberian peninsula is not optimal while its approximation by Physarum is optimal.*

The statement is validated as follows. Let us consider the intersection of the motorway graph and proximity graphs. The intersection of **H** and **RNG** has one isolated node (Granada). Five edges of **RNG** are not represented in

Node	ST length
1. A Coruna	1.04
2. Gijon–Oviedo	1.02
3. Santander	1.08
4. Bilbao	1.04
5. San Sebastian	1.05
6. Vigo	1.10
7. Valladolid	1.08
8. Zaragoza	1.02
9. Porto	1.09
10. Tarragona	1.06
11. Barcelona	1.06
12. Pombal	1.09
13. Madrid	1.10
14. Valencia	1.05
15. Lisbon	1.02
16. Alicante	1.00
17. Cordoba	1.09
18. Murcia	1.05
19. Sevilla	1.12
20. Faro	1.10
21. Granada	1.07
22. Malaga–Marbella	1.06
23. Cadiz	1.05

Fig. 14.9 Ratios of lengths of spanning trees rooted in nodes of **U** to the length of minimum spanning tree (rooted in Alicante).

H: A Coruna–Gijon–Oviedo, Zaragoza–Tarragona, Murcia–Granada, Granada–Malaga–Marbella and Faro–Cadiz. There are three isolated nodes: Granada, Murcia and Faro in the intersection of **H** and **MST**. Let **w** = Santander-Valladolid and **L** = Santander–Valladolid, Valladolid–Porto, Madrid–Cordoba, Madrid–Zaragoza. The relative neighbourhood graph is almost a subgraph of Physarum graphs because $\mathbf{RNG} - \mathbf{w} \subset \mathbf{P}(0)$ and $\mathbf{RNG} - \mathbf{L} \subset \mathbf{P}(\frac{5}{30})$. Moreover, we see that $\mathbf{MST} - \mathbf{w} \subset \mathbf{P}(0)$ and $\mathbf{MST} - \mathbf{L} \subset \mathbf{P}(\frac{16}{30})$.

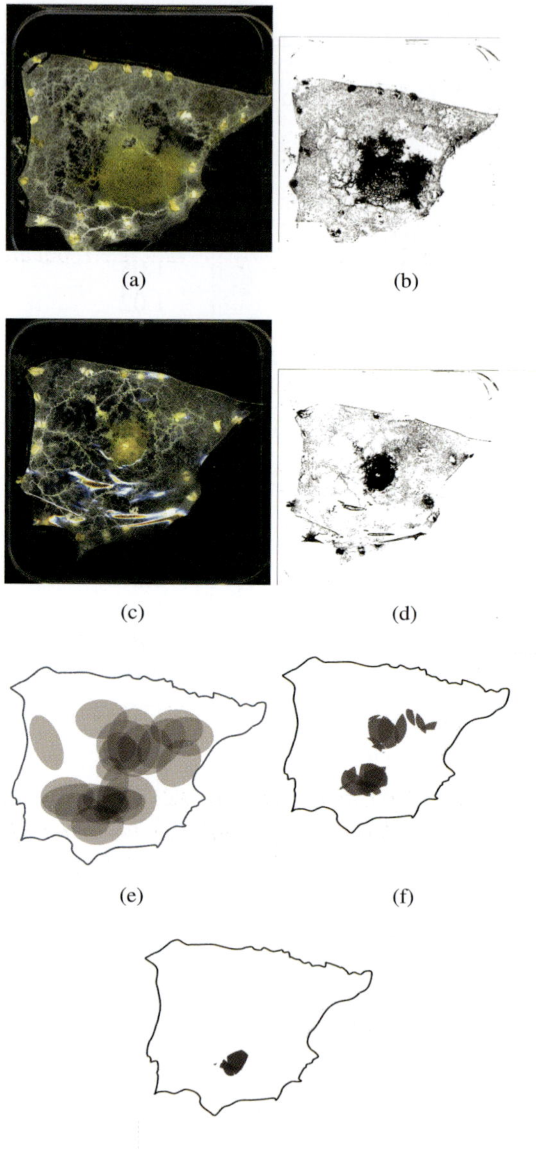

Fig. 14.10 Imitating collapse of infrastructure with *Physarum polycephalum*: (a–d) two examples of dried substrate with sclerotium formed. Scanned images of experiments are shown in (a) and (c), their binarised versions in (b) and (d). Positions of sclerotium are overlapped from 20 experiments. All outcomes of sclerotinisation are displayed in (e). Domains occupied by sclerotia in 50% of experiments (f) and 80% of experiments (g).

14.3 Collapse of infrastructure

What would happen with transport networks in the Iberian peninsula if the whole infrastructure collapses? We tried to simulate a fairly real and frequent kind of motorway crisis: that of the relatively frequent traffic interruption caused by very heavy and intense storms which happen on the east coast of Spain. These punctual episodes, associated with very heavy rain in a short time, generate huge floods. The disastrous effects are induced not only by the rain itself, but also by the lack of vegetative cover (due to deforestation) and the occupation by human urbanisation of many of the ways of natural water evacuation [Bull et al. (2000); Camaras Belmonte and Segua Beltrun (2001)].

To imitate possible scenarios of infrastructure breakdown, we deprived the Physarum transport network from further supply of nutrients and made its condition harsh by allowing the substrate to dry. When the substrate dries the plasmodium abandons its protoplasmic tubes and migrates to a single domain of the substrate, where it forms a hardened protoplasm (sclerotium). In natural conditions sclerotium can survive in such hibernation for a very long period of time; it can then come back to life when moisturised.

Two examples of sclerotium are shown in Fig. 14.10a–d. In the scanned images of dried agar gel sclerotia are visible as brownish concentrations (Fig. 14.10a and c), while in the binarised images the sclerotia are represented by areas with high density of black pixels (Fig. 14.10b and d).

Finding 99. *In the case of infrastructure breakdown in the Iberian peninsula the following scenarios may develop. Portugal may become isolated. Spain will concentrate to Madrid, Castilla La Mancha and Andalusia with probability 0.5 and concentrate to a small region at the boundary between Castilla La Mancha and Andalusia with probability 0.8.*

We undertook 20 experiments; for each experiment we logged positions and areas of sclerotia and superposed the sclerotia, represented by ellipses, onto one image (Fig. 14.10e). From this image we extracted domains occupied by sclerotia in 50% (Fig. 14.10f) and 80% (Fig. 14.10g) of experiments.

14.4 Discovering ancient roads

To evaluate Physarum's performance in imitating road development, we also compared Physarum graphs with road networks as they were in the Iberian peninsula in maps of the Iberian peninsula in the year 125 [Nacu (2012)]. The following

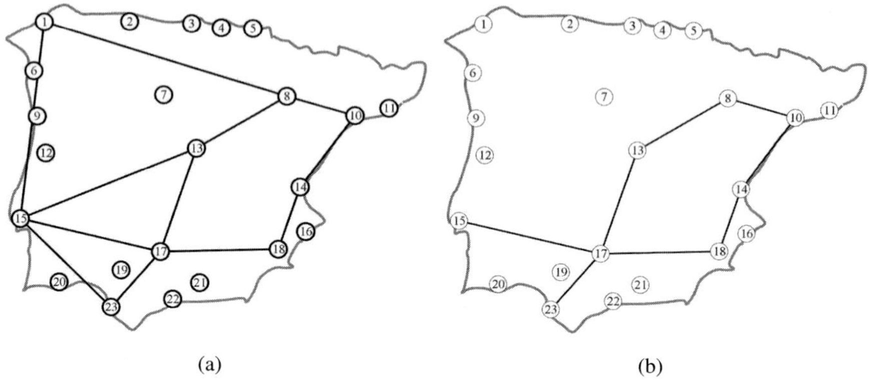

Fig. 14.11 Motorway graph of roads in Iberian peninsula in year 125 (a) and its intersection with Physarum graph $\mathbf{P}(0)$ (b). Node numbering is as follows: (1) Brigantium, (8) Caesaraugusta, (10) Tarraco, (13) Toletum, (14), Valentia, (15) Felicitas Iulia, (17) Corduba, (18) Nova Carthago, (23) Gades. Other nodes are shown for completeness.

Iberian settlements were taken into account: Tarraco (Tarragona), Valentia (Valencia), Caesaraugusta (Zaragoza), Felicitas Iulia (Lisbon), Corduba (Cordoba), Gades (Cadiz). We assumed that due to the proximity of Brigantium (Betanzos) to A Coruna, Tolentum (Toledo) to Madrid and Nova Carthago (Cartagena) to Murcia, they can be identified with these urban areas. We did not incorporate Castra Legionis (Leon) and August Emerita (Merida) (Fig. 14.11a).

We found that the Physarum protoplasmic network represents 7 of 11 major roads (Fig. 14.11b), including Via Hercúlea and Via del Norte. Via del Atlantico is not represented in a straightforward way. However, by allowing intermediary cities the representation could be achieved.

Incorporating mountains in experiments would change the resultant transport networks developed by plasmodium of *Physarum polycephalum*. This issue (that will be scrutinised in a further study) is particularly important in the case of the Iberian peninsula, whose transport network has been notably conditioned by important orographic accidents. Incidentally, although the land-based boundaries between Portugal and Spain are mostly not defined by geomorphological entities (e.g. mountains and/or rivers), they have been notably stable in history [History of Portugal (2012)].

14.5 Summary

Is there a match between man-made motorways in Spain and Portugal and their biological analogues developed by slime mould? To get an answer, we employed a promising biological computing substrate [Adamatzky (2010)b] — plasmodium of *Physarum polycephalum*. We represented major urban regions in the the Iberian peninsula with oat flakes, inoculated the plasmodium in Madrid and recorded formation of protoplasmic networks. Then we compared the Physarum transport graph with man-made motorways and abstract proximity graphs.

Experimenting with *Physarum polycephalum*, we got more than we bargained for. Contrary to our expectations, the plasmodium did not build a spanning tree rooted in Madrid, but instead spanned major Iberian cities by propagating (anti)clockwise along the shore. We found that the plasmodium likes to segregate transport networks of Spain and Portugal and prefers not to integrate Madrid (13) and Valladolid (7) into the united transport network of Spain.

Overall, we can claim that plasmodium offers an alternative yet optimal transport network different from the existing motorway network. Additional indications on optimality of the Physarum network of protoplasmic tubes are that the minimum spanning tree is matched better by plasmodium than by the motorway network, and that the relative neighbourhood graph is a subgraph of the Physarum graph, while the motorway graph is not.

There are two traits in the Iberian peninsula that notably distinguish it from previously studied countries. First, contrary to a fairly universal rule, the most populated area (Madrid) in the Iberian peninsula is not located near the coast. Moreover, Madrid is located roughly in the geographical centre. Second, the considerable depopulation of the central region (besides Madrid), which reaches extremely low population densities in some regions, dramatically contrasts with very high population densities on the coast, coming from both resident inhabitants and tourism [Desequilibrios demográficos (2012)].

Chapter 15

United Kingdom road planning with slime mould

Andrew Adamatzky and Jeff Jones

Plasmodium of *P. polycephalum* approximates a shortest path, builds proximity graphs and grows spanning trees. Does the plasmodium emulate road networks? Given cities represented by oat flakes and plasmodium of *P. polycephalum* inoculated in one of the cities, will the plasmodium develop a protoplasmic network connecting the oat flakes that matches the network of roads connecting the cities?

We decided to concentrate on only a few major urban areas (Fig. 15.1) and consider the whole United Kingdom as the plasmodium growth domain. The areas are projected onto gel or filter paper and oat flakes are placed in the positions of the urban areas (Fig. 15.2).

We consider the 10 most populous urban areas in the United Kingdom (Fig. 15.1):

- Greater London,
- Bristol,
- Sheffield,
- Nottingham,
- Liverpool,
- Tyneside,
- Greater Glasgow,
- West Yorkshire,
- Greater Manchester,
- West Midlands,

Fig. 15.1 Schematic map of 10 most populous urban areas. Adapted from [Pointer (2005)].

as per the 2001 Census[1], see details and boundaries of the areas in [Pointer (2005)].

We try to match size of urban areas with size of oat flakes. At the beginning of each experiment, plasmodium is inoculated in the centre of the Greater London urban area.

[1]Office for National Statistics, General Register Office for Scotland and Northern Ireland Statistics and Research Agency.

Fig. 15.2 Typical experimental setup: urban areas are represented by oat flakes, plasmodium is inoculated in London, the plasmodium spans the oat flakes by protoplasmic transport network. Map of urban areas is adapted from [Pointer (2005)].

15.1 From London to Bristol and Glasgow

Being placed in the centre of the Greater London urban area, plasmodium typically consumes some nutrients from its nearest (London) oat flake and starts propagating north, north-west or west (Fig. 15.3a). Birmingham (Fig. 15.3a) and Bristol (Fig. 15.4c) are the usual candidates which are spanned by the London-originated plasmodium. When urban areas in the Midlands are colonised by the plasmodium and linked by protoplasmic tubes, the plasmodium heads towards the Tyneside urban area (Fig. 15.3b).

After taking in Tyneside, the plasmodium propagates north, crosses the Scottish boundary and finally reaches the Glasgow urban area (Fig. 15.3c). Then the

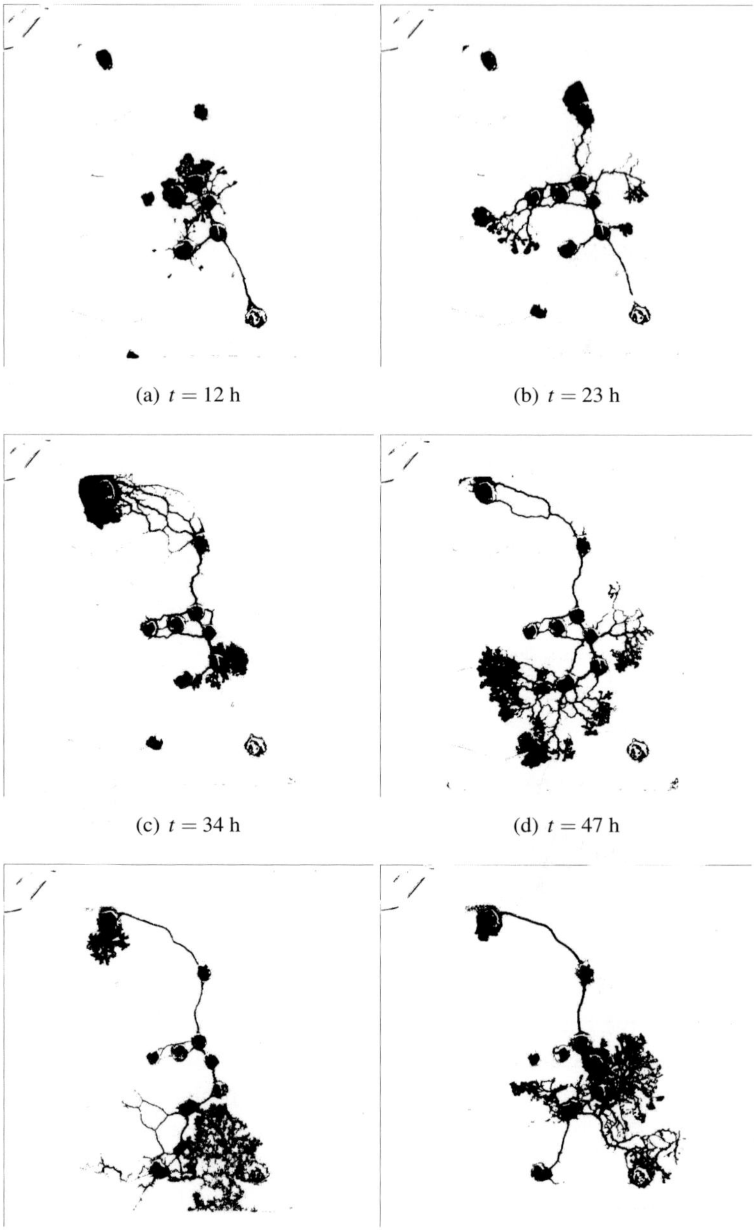

(a) $t = 12$ h (b) $t = 23$ h

(c) $t = 34$ h (d) $t = 47$ h

(e) $t = 69$ h (f) $t = 80$ h

Fig. 15.3 Typical plasmodium development.

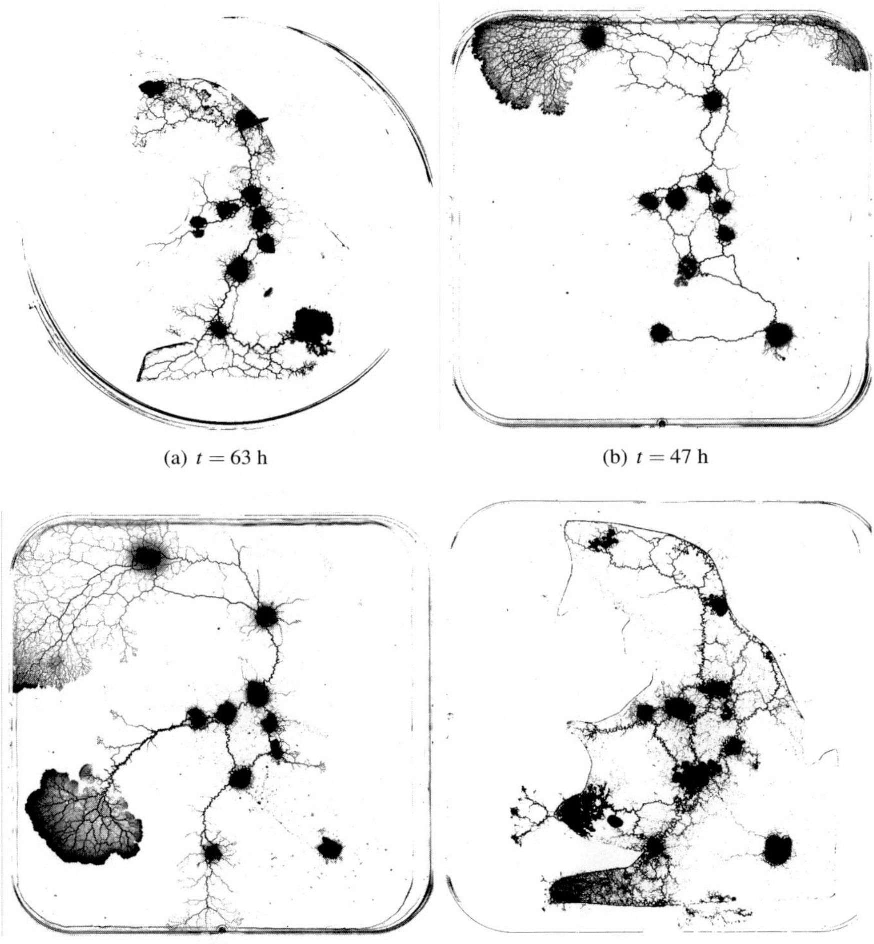

(a) $t = 63$ h

(b) $t = 47$ h

(c) $t = 53$ h

(d) $t = 46$ h

Fig. 15.4 Examples of protoplasmic networks occurring in experiments. Times passed after inoculation of plasmodium in London are shown in subfigure captions.

plasmodium continues colonisation of the substrate until all urban areas (sources of nutrients) are colonised (Fig. 15.3d–f).

Examples of plasmodium networks connecting urban areas are shown in Fig. 15.4. The figure demonstrates that, in general, the structure of the network does not depend significantly on size and shape of the substrate but mainly on the configuration of sources of nutrients. We demonstrated this by undertaking experiments in a 90 mm diameter Petri dish (Fig. 15.4a), 120 mm square dishes fully covered with agar gel (Fig. 15.4b–c) and the shape of the UK island cut out of an agar gel plate in a 120 mm square Petri dish (Fig. 15.4d).

The plasmodium does not stop its foraging activity even when all sources of nutrients are colonised. It propagates away from the 'designated' area (Fig. 15.4c) unless stopped by an unfriendly substrate (like the bottom of a plastic Petri dish not covered by gel, Fig. 15.4d (bottom)).

The plasmodium does not always keep all sources of nutrients spanned by its protoplasmic tubes. Sometimes, a few tubes are abandoned during colonisation. Thus, in Fig. 15.5, we see that at the beginning of its development plasmodium links London and Nottingham (Fig. 15.5a). When urban areas in the Midlands are colonised and linked by protoplasmic tubes to the Tyneside and Glasgow urban areas, the plasmodium abandons its tube connecting the Greater London and Nottingham urban areas (Fig. 15.5b).

Even a limited number of examples (e.g. Figs. 15.4 and 15.5) demonstrate that plasmodium of *P. polycephalum* is a very dynamic system, whose morphology is

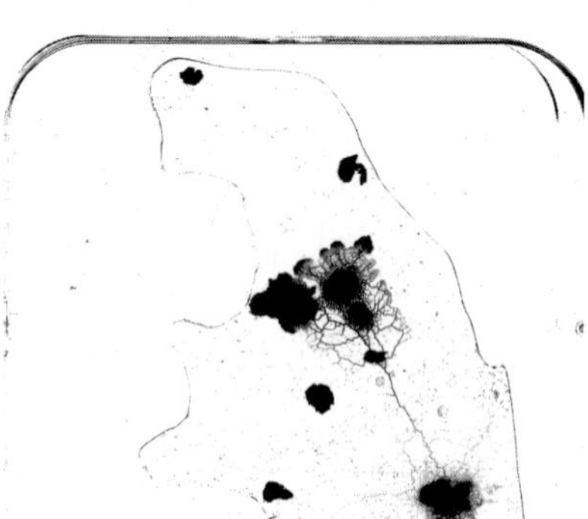

(a) $t = 8$ h

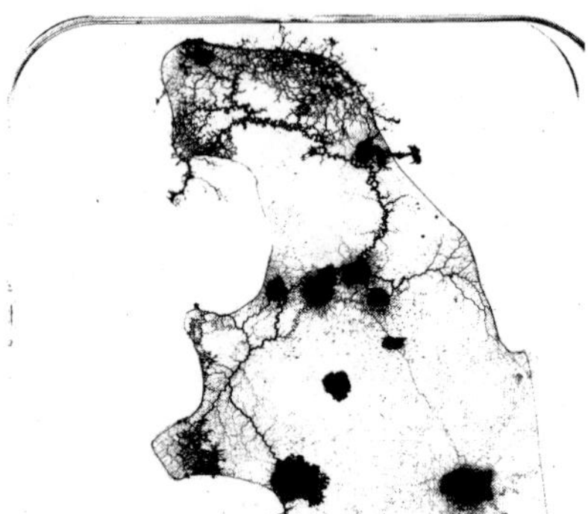

(b) $t = 31$ h

Fig. 15.5 Examples of reconfigurations of protoplasmic network.

continuously changing and whose spatiotemporal dynamics rarely reaches a fixed stable point (unless humidity decreases and the plasmodium enters the sclerotium stage of its life cycle). There is no such thing as a stationary configuration of a protoplasmic network; therefore, when extracting a graph of transport links from plasmodium experiments we can only rely on the Physarum graph **P**.

Physarum graphs extracted from 25 laboratory experiments are shown in Fig. 15.6; the maximum edge weight is 22. The graph becomes planar when we remove edges with weights below 6 (Fig. 15.6b). The graph is acyclic, or a tree, when only edges appearing in over 40% of experiments are shown (Fig. 15.6d). If we increase the cutoff value to 14, the graph becomes disconnected, and the node corresponding to the Bristol urban area becomes isolated.

15.2 Physarum vs Department for Transport

The graph **M** is extracted from the motorway network as shown in `maps.google.com` and `www.openstreetmap.org`. The motorway graph **M** is shown in Fig. 15.7. By comparing **M** (Fig. 15.7) and the Physarum graph **P** and its subgraphs (Fig. 15.6), we found the following.

Finding 100. *The motorway graph* **M** *is a subgraph of the Physarum graph* **P**.

This shows that distributed 'logic' underpinning plasmodium's decision-making routines corresponds to human logic behind road-planning decisions. However, the graph **P** is not planar and thus is of little practical importance. The motorway link M6/M74 connecting the Greater Manchester and Greater Glasgow urban areas is represented by plasmodium only in three of 25 experiments. The corresponding edge does not appear in the graphs $\mathbf{P}_5$, $\mathbf{P}_{10}$ and $\mathbf{P}_{12}$ (Fig. 15.6b–d). The motorway M4, linking the Greater London and Bristol urban areas, is represented by protoplasmic tubes of *P. polycephalum* only in 20% of experiments, the graph $\mathbf{P}_5$ in Fig. 15.6b.

Finding 101. P. polycephalum *satisfactorily approximates the motorway network linking the 10 most populous urban areas in the United Kingdom except for the motorway link M6/M74 connecting the Greater Manchester and Greater Glasgow urban areas.*

15.3 Linking Newcastle to Glasgow

Is there a rationale behind plasmodium's behaviour?

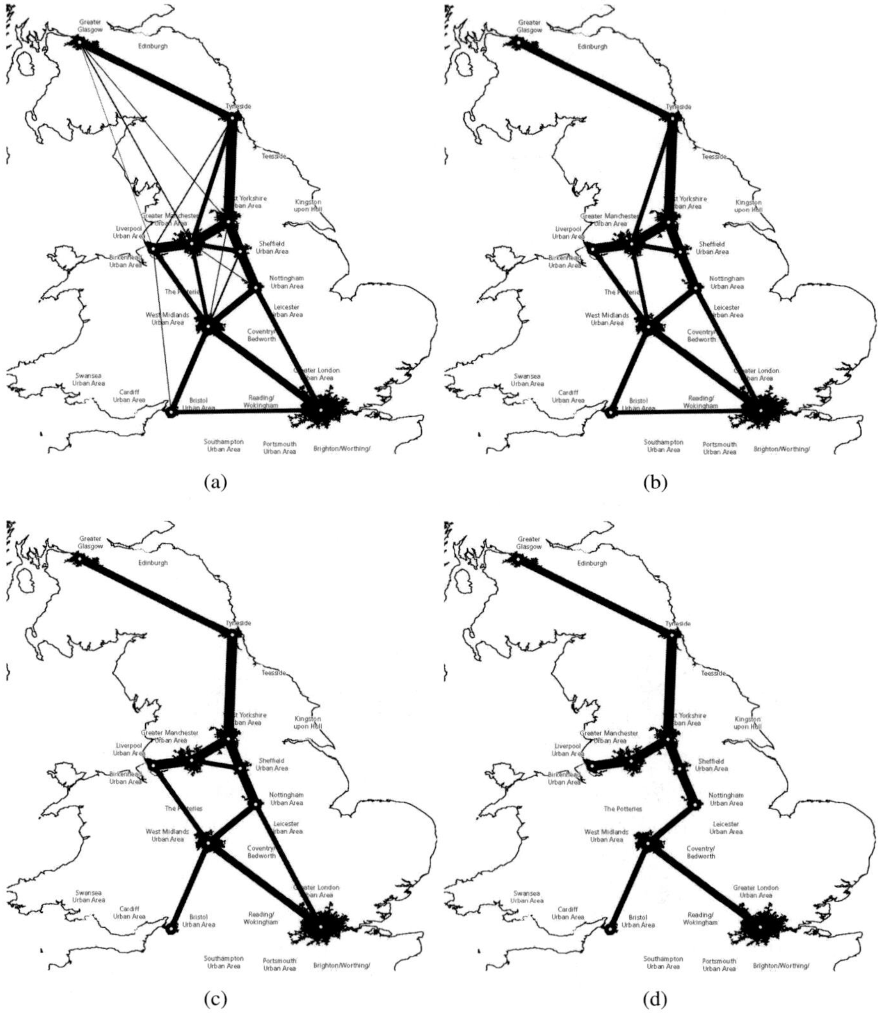

Fig. 15.6 Physarum graphs for various values of edge weights: (a) all edges of Physarum graph are shown, thickness of each edge is proportional to the edge's weight, (a–d) only edges with weights exceeding 5 (b), 10 (c) and 12 (d) are shown.

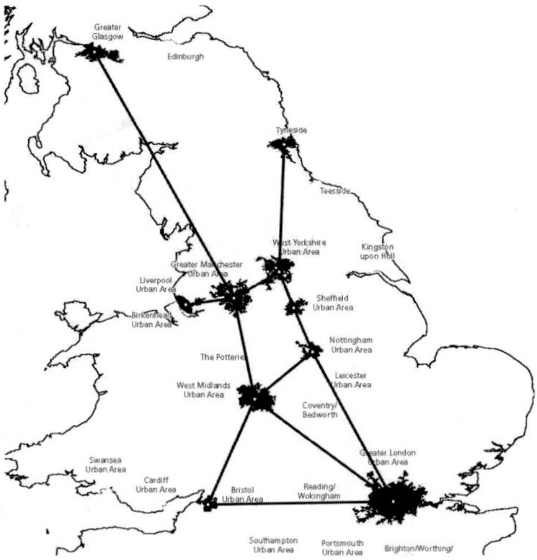

Fig. 15.7 Graph **M** of man-made motorway network connecting the 10 most populous urban areas.

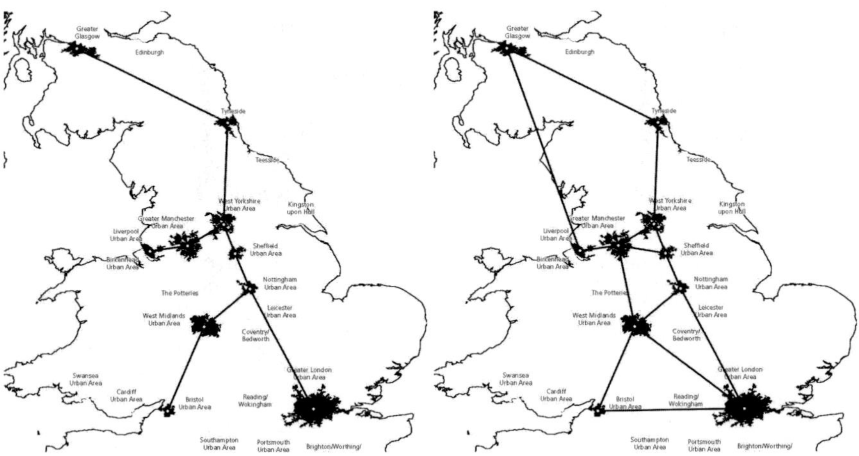

(a) (b)

Fig. 15.8 Two proximity graphs constructed on urban areas **U**: (a) relative neighbourhood graph **RNG**, (b) Gabriel graph **GG**.

Let us have a look at the two most popular planar proximity graphs, the relative neighbourhood graph **RNG** (Fig. 15.8a) and the Gabriel graph **GG** (Fig. 15.8b) constructed over nodes corresponding to centres of urban areas **U**.

Our experiments have shown that

Finding 102. $\mathbf{P}_{10} \subset \mathbf{GG}$ *and* $\mathbf{P}_{12} = \mathbf{RNG}$.

Moreover, $\mathbf{P}_{12}$ is a minimum spanning tree (**MST**) over **U**. We know that $\mathbf{MST} \subseteq \mathbf{RNG}$ but, in the particular case of urban areas' configurations **U**, we even have $\mathbf{MST}(\mathbf{U}) = \mathbf{RNG}(\mathbf{U})$.

With regard to relations between the motorway graph and the proximity graphs, we see that **M** is neither a sub- nor a supergraph of **RNG** and **GG**. To transform **M** to **RNG**, one needs to remove two edges from and relocate one edge in **M**, while to transform **M** to **GG** no edges should be removed from but three edges added to **M**. The edge connecting Tyneside and Greater Glasgow is present in **RNG** and **GG** but absent in **M**.

Finding 103. *Experiments with plasmodium of* P. polycephalum *show that the M6/M74 motorway is not optimally positioned and should be rerouted from Newcastle to Glasgow. Alternatively, the M6/M74 motorway may remain intact but a new Newcastle–Glasgow motorway must be built.*

15.4 Salt in Leeds

Here we consider three examples: disaster in and contamination of the West Yorkshire urban area (Fig. 15.9, wet filter paper substrate, and Fig. 15.10, agar gel substrate) and the Tyneside urban area (Fig. 15.11, agar gel substrate).

In the experiment shown in Fig. 15.9, a plasmodium network spanning most urban areas is formed 32 h after inoculation of the Greater London area with plasmodium. We place a salt crystal in Leeds (West Yorkshire urban area) (Fig. 15.9a). In response to the increased concentration of sodium chloride the plasmodium abandons the West Yorkshire area and disconnects the area from the neighbouring Tyneside, Greater Manchester and Sheffield urban areas (Fig. 15.9b). At the same time the plasmodium starts mass exploration of Scotland and North Wales, and restores the previously abandoned transport link with London (Fig. 15.9b). In around 27–30 h after the 'disaster' in West Yorkshire the plasmodium completes its evacuation from the Midlands and northern England and regroups itself in northern Scotland and southern England (Fig. 15.9c).

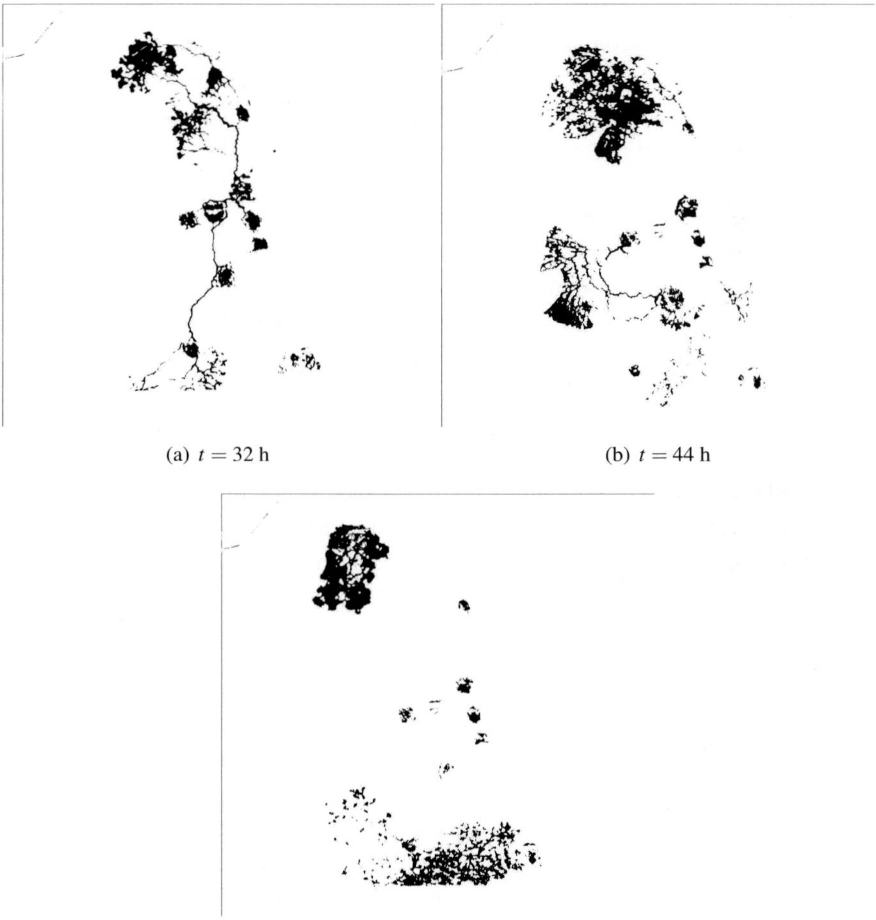

(a) $t = 32$ h

(b) $t = 44$ h

(c) $t = 59$ h

Fig. 15.9 Plasmodium's response to disaster in and subsequent contamination of West Yorkshire area, filter paper substrate. Grain of salt placed in West Yorkshire urban area at 32 h of plasmodium development.

Due to the lower rate of sodium chloride diffusion in agar gel (compared to wet filter paper), the plasmodium's response to imitated contamination of agar gel is less dramatic.

As in the previous experiment, we wait until plasmodium forms a well-established protoplasmic network and then place a salt crystal in Leeds (Fig. 15.10a). The plasmodium temporarily breaks all transport links leading to the contaminated zone (West Yorkshire), and increases exploratory activity in Wales (even developing pronounced protoplasmic routes in west Wales) and the south-west of England (Fig. 15.10b). One and a half days after contamination of West Yorkshire the plasmodium restores transport links with Leeds (Fig. 15.10c) and decreases its activity in Wales and the south-west.

In the example shown in Fig. 15.11, we strike an urban area with contaminant before any transport link leading to the area is established. Plasmodium inoculated in the Greater London area spans urban areas in the Midlands with a protoplasmic network (Fig. 15.11a). When the plasmodium's active zone approaches Tyneside, we place a salt crystal in Newcastle (Fig. 15.11a). In response to the contamination, the plasmodium abandons its attempt to colonise northern England and Scotland but heads its foraging activity west and south, and 9 h after contamination of Tyneside the plasmodium reaches Bristol (Fig. 15.11b). The plasmodium increases its exploration of Wales and the south-west of England (Fig. 15.11c). In 32 h after the disaster strikes Tyneside the plasmodium restores the transport network between the Midlands and London, this time via Bristol (Fig. 15.11d).

Finding 104. *Contamination of a single urban area stimulates exploration of uncolonised areas and leads to restoration of previously abandoned transport links.*

15.5 Summary

We represented the 10 most populated UK urban areas by sources of nutrients, inoculated plasmodium of *P. polycephalum* in one of the areas and analysed dynamics of colonisation of the areas by the plasmodium. We studied space–time dynamics of spanning the urban areas by the plasmodium's network of protoplasmic tubes and demonstrated that the plasmodium transport network sufficiently well matches the topology of the existing man-made motorway networks. We found two discrepancies.

The M4 motorway (Bristol–London) rarely occurs in plasmodium networks. The route M6/M74 is absent.

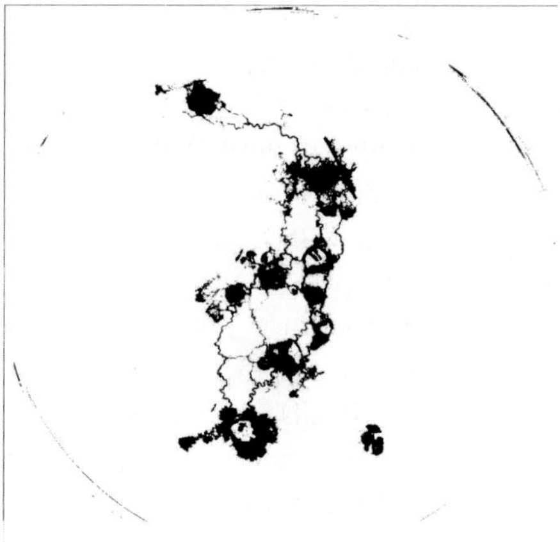

(a) $t = 37$ h

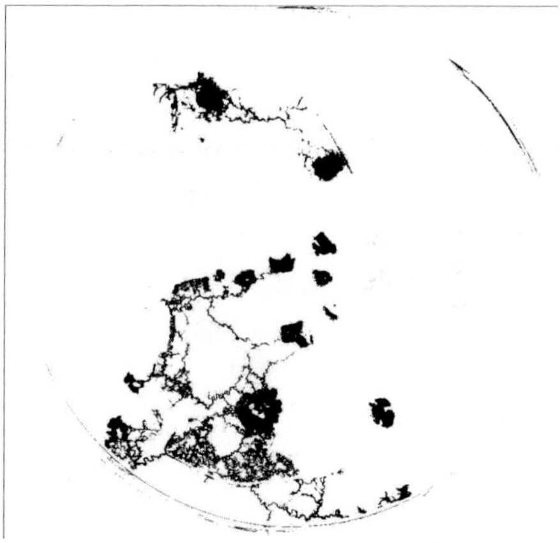

(b) $t = 51$ h

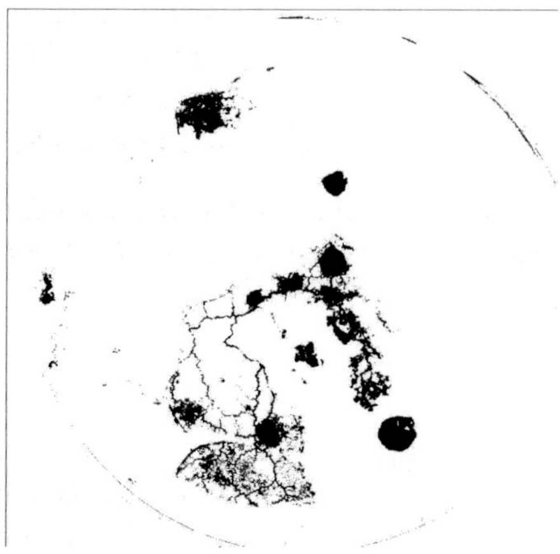

(c) $t = 85$ h

Fig. 15.10 Plasmodium's response to disaster in and subsequent contamination of West Yorkshire area, agar gel substrate. A grain of salt placed in West Yorkshire urban area at 37 h of plasmodium development.

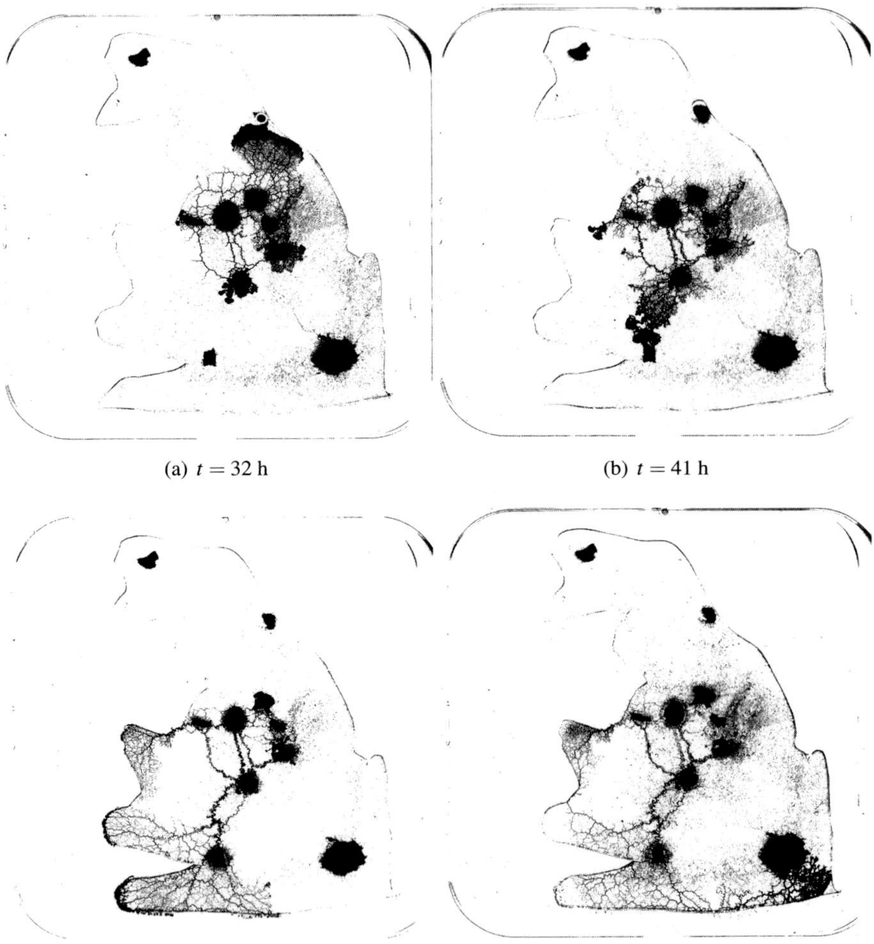

(a) $t = 32$ h (b) $t = 41$ h

(c) $t = 52$ h (d) $t = 64$ h

Fig. 15.11 Plasmodium's response to disaster in and subsequent contamination of Tyneside area, agar gel substrate. A grain of salt placed in Tyneside urban area at 32 h of plasmodium development.

As a 'byproduct' of the experiments, we provided an insight into bioinspired response to disastrous contamination of urban areas. Two main components of the response are exploring uncolonised territories and restoration of abandoned transport links.

Our experiments did not take terrain into account. This may explain the 'anomalous' situation with plasmodium not imitating the route M6/M74 but developing a transport link directly from Newcastle to Glasgow. Terrain-based experiments on Physarum road planning may form a subject for further studies.

Chapter 16

Slimy interstates in the USA

Andrew Adamatzky and Andrew Ilachinski

The USA interstate system, created by Dwight D. Eisenhower, is considered to be an example of a simple yet highly efficient transport that changed the lifestyle and economy of the country [Lewis (1997); McNichol (2005); Kames (2009)]. The interstate system is an ideal example of a transport system aimed to maximise spanning of a territory with a minimal (but still fault-tolerant) number of links. The interstate system was designed by rules of human logic. An interesting question to ask is whether this logic is consistent with that of such primitive living creatures as slime mould?

We consider the 20 most populated urban areas **U** of the USA (Fig. 16.1a):

(1) New York area, including Philadelphia, Baltimore and Washington,
(2) Boston,
(3) Charlotte, NC,
(4) Atlanta,
(5) Jacksonville, FL,
(6) Chicago area, including Detroit, Indianapolis, Columbus and Louisville,
(7) Milwaukee,
(8) Nashville,
(9) Memphis,
(10) Kansas City,
(11) Denver,
(12) Oklahoma City,
(13) Dallas–Fort Worth,
(14) Houston area, including San Antonio and Austin,

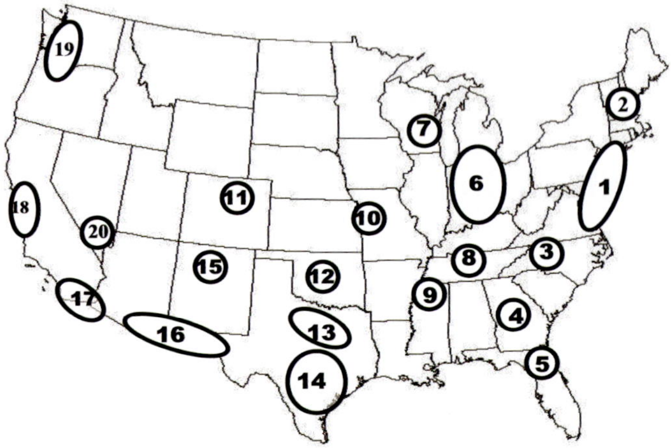

(e)

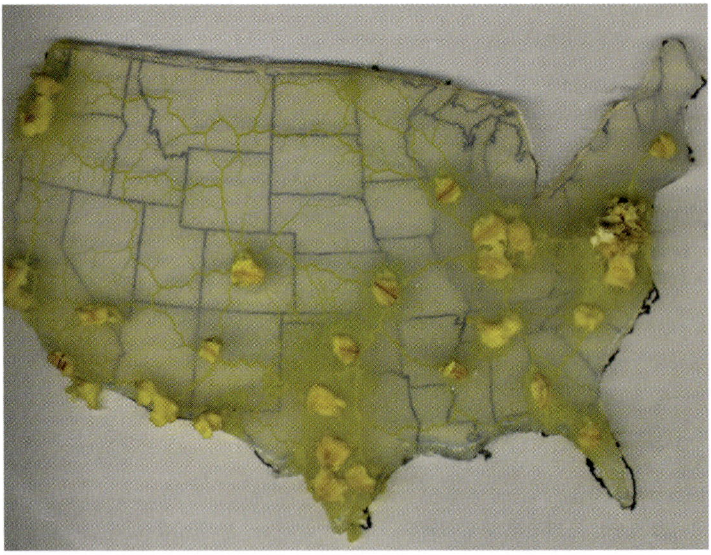

(f)

Fig. 16.1 Experimental setup: (a) urban areas to be represented by oat flakes, (b) snapshot of protoplasmic transport network developed by *P. polycephalum*; the snapshot is made on a state map.

(15) Albuquerque,
(16) Phoenix area, including El Paso and Tucson,

(17) Los Angeles area, including San Diego,
(18) San Jose area, including San Francisco,
(19) Seattle area, including Portland,
(20) Las Vegas.

To project regions of **U** onto agar gel, we placed oat flakes in the positions of the regions of **U** (Fig. 16.1). At the beginning of each experiment a piece of plasmodium, usually already attached to an oat flake in a cultivation box, was placed in the New York area (region 1 in Fig. 16.1a).

16.1 Physarum and Eisenhower

In a few hours after being inoculated at the site of the New York area plasmodium of *P. polycephalum* begins its colonisation of the substrate. In the example illustrated in Fig. 16.2, we see that in the first 12 h the plasmodium spans the New York area with Boston and Charlotte and starts colonising Atlanta and Nashville (Fig. 16.2a). In the next 24 h the plasmodium propagates to and colonises Milwaukee in the north, Atlanta and Jacksonville in the south and the Chicago area, the Kansas area and Oklahoma in the west (Fig. 16.2a). At early stages of colonisation a fine network of protoplasmic tubes is formed. The network is roughened up; some tubes are abandoned and others increase in size later on. The rest of the urban areas of **U** are colonised in a further 24 h.

Plasmodium of *P. polycephalum* rarely repeats its foraging pattern, and almost never builds *exactly* the same protoplasmic network twice (Fig. 16.3). To generalise our experimental results, we constructed a Physarum graph with weighted edges.

A Physarum graph $\mathbf{P}(0)$ is shown in Fig. 16.4, all edges which represented at least once in experiments are shown.

Finding 105. *The Physarum graph is planar for* $\theta \geq \frac{5}{29}$ *and disconnected for* $\theta \geq \frac{18}{29}$.

When θ increases from $\frac{17}{29}$ (Fig. 16.4c) to $\frac{18}{29}$ (Fig. 16.4d), the graph loses its connectivity due to removal of the link Phoenix area to Dallas area. The link roughly corresponds to Interstates 10 and 20. Also the links, yet not responsible for keeping connectivity of the transport network, Seattle–Portland to Las Vegas and Seattle–Portland to Denver, are removed in the transition $\mathbf{P}(\frac{17}{29}) \rightarrow \mathbf{P}(\frac{17}{29})$.

We call a link strong if it appears in over 75% of laboratory experiments (Fig. 16.4e) and superstrong if it appears in over 95% of experiments (Fig. 16.4f).

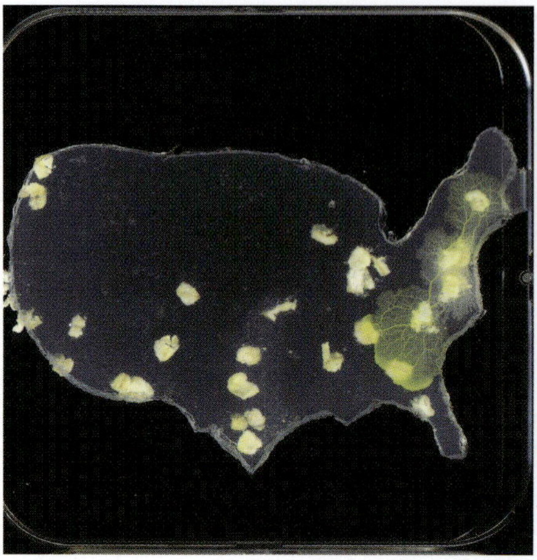

(a) 12 h

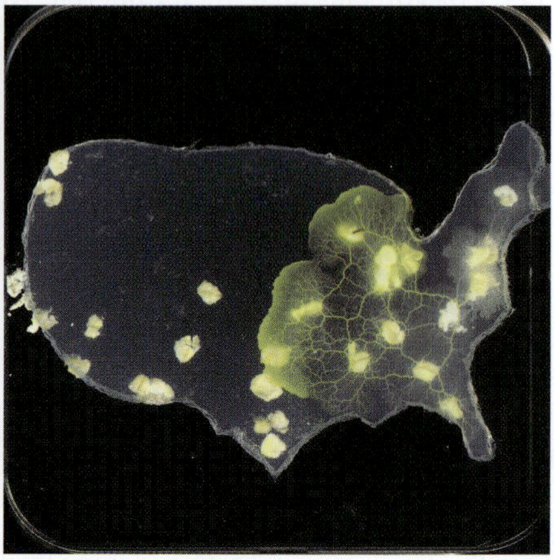

(b) 36 h

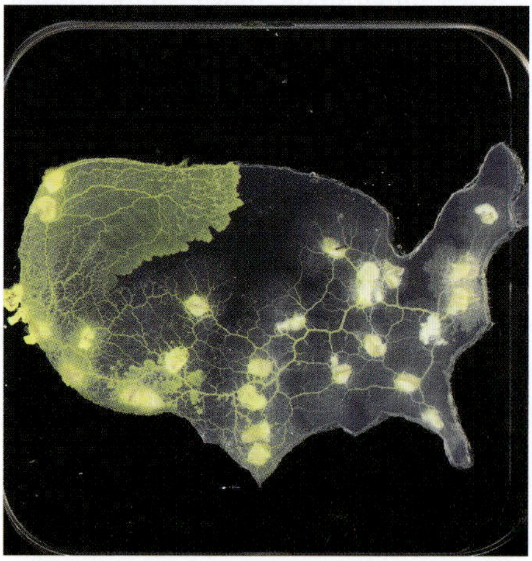

(c) 60 h

Fig. 16.2 Example of plasmodium colonising experimental area. Images are recorded at (a) 12 h, (b) 36 h and (c) 60 h inoculation.

Finding 106. *Physarum imitation of the USA interstate transport network has strong components:*

- *the route from San Jose area to Las Vegas,*
- *the chain of links connecting Denver to Albuquerque to Phoenix area to Los Angeles area,*

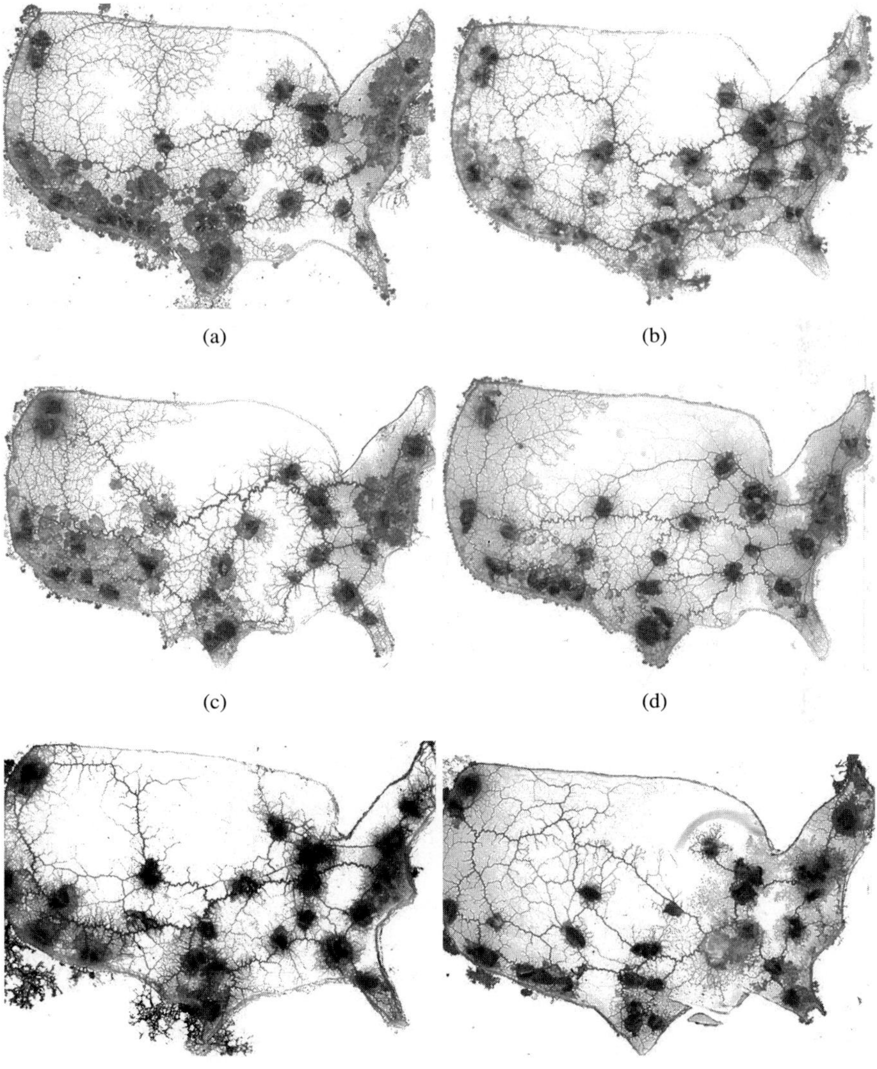

(a) (b) (c) (d) (e) (f)

Fig. 16.3 Grey-scale images of protoplasmic networks developed in laboratory experiments. The images are recorded approx. 48 h after inoculation of plasmodium in the New York area.

- *the chain of links connecting Kansas City to Oklahoma City to Dallas area to Houston area,*

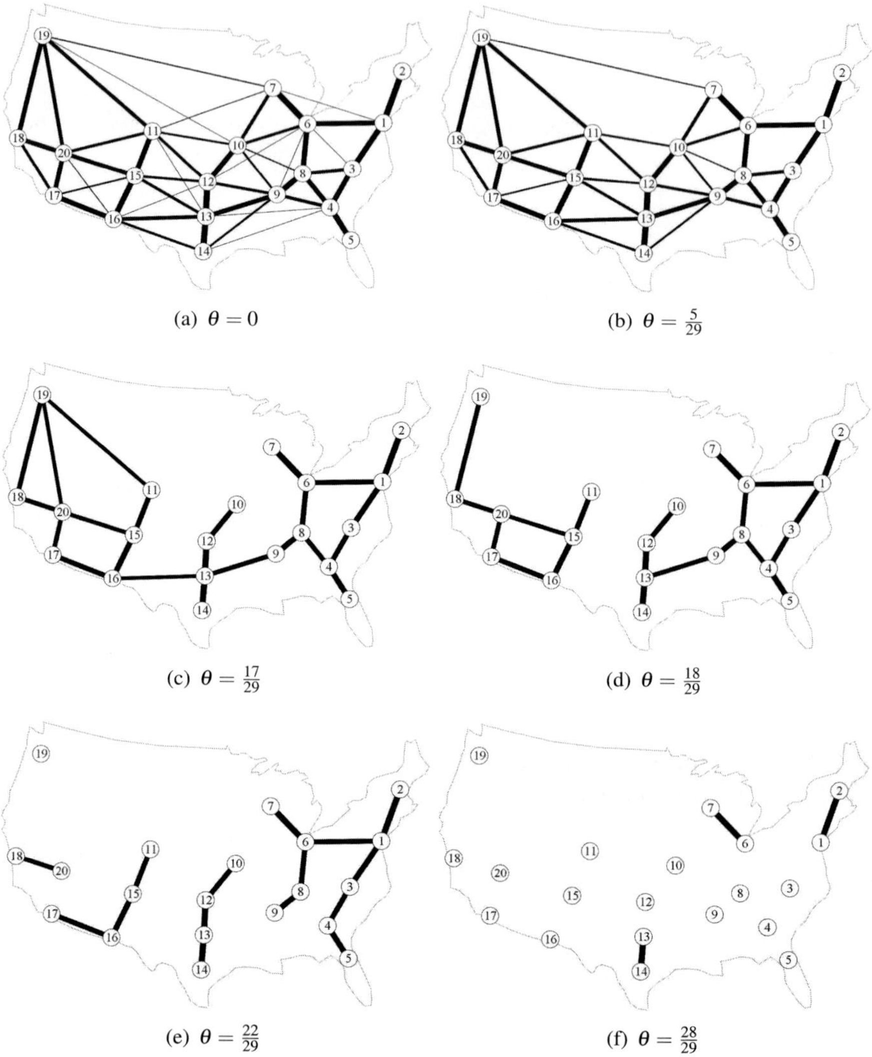

(a) $\theta = 0$

(b) $\theta = \frac{5}{29}$

(c) $\theta = \frac{17}{29}$

(d) $\theta = \frac{18}{29}$

(e) $\theta = \frac{22}{29}$

(f) $\theta = \frac{28}{29}$

Fig. 16.4 Configurations of Physarum graph $\mathbf{P}(\theta)$ for various cutoff values of θ. Thickness of each edge is proportional to the edge's weight.

- *two chains: Milwaukee to Chicago area to Nashville to Memphis and Boston to New York area to Charlotte to Atlanta to Jacksonville, bridged by a link from Chicago area to New York area.*

Finding 107. *The links Dallas area to Houston area, Chicago area to Milwaukee and New York area to Boston are superstrong links of the USA transport network from a perspective of slime mould's foraging behaviour.*

The superstrong links of the Physarum graph are shown in Fig. 16.4f. They roughly correspond to Interstate 35 (Dallas–Houston) and Interstate 96 (New York area–Boston).

16.2 Redundancy of Physarum

The interstate graph **H** shown in Fig. 16.5b is extracted from a scheme of the USA interstate network (Fig. 16.5a). Intersections of the Physarum graph $\mathbf{P}(\theta)$ and **H** are shown in Fig. 16.6.

Finding 108. *The Physarum graph* $\mathbf{P}(0)$ *approximates 36 of 41 edges of the interstate graph* **H**.

The following transport routes of **H** are not represented in protoplasmic networks (Fig. 16.6a):

- Denver to San Jose area,
- Houston area to Albuquerque,
- Houston area to Jacksonville,
- New York area to Nashville,
- Boston to Chicago area.

Finding 109. *Slime mould overdoes the interstate network by producing additional links:*

- *San Jose area to Las Vegas,*
- *Denver to Seattle area,*
- *Albuquerque to Las Vegas.*

This is obtained by direct comparison of $\mathbf{P}(\frac{17}{29})$, $\mathbf{P}(\frac{18}{29})$ and $\mathbf{P}(\frac{22}{29})$ (Fig. 16.4) and intersections of these graphs with the interstate graph **H** (Fig. 16.6).

16.3 New York to Chicago is the strongest link

Finding 110. *Slime mould's protoplasmic network includes minimum spanning of the urban areas* **U**.

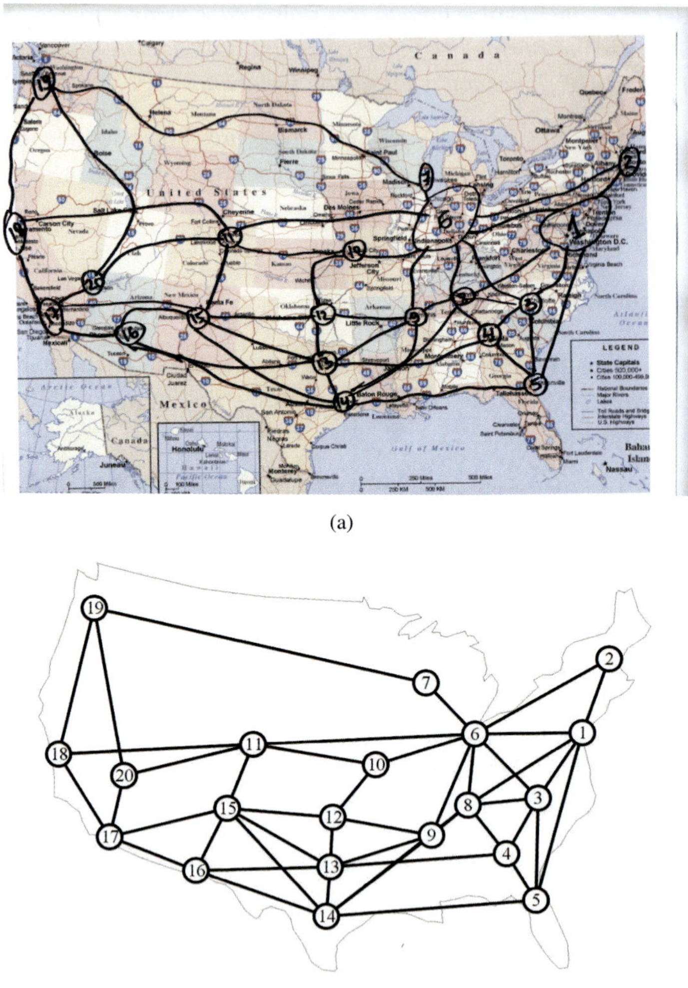

(a)

(b)

Fig. 16.5 Transport network of the USA: (a) scheme of interstates, (b) interstate graph **H**.

This is because $\mathbf{P}(\frac{0}{29}) \bigcap \mathbf{MST} = \mathbf{MST}$ (Fig. 16.8a). The minimum spanning component of slime mould's network is stable: $\mathbf{P}(\theta') \bigcap \mathbf{MST} = \mathbf{P}(\theta'') \bigcap \mathbf{MST}$ for any $0 \leq \theta', \theta'' \leq \frac{19}{29}$.

Finding 111. *The transport link connecting the New York area and the Chicago area is the only strong link not present in the minimum spanning tree* **MST**.

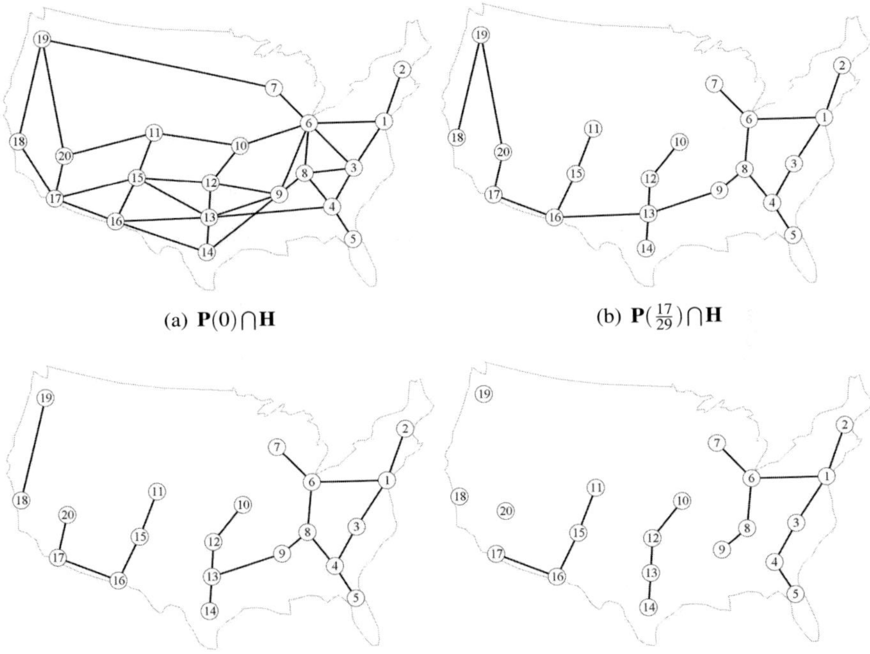

(a) $\mathbf{P}(0) \bigcap \mathbf{H}$ (b) $\mathbf{P}(\frac{17}{29}) \bigcap \mathbf{H}$

(c) $\mathbf{P}(\frac{18}{29}) \bigcap \mathbf{H}$ (d) $\mathbf{P}(\frac{22}{29}) \bigcap \mathbf{H}$

Fig. 16.6 Intersections of Physarum graph $\mathbf{P}(\theta)$ with interstate graph $\mathbf{H}$ for $\theta = 0, \frac{8}{23}, \frac{17}{23}, \frac{18}{23}, \frac{22}{23}$.

This is because $\mathbf{P}(\frac{28}{29}) \bigcap \mathbf{MST} = \mathbf{P}(\frac{28}{29})$−New York–Chicago (Fig. 16.8b); see also Finding 106.

Finding 112. *Superstrong links are edges of the minimum spanning tree on* **MST**.

As we see in Fig. 16.8c, $\mathbf{P}(\frac{28}{29}) \bigcap \mathbf{MST} = \mathbf{P}(\frac{28}{29})$, which are indeed the links described in Finding 107.

Finding 113. *The Physarum graph* $\mathbf{P}(0)$ *includes all the proximity graphs considered:* $\mathbf{MST} \subseteq \mathbf{RNG} \subseteq \mathbf{GG} \subseteq \mathbf{P}(0)$.

This is in agreement with our previous studies on relationships between slime mould's protoplasmic networks and proximity graphs, see e.g. [Adamatzky (2010)b]; see also Fig. 16.8a, d and e.

Finding 114. *The Physarum graph* $\mathbf{P}(\theta)$ *is a subgraph of the Gabriel graph* **GG** *for* $\theta \geq 17$.

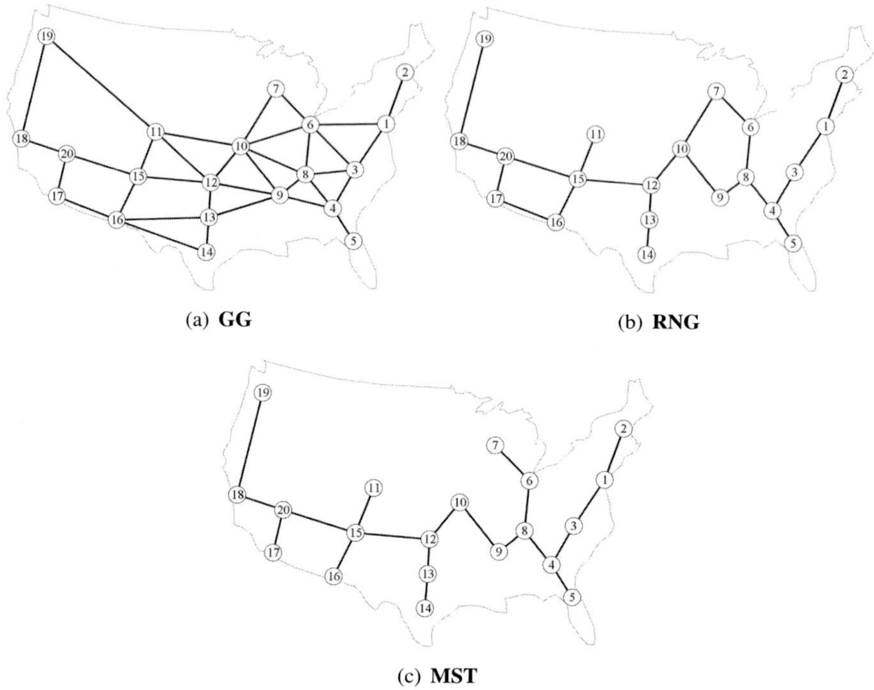

Fig. 16.7 Principal proximity graphs computed on a set **U** of urban ares: (a) Gabriel graph, (b) relative neighbourhood graph, (c) minimum spanning tree.

See examples in Fig. 16.8f and g.

Finding 115. *The minimum spanning tree* **MST** *is almost a subgraph of the interstate graph* **H**.

The following links are missing from $\mathbf{H} \bigcap \mathbf{MST}$ (Fig. 16.8h): Memphis to Kansas City, Las Vegas to Albuquerque and San Jose area to Las Vegas.

16.4 Reconfiguration in disasters

To imitate a large-scale disaster, we placed a crystal of sodium chloride in the place of an agar plate corresponding to Palo Verde nuclear generating station (Fig. 16.9a–c). As an immediate response to diffusing chloride, plasmodium withdraws from the zone immediate to the epicentre of contamination (Fig. 16.10). Further response usually fits into two types: hyperactivation of transport as an

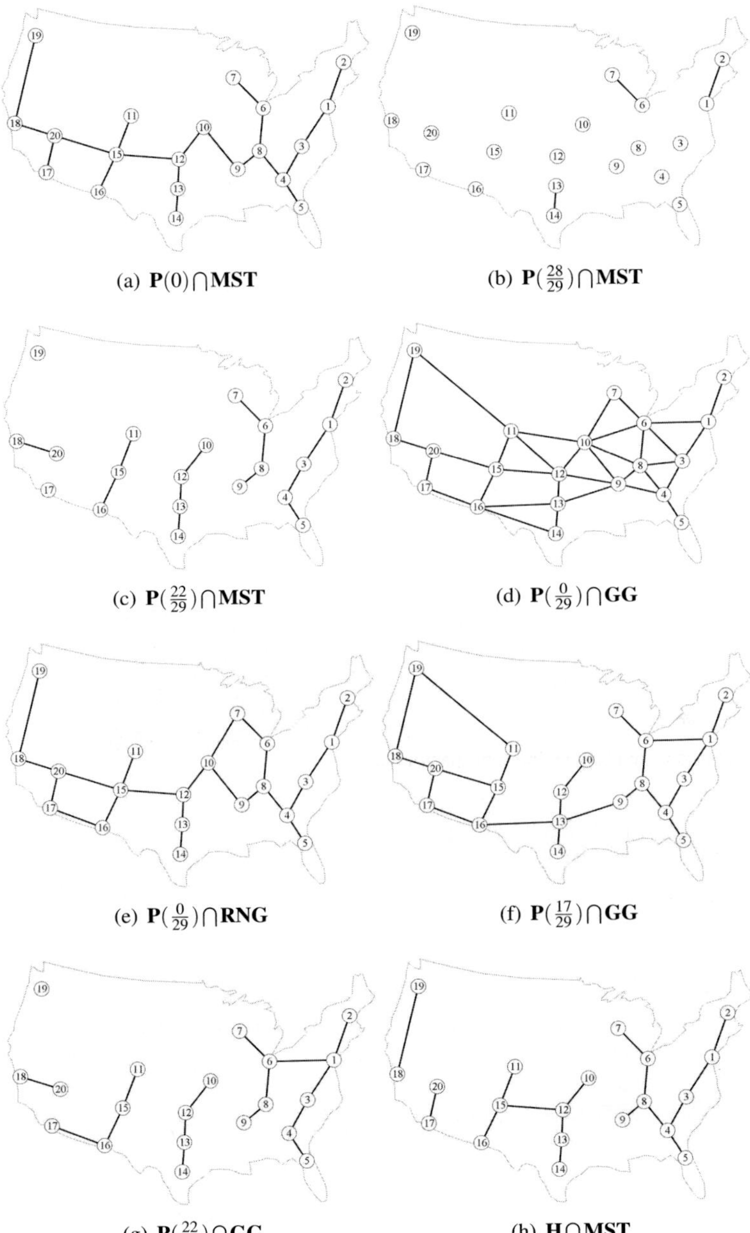

(a) $\mathbf{P}(0) \bigcap \mathbf{MST}$

(b) $\mathbf{P}(\frac{28}{29}) \bigcap \mathbf{MST}$

(c) $\mathbf{P}(\frac{22}{29}) \bigcap \mathbf{MST}$

(d) $\mathbf{P}(\frac{0}{29}) \bigcap \mathbf{GG}$

(e) $\mathbf{P}(\frac{0}{29}) \bigcap \mathbf{RNG}$

(f) $\mathbf{P}(\frac{17}{29}) \bigcap \mathbf{GG}$

(g) $\mathbf{P}(\frac{22}{29}) \bigcap \mathbf{GG}$

(h) $\mathbf{H} \bigcap \mathbf{MST}$

Fig. 16.8 Intersections of (e–g) Physarum graphs $\mathbf{P}(\theta)$ with proximity graph and (h) the interstate graph **H** with minimum spanning tree **MST**. See original proximity graphs in Fig. 16.7.

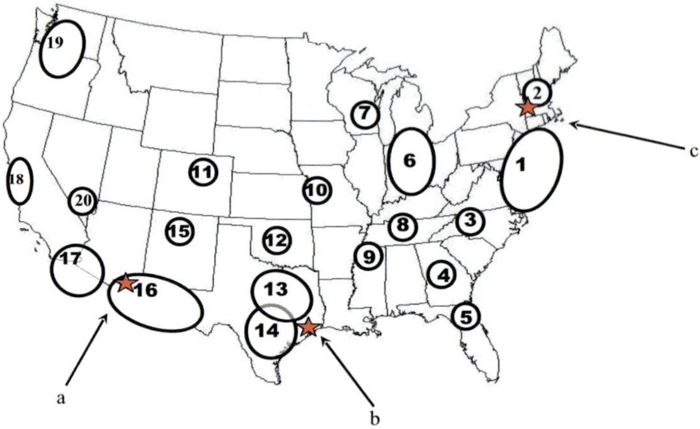

Fig. 16.9 Sites of contamination epicentres are shown by stars and arrows: (a) Palo Verde nuclear generating station, (b) South Texas nuclear generating station, (c) Vermont Yankee nuclear power plant.

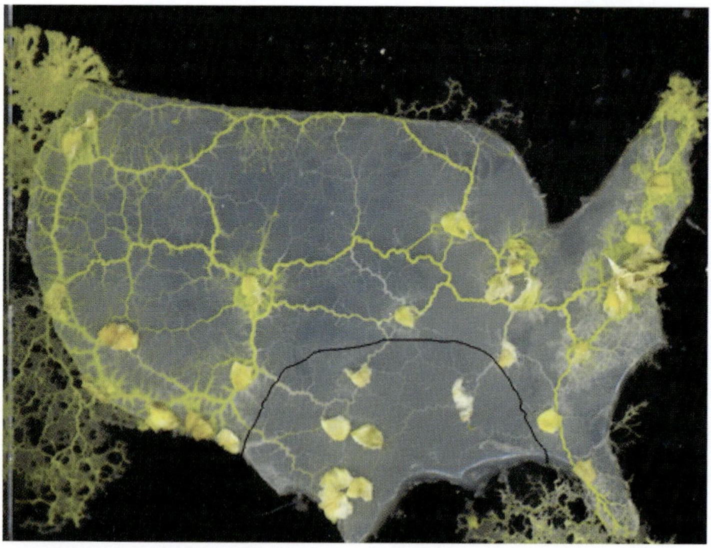

Fig. 16.10 Early stages (8 h) of the plasmodium's response to contamination with epicentre in South Texas nuclear generating station. Boundary of contaminated area is marked by a line.

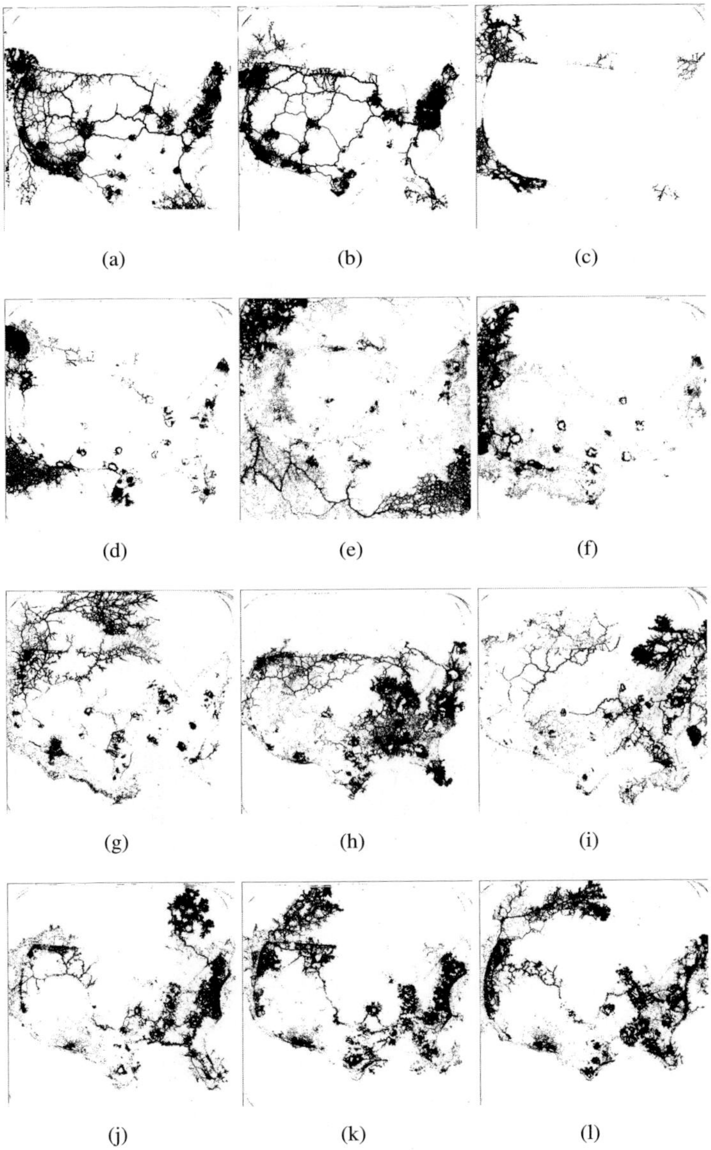

Fig. 16.11 Examples of plasmodium response to large-scale contamination of USA, with epicentres in (a–c) South Texas nuclear generating station, (d–f) Vermont nuclear power plant and (g–l) Palo Verde nuclear generating station. Images are scanned in 15–24 h after start of contamination and binarised by the following procedure. Only pixels from original scans of experimental dishes whose red and green components exceed 100 and blue component is less than 100 are drawn as black pixels, otherwise white.

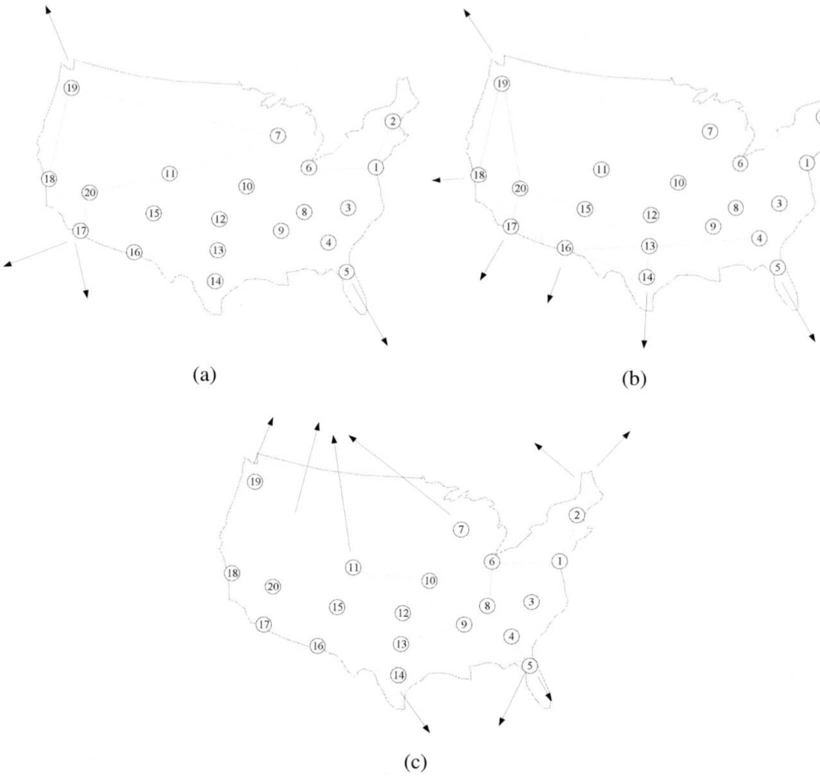

Fig. 16.12 Schematics of plasmodium's response to large-scale contamination with epicentres in (a) South Texas nuclear generating station, (b) Vermont nuclear power plant and (c) Palo Verde nuclear generating station. Transport routes enhanced in a response to contamination are shown by lines. Routes of attempted emigration are shown by arrows.

attempt to deal with the situation and migration away from contamination and even beyond the growth substrate as an attempt to completely avoid the situation (Figs. 16.11 and 16.12). The following scenarios are observed in laboratory experiments.

16.4.0.1 *Epicentre of contamination in Palo Verde nuclear generating station*

Normal activities in areas as far from Palo Verde as San Jose and San Francisco, Seattle–Portland, Albuquerque and Oklahoma City are totally disrupted. Plenty of abandoned protoplasmic tubes can be found in the contaminated zone. Transport links at the boundary between contaminated and clean areas are substantially enhanced. A huge increase in diameter, equivalent to throughput, of protoplasmic tubes is observed in the routes from Seattle–Portland to Denver, from Denver to Kansas City and from Kansas City to Milwaukee and Memphis. A slight but yet clearly visible increase of transport links is also recorded in routes from Dallas and Fort Worth to Memphis and from the Houston area to Memphis. Sometimes we can also note that tubes connecting the Chicago area with Nashville and the New York area, and from Atlanta to Charlotte and from the New York area to Boston, become hypertrophic (Figs. 16.11g–l and 16.12d). At the same we observe the plasmodium migrating in the following directions:

- from Jacksonville to Miami and then towards Cuba,
- from Houston, San Antonio and Austin, across the Mexican border towards Monterrey and Ciudad,
- from Vermont, New Hampshire and Maine across the Canadian boundary towards New Brunswick and Québec,
- the greatest wave of migration is directed from the USA to Calgary, Edmonton and Saskatoon in Canada: directed links of migrating plasmodium are observed propagating from Milwaukee, Denver, Seattle and Portland towards and across the Canadian boundary.

16.4.0.2 *Epicentre of contamination in Texas nuclear generating station*

Three types of migratory response to contamination spreading from the Texas nuclear generating station are observed in laboratory experiments:

- from Los Angeles and San Diego to Tijuana and Ensenada,
- from Jacksonville to Miami and then towards Cuba,
- from Seattle and Portland towards Vancouver and further in British Columbia.

Transport links from Los Angeles and San Diego via San Jose and San Francisco to Seattle and Portland, and from Seattle–Portland to the Houston area and Milwaukee, are substantially increased in their size. An infrequent hypertrophic reaction is also observed in routes from Milwaukee to the Chicago area, and from the New York area to Boston and to the Chicago area (Figs. 16.11a–c and 16.12a).

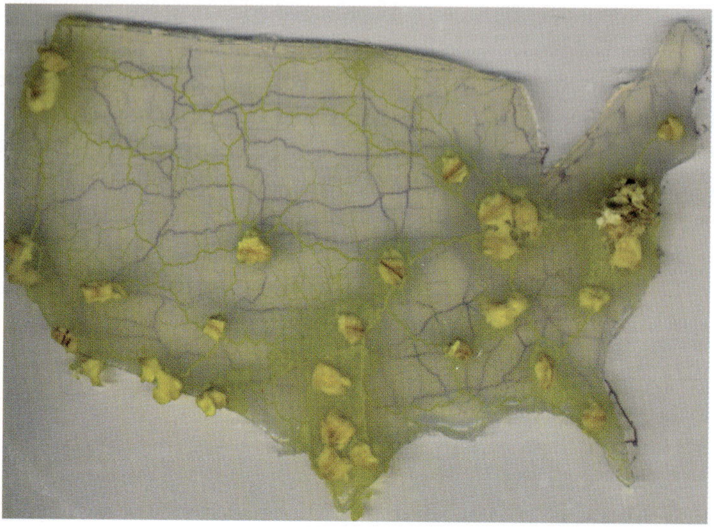

(a)

Fig. 16.13 Examples of protoplasmic network on the interstate map of the USA.

16.4.0.3 *Epicentre of contamination in Vermont nuclear power plant*

In the situation when the epicentre of contamination is located in the Vermont nuclear power plant, we observe migration and shift of plasmodium's activity towards the southern and south-western parts of the USA. Namely, hypertrophy of the following transport routes is recorded in experiments (Figs. 16.11d–f and 16.12b):

- Seattle–Portland to San Jose and San Francisco and to Las Vegas,
- Los Angeles area to Phoenix area to Dallas area and Houston area,
- Las Vegas to Albuquerque to Dallas area and Forth Worth,
- Dallas area to Memphis to Jacksonville.

Attempted migration of plasmodium outside the USA boundaries is observed almost everywhere along the southern and south-western boundaries. The most pronounced sites of migration are Houston area, Phoenix area, Los Angeles and San Diego, San Jose and San Francisco, and Seattle and Portland.

(a)

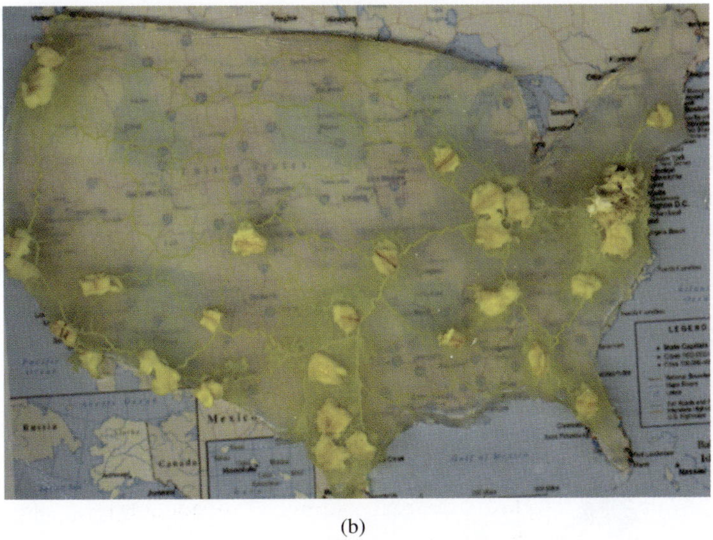

(b)

Fig. 16.14 Examples of protoplasmic network on the map of the USA.

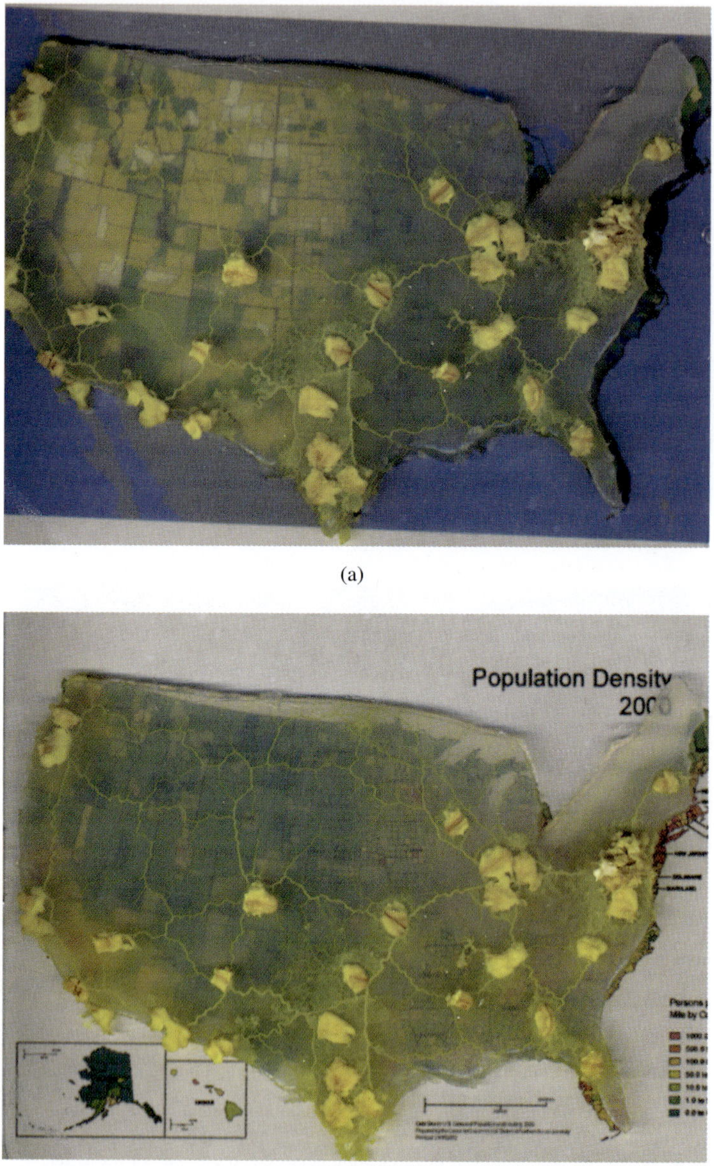

(a)

(b)

Fig. 16.15 Examples of protoplasmic network on the map of population density of the USA.

16.5 Summary

We represented 20 major areas of the USA with oat flakes placed on an agar plate in the shape of the USA, inoculated plasmodium of *P. polycephalum* on the oat flake representing the New York area and analysed networks of plasmodium's protoplasmic tubes spanning the oat flakes (Fig. 16.13). We found that the slime mould approximates almost all interstates apart from routes directly connecting Denver to the San Jose area, the Houston area to Albuquerque, the Houston area to Jacksonville, the New York area to Nashville and Boston to the Chicago area. We discovered that the slime mould's network includes the Gabriel graph, the relative neighbourhood graph and has a strong core corresponding to a minimum spanning tree on major urban areas. We identified that Interstates 10 and 20 are responsible, at least from the slime mould's 'point of view', for connectivity of the USA transport network: when these Interstates are removed the network becomes separated into western and eastern components. In laboratory experiments we imitated an abstract major disaster leading to spreading contamination and outlined several possible scenarios of how the transport network will restructure in response to such disasters. Despite a range of impressive findings (Fig. 16.14), we must admit that our experimental approach suffers from low resolution and also does not take into account terrain and other geographical conditions; it does however take into account the population density of the country (Fig. 16.15).

Chapter 17

World colonisation and trade route formation

Andrew Adamatzky

In the present chapter we discuss an experimental laboratory exercise on playing a hypothetical scenario of the colonisation of the world by a large-scale amorphous substrate and formation of world-wide transportation routes crossing countries and linking continents. The resultant protoplasmic networks were compared, at the abstract level of graphs, with an ancient road network — the Silk Road — and a future road network — the Asian Highways. For experiments we use Petri dishes with 2% agar gel and a 15 cm globe coated with agar. Agar plates in the dishes and the globe are cut in shapes of continents.

We considered 24 metropolitan areas as follows (Fig. 17.1a):

(1) Beijing,
(2) Seoul,
(3) Tokyo,
(4) Hong Kong,
(5) Ha Noi,
(6) Ho Chi Minh,
(7) Jakarta,
(8) Kolkata,
(9) Mumbai,
(10) Delhi,
(11) Karachi,
(12) Tehran,
(13) Moscow,
(14) Istanbul,

(a)

(b)

Fig. 17.1 Experimental setup: (a) configuration of points, representing selected metropolitan or urban areas **U**, (b) slime mould *P. polycephalum* occupies oat flakes representing urban areas **U**.

(15) London,
(16) Lagos,

Fig. 17.2 Experimental setup. Globe covered with agar gel is colonised by slime mould *P. polycephalum*. Oat flakes represent areas of **U**.

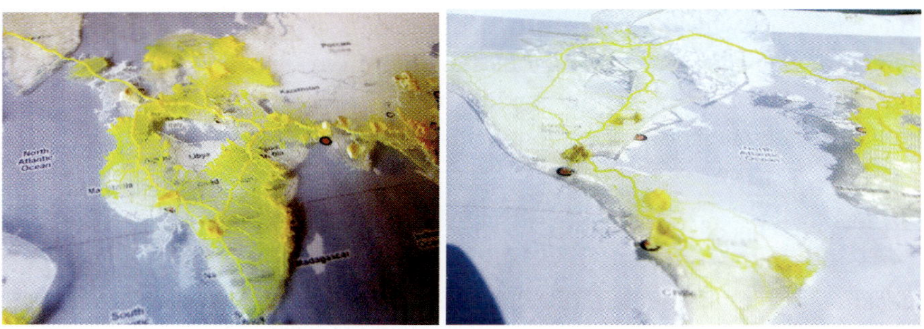

(a) (b)

Fig. 17.3 Plasmodium crosses bare space between the continent-shaped agar plates.

(17) Kinshasa,
(18) New York,
(19) Mexico City,
(20) Bogota,
(21) Lima,
(22) São Paulo,
(23) Canberra,
(24) Wellington.

The elements of **U** were selected based on their population size, proximity to other urban/metropolitan areas and representativeness of a continent/country. The following areas were not chosen for experiments, despite being amongst the largest metropolitan areas: Manila is not included due to its proximity to Jakarta; Shanghai due to its proximity to Beijing; Buenos Aires due to its proximity to São Paulo; Osaka–Kobe–Kyoto due their proximity to Tokyo; Dhaka due to its proximity to Kolkata. The capital cities Canberra and Wellington are by no means in the list of largest cities or metropolitan areas; however, they were included to give slime mould a chance to move to these countries. To represent areas of **U**, we placed oat flakes in the positions of the agar plate corresponding to the areas (Figs. 17.1b and 17.2). At the beginning of each experiment an oat flake colonised by plasmodium (25–30 mg plasmodial weight) was placed in the Beijing area. Reasons for choosing Beijing as a starting point for the slime mould are because it is amongst the most populated cities in the world [Beijing (2011)] and also because the Chinese civilisation is amongst the earliest civilisations [Makeham and Buell (2008)].

We undertook 38 experiments in total: eight experiments with the globe and 30 experiments with the Petri dish. As exemplified in Fig. 17.3, absence of a humid growing substrate prevented plasmodium from spreading 'uncontrollably' into parts of the substrate corresponding to oceans and seas yet allows the plasmodium to migrate between the continents when necessary.

17.1 Scenarios of colonisation

Scenarios of the plasmodium development in Petri dishes and on the globe are strikingly similar. Here we consider one example of colonisation of **U** in the Petri dish and three examples of colonisation on the globe. In the first day after inoculation in the Petri dish, the plasmodium propagates from Beijing to Seoul and Tokyo in the east; from Beijing to Hong Kong, Ha Noi and Ho Chi Minh in the south; and from Beijing to Kolkata, Mumbai, Delhi and Karachi in the west (Fig. 17.4a).

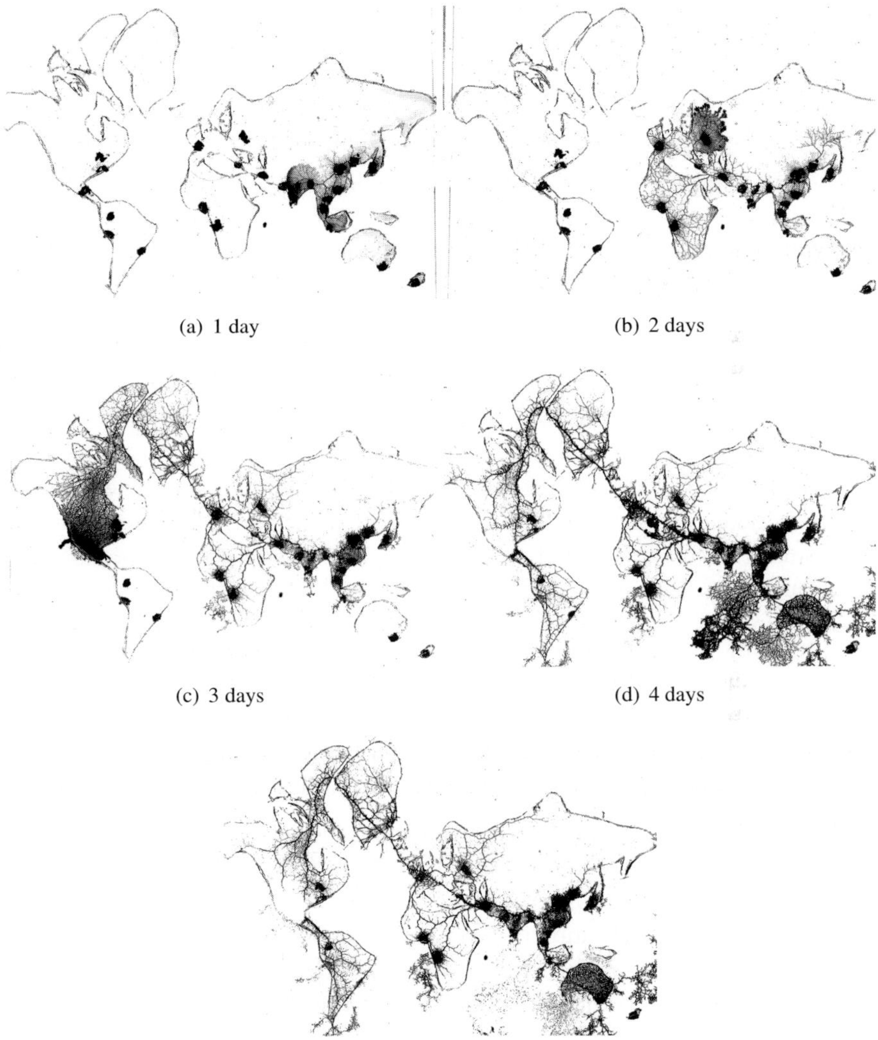

(a) 1 day (b) 2 days

(c) 3 days (d) 4 days

(e) 5 days

Fig. 17.4 Example of plasmodium development on a configuration of continent-shaped agar plates.

On the second day the plasmodium explores the space north of Beijing and links Karachi and Tehran, and Tehran and Istanbul, with protoplasmic tubes. It also grows from Istanbul to Moscow and from Moscow to London, and from Tehran to Lagos and Kinshasa (Fig. 17.4b). The plasmodium grows from London to Iceland

and then to Greenland. It propagates from Greenland to Canada and eventually reaches New York and Mexico City on the fifth day of the experiment (Fig. 17.4c). The final stage of spanning **U** takes place on the sixth day of the experiment: the plasmodium propagates from Mexico City to Bogota and from Bogota to Lima and São Paulo. At the same time the plasmodium grows from Ho Chi Minh to Jakarta and from Jakarta to Canberra and Wellington (Fig. 17.4d). As soon as all sources of nutrients, representing the areas of **U**, are spanned by the plasmodium, the plasmodium remains in its configuration for a couple of days (Fig. 17.4c). By that time the nutrients become depleted, the substrate becomes contaminated with products of metabolism and the plasmodium tries to migrate to other areas and/or form sclerotium.

Two examples shown in Fig. 17.5 demonstrate that colonisations of East and South Asia can go by different scenarios; however, propagations towards Western Europe and the Americas have matching trajectories. Let us consider the scenario in Fig. 17.5a. In the second day after inoculation plasmodium propagates from Beijing to Seoul and then to Tokyo, and from Beijing to Delhi and to Kolkata, and from Delhi to Karachi and to Mumbai. On the third day, the plasmodium propagates from Karachi to Tehran to Istanbul to Moscow (Fig. 17.6a). It then continues to explore the space north-east of Moscow. On the fourth day the plasmodium develops protoplasmic links from Delhi to Kolkata, from Kolkata to Ha Noi and Ho Chi Minh, and from Ho Chi Minh to Jakarta; and completes the route from Moscow to Beijing. It also explores the space east of Jakarta (Fig. 17.6b). On the fifth, sixth and seventh days slime mould propagates from Jakarta to Canberra (Fig. 17.6c) and from Istanbul to London. The plasmodium reaches New York from London via Iceland, Greenland and Canada on the ninth day after inoculation in Beijing. Then it propagates from New York to Mexico City, Bogota, Lima and São Paulo (Fig. 17.5a).

Early stages of colonisation (Fig. 17.5a) show the following priorities of the plasmodium's development: east then west then south. The scenario with early development: east then south then west, is shown in Fig. 17.5b and illustrated in Fig. 17.7.

In the first two days after inoculation in Beijing the plasmodium propagates to Seoul and Tokyo in the east, to Delhi in the south-west and Hong Kong in the south-east. It also ventures north to Siberia. On the third day the plasmodium grows from Ha Noi to Kolkata and Ho Chi Minh, and from Ho Chi Minh to Jakarta and Canberra (Fig. 17.7a). On the fourth day the plasmodium propagates from Delhi to Karachi, Tehran, Istanbul, Moscow and London, and ventures from London to Iceland (Fig. 17.7b). Protoplasmic links between London and Lagos and Kinshasa are developed by the slime mould at the fifth day of the experiment.

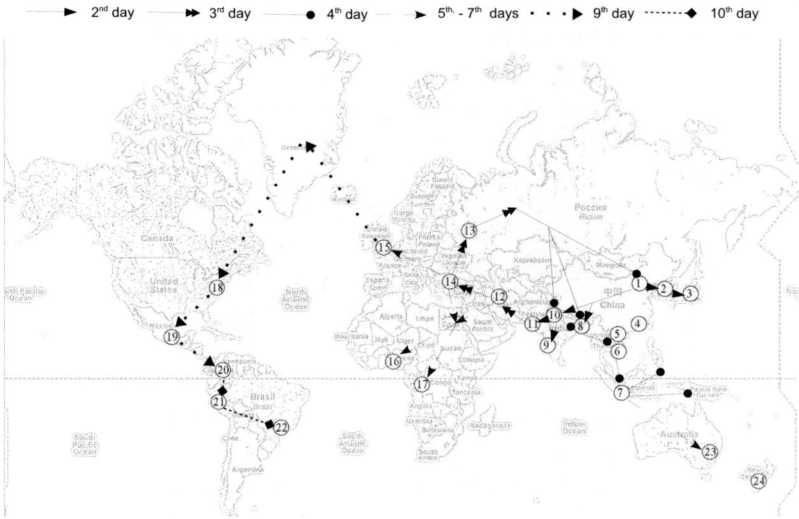

(a)

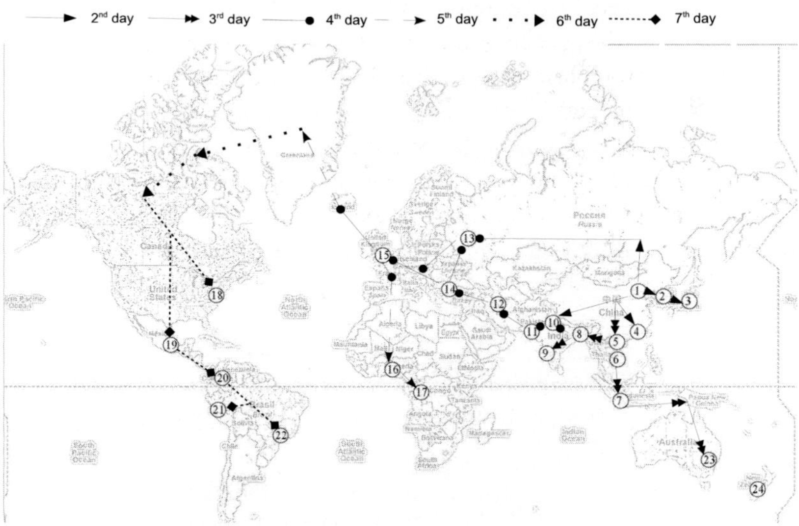

(b)

Fig. 17.5 Two scenarios of plasmodial active zones propagating on the globe.

On the sixth and seventh days the plasmodium crosses Greenland and Canada towards the USA. It firstly reaches New York and then propagates towards Mexico City, from where it spans Bogota, Lima and São Paulo (Fig. 17.7b). This scenario

(a)

(b)

(c)

Fig. 17.6 Experimental examples of scenario of Fig. 17.5a: (a) propagation of slime mould from Karachi to Tehran to Istanbul to Moscow on the third day after inoculation, (b) on the fourth day of the experiment plasmodium links Ha Noi to Ho Chi Minh to Jakarta with its protoplasmic tubes and explores the space east of Jakarta, (c) plasmodium's active zone enters Australia at the port of Derby.

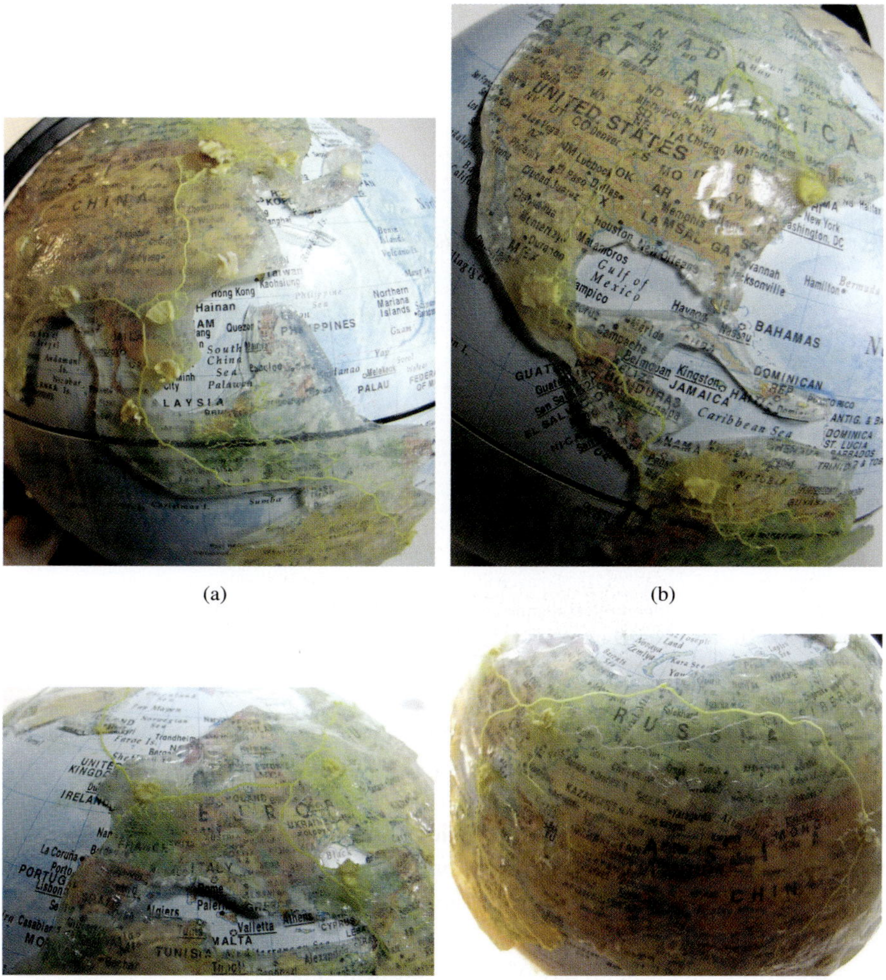

(a) (b)

(c) (d)

Fig. 17.7 Illustrations of the scenario of Fig. 17.5b. (a) Plasmodium propagates from South-East Asia to Australia. (c) Plasmodium propagates west and north-west. (b) Plasmodium reaches USA via Iceland, Greenland and Canada and then spans Latin America. (d) Transport link between Beijing and London.

also displays a pronounced transport link between Beijing and London (Fig. 17.7c and d).

An example of incomplete spanning is shown in Fig. 17.8. Transport links between Beijing, Seoul and Tokyo, and Beijing and Hong Kong, are established in

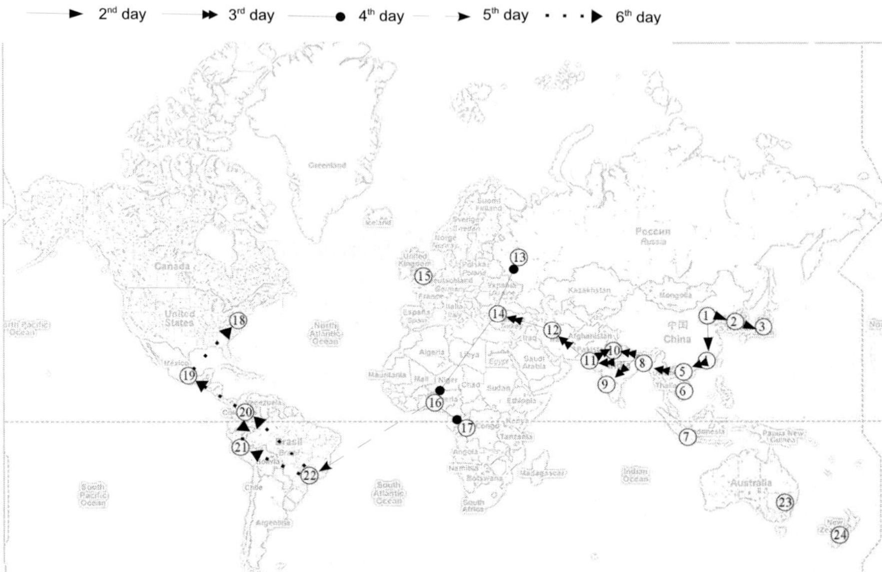

Fig. 17.8 Examples of incomplete spanning of urban areas. Trajectories of plasmodial active zone propagation in 10 days of experiment.

two days after inoculation (Fig. 17.9a). On the third day the plasmodium connects the following areas by the chain of protoplasmic tubes: Hong Kong–Ha Noi–Kolkata–Delhi–Karachi–Tehran–Istanbul (Fig. 17.9b). Moscow and Lagos and Kinshasa are spanned by the protoplasmic network on the fourth day of colonisation (Fig. 17.9c). It takes slime mould one more day to cross the South Atlantic to make the link between Lagos and São Paulo (Fig. 17.9a). Protoplasmic links São Paulo–Lima, São Paulo–Bogota, Bogota–Mexico City and Mexico City–New York are grown on the sixth day of the experiment (Fig. 17.8).

Physarum graphs derived from experiments in Petri dishes (let us call them P-graphs) and the globe (G-graphs) are shown in Fig. 17.10. P-graphs represent G-graphs sufficiently. Only three edges (albeit represented in one experiment each) of P-graphs are not represented by G-graphs: Kinshasa–São Paulo, Moscow–Istanbul) and Beijing–Hong Kong. In contrast, eight edges of P-graphs are not represented by G-graphs: Hong Kong–Jakarta, Ha Noi–Ho Chi Minh, Jakarta–Kolkata, Tehran–Lagos, Istanbul–Kinshasa, London–Mexico City, Mexico City–Bogota and Lagos–São Paulo. Further on, we consider Physarum graphs as a union of P- and G-graphs.

Physarum graphs for critical values of θ are shown in Fig. 17.10. Physarum

Fig. 17.9 Illustration of the scenario of Fig. 17.8. (a) Plasmodium crosses South Atlantic from West Africa to South America. (c) Protoplasmic tube connecting Istanbul to Lagos develops on the fourth day of experiment. (d) Transport links Beijing, Seoul and Tokyo, and Beijing and Hong Kong, are established in two days after inoculation. (b) Chain Hong Kong–Ha Noi–Kolkata–Delhi–Karachi–Tehran–Istanbul is developed on the third day of experiment.

graphs $\mathbf{P}(\theta)$ are subgraphs of P- and G-graphs for $\theta \geq \frac{14}{38}$. Physarum graphs are planar for $\theta \geq \frac{6}{38}$; however, with acquiring planarity the Physarum graph $\mathbf{P}(\frac{6}{38})$ loses connectivity and Wellington becomes isolated (Fig. 17.10b). The highest

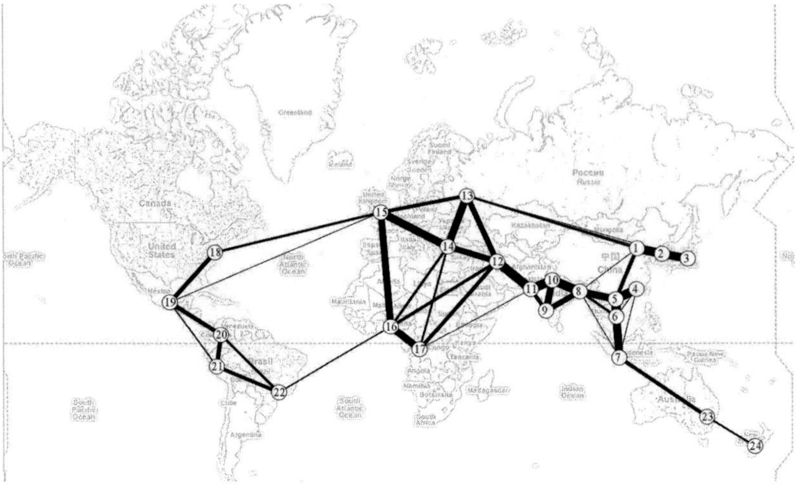

(a)

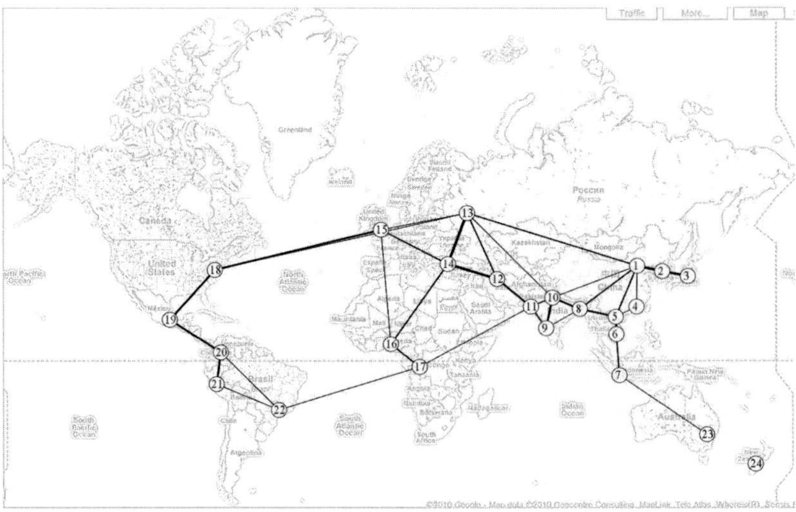

(b)

Fig. 17.10 Physarum graphs obtained from experiments in (a) the Petri dishes, P-graph, and (b) the globe, G-graph. Thickness of a line is proportional to the line's weight.

value of θ for which the graph $\mathbf{P}(\theta) - \{\text{Wellington }\}$ remains connected is $\frac{14}{38}$ (Fig. 17.10c). When θ increases to $\frac{15}{38}$, the Americas become disconnected from Eurasia and Africa, and American transport pathways become a single chain New

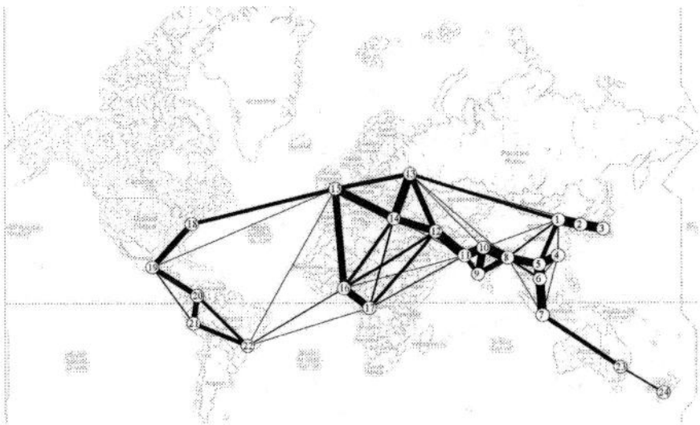

(a) $\theta = \frac{1}{38}$

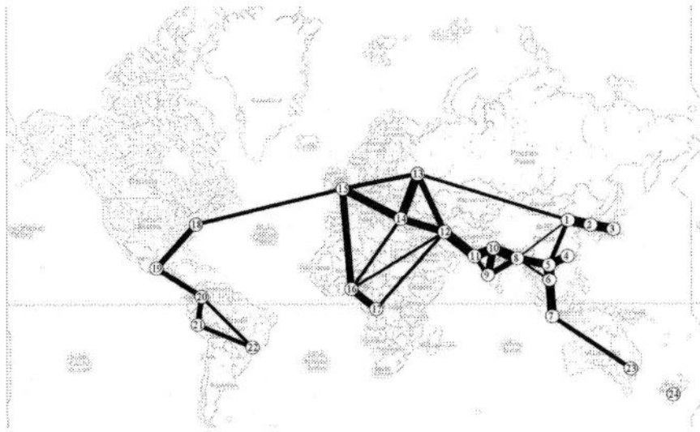

(b) $\theta = \frac{6}{38}$

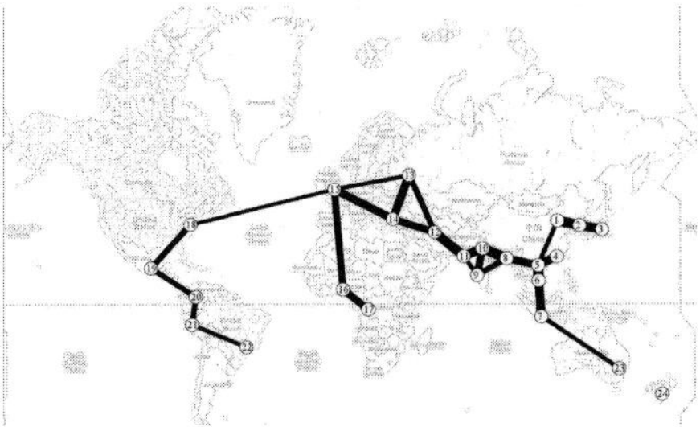

(c) $\theta = \frac{14}{38}$

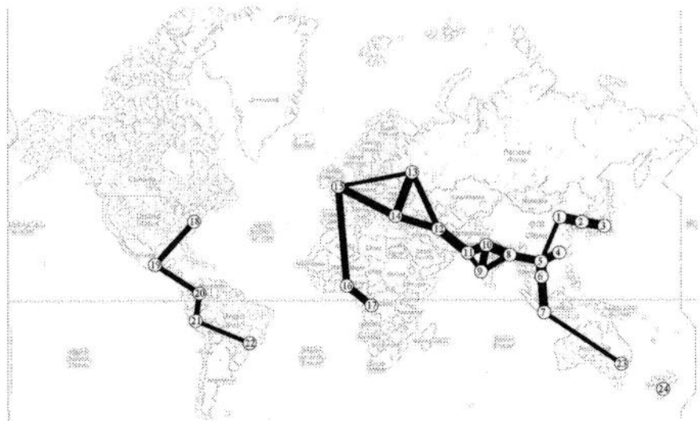

(d) $\theta = \frac{15}{38}$

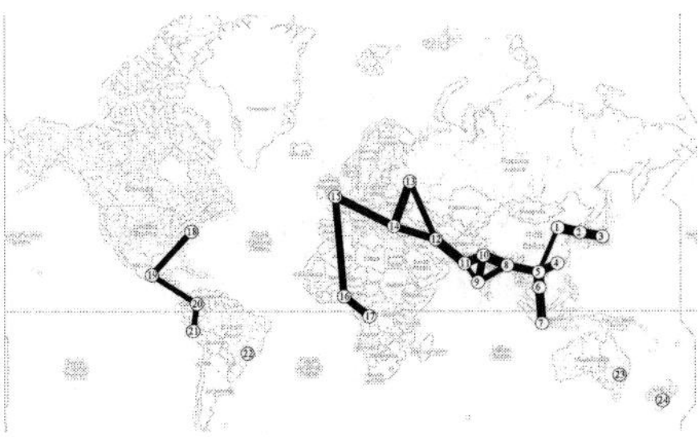

(e) $\theta = \frac{16}{38}$

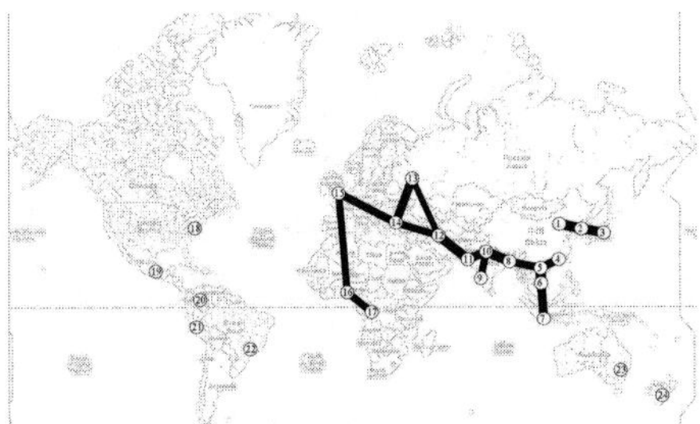

(f) $\theta = \frac{22}{38}$

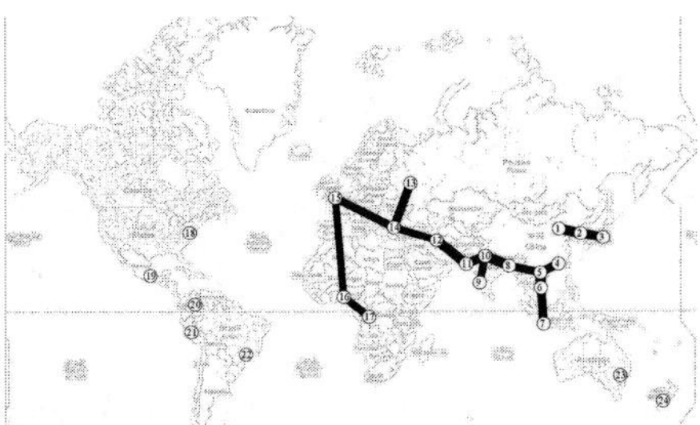

(g) $\theta = \frac{23}{38}$

Fig. 17.11 Physarum graphs $\mathbf{P}(\theta)$ for selected values of θ.

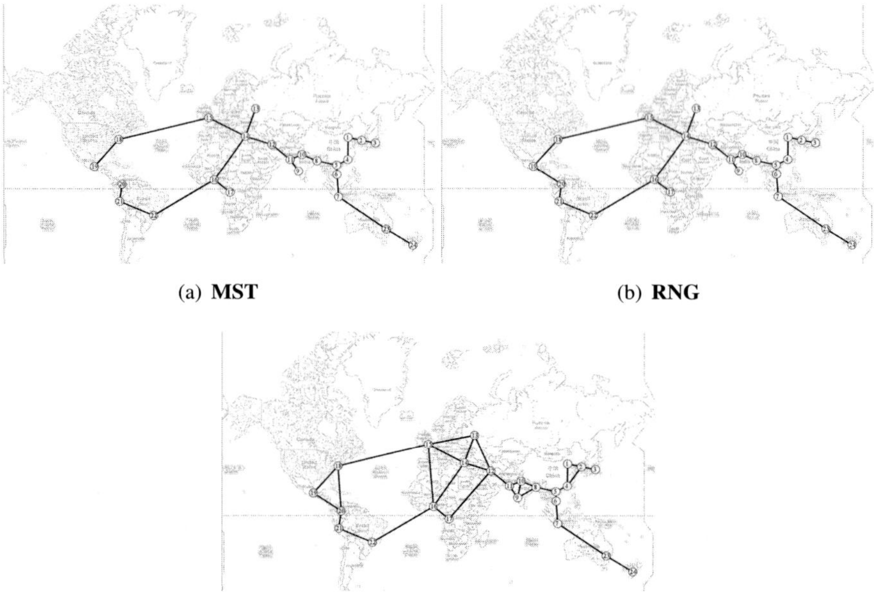

(a) **MST** (b) **RNG**

(c) **GG**

Fig. 17.12 Relative neighbourhood graph (a) and Gabriel graph (b) constructed on urban areas of **U**.

York– Mexico City–Bogota–Lima–São Paulo (Fig. 17.10d). Further increase of θ to $\frac{16}{38}$ leads to isolation of São Paulo and Canberra (Fig. 17.10e).

For $\theta = \frac{22}{38}$, all the American areas of **U** become isolated and the transport network at this continent virtually disappears. The graph $\mathbf{P}(\frac{22}{38})$ has two components of connectivity:

- the chain Beijing–Seoul–Tokyo,
- the component consisting of the chain Jakarta–Ho Chi Minh– Kolkata–Delhi–Tehran–Istanbul–London–Lagos– Kinshasa with two branches Ha Noi–Hong Kong and Delhi–Mumbai attached and the cycle Tehran–Moscow–Istanbul– Tehran (Fig. 17.10f).

The Physarum graph is transformed to the acyclic graph at $\theta = \frac{23}{38}$ when the edge Tehran–Moscow becomes cut off (Fig. 17.10g.)

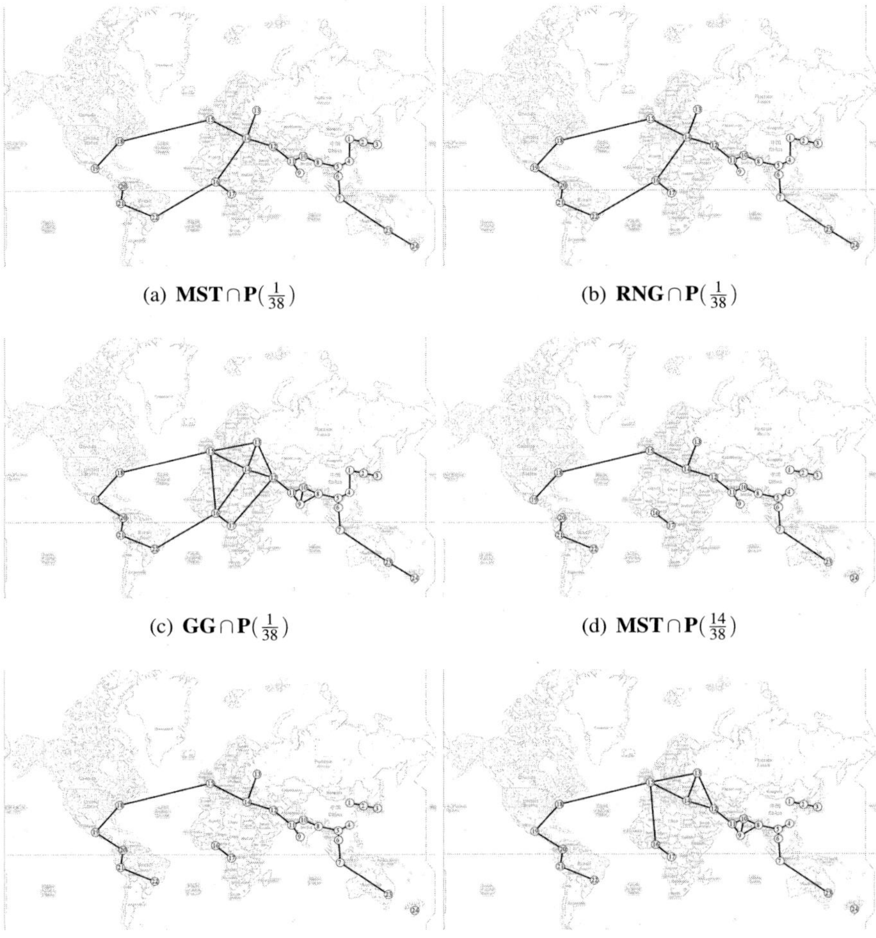

(a) $\mathbf{MST} \cap \mathbf{P}(\frac{1}{38})$ (b) $\mathbf{RNG} \cap \mathbf{P}(\frac{1}{38})$

(c) $\mathbf{GG} \cap \mathbf{P}(\frac{1}{38})$ (d) $\mathbf{MST} \cap \mathbf{P}(\frac{14}{38})$

(e) $\mathbf{RNG} \cap \mathbf{P}(\frac{14}{38})$ (f) $\mathbf{GG} \cap \mathbf{P}(\frac{14}{38})$

Fig. 17.13 Intersections of Physarum graphs (a–c) $\mathbf{P}(\frac{1}{38})$ and (d–f) $\mathbf{P}(\frac{14}{38})$ with proximity graphs: (a, d) minimum spanning tree **MST**, (b, e) relative neighbourhood graph **RNG**, (c, f) Gabriel graph **GG**.

17.2 Physarum includes spanning tree

Intersections of Physarum graphs with the proximity graphs are shown in Fig. 17.13.

Finding 116. *The minimum spanning tree and the relative neighbourhood graph* *are subgraphs of* $\mathbf{P}(\frac{1}{38})$. *The Gabriel graph would be a subgraph of* $\mathbf{P}(\frac{1}{38})$ *if the Physarum graph had the edges Hong Kong–Seoul and New York–Bogota.*

MST differs from **RNG** only by absence of the single edge Mexico City–Bogota; the rest is the same and is matched by edges of the Physarum graph $\mathbf{P}(\frac{1}{38})$ (Fig. 17.13a and b). The fact that $\mathbf{GG} \subseteq \mathbf{P}(\frac{14}{38})\cup$(Hong Kong, Seoul)$\cup$(Mexico City–Bogota) (Fig. 17.13c) plays against the geographical validity of the Gabriel graph. The only physically possible routes from New York to Bogota and from Hong Kong to Seoul are maritime routes; the overland route from New York to Bogota is passing via Mexico City and from Hong Kong to Seoul via Beijing.

Finding 117. *The longest chain in* $\mathbf{MST} \cap \mathbf{P}(\frac{14}{38})$ *is* C_1=*Canberra–Jakarta–Ho Chi Minh–Ha Noi–Kolkata–Delhi–Karachi–Tehran–Istanbul–London–New York–Mexico City.*

Two fragments of the chain C_1: Jakarta–Ho Chi Minh–Ha Noi and Kolkata to London (Fig. 17.13d) match the Silk Road. The chain C_1 is elongated by the segment Bogota–Lima–São Paulo in $\mathbf{RNG} \cap \mathbf{P}(\frac{14}{38})$ (Fig. 17.13e) and $\mathbf{GG} \cap \mathbf{P}(\frac{14}{38})$ (Fig. 17.13f).

17.3 The Silk Road and the Asian Highways

To evaluate how well plasmodial networks match large-scale transportation systems, we compare Physarum graphs with graphs derived from the Silk Road and the Asian Highway network.

The Silk Road is a long-distance trade route across Asia developed in the first millennium to transport goods between China and Western Asia and Northern Europe [Thubron (2008)]. Commonly, the Silk Road is composed of the main overland route from Luoyang to Seleucia/Clesiphon and associated overland and maritime routes reaching as far as Japan in the east and as far as Colchester and Holborough in the west [Scarre (1996); Liu (2010)].

A graph of the Silk Road is shown in Fig. 17.14a. When constructing the Silk road graph we tracked geographical positions of modern cities along the historical Silk Road and associated trade routes, with minor assumptions. Thus, the Silk Road is represented by the edge Beijing–Tehran. The road actually started at Luoyang and passed via Hecatompylos and ended in Ctesiphon. We assumed that Hecatompylos was in the proximity of Tehran and Beijing is in the proximity of Luoyang. Similarly, we assume that the edge Delhi–Kolkata corresponds to the trade route connecting Mathura (proximity of Delhi) and Tampluk via Patalipu-

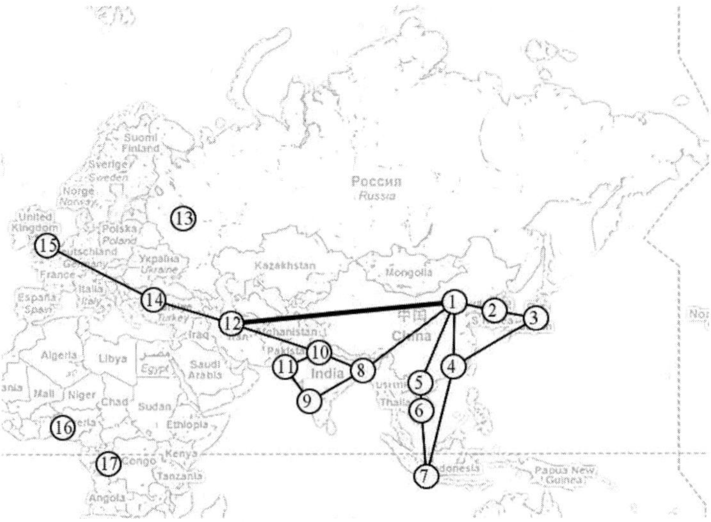

(a)

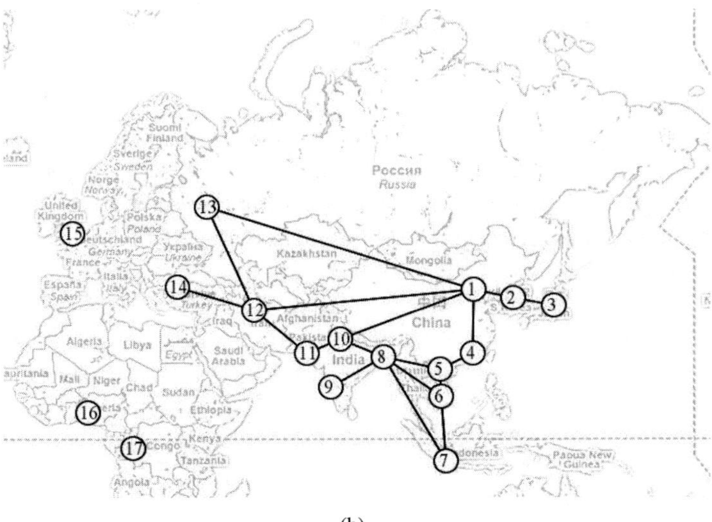

(b)

Fig. 17.14 Graphs of the Silk Road and the Asian Highways: (a) the Silk Road graph, comprising the Silk Road (thick line) and associated trade routes (thin lines); the graph is derived from the scheme of trans-Asian trade 500 B.C.–A.D. 750 [Scarre (1996)], (b) the Asian Highway network graph; the graph is derived from the Asian Highway network map [Priority Investment Needs (2006)].

tra (proximity of Kolkata). The link Tokyo–Hong Kong corresponds to a maritime route from the south of Japan to Oc Eo via Guangzhou and passing between continental China and Hong Kong island. The edge Tehran–Istanbul represents the land route from Hecatompylos to Dura Europos and Antioch combined with the maritime trade route from Antioch to Constantinople. The edge Istanbul–London represents the maritime route Constantinople–Athens–Rome–Massilia–Colchester.

The Asian Highway network is a network of roads selected for regional transport cooperation: "initiative aimed at enhancing efficiency and development of the road transport infrastructure in Asia" [Priority Investment Needs (2006)]. The Asian Highway network consists of 141,000 km of roads running across 32 member states. When constructing the Asian Highway graph (Fig. 17.14b), we adopted the following match between the graph's edges and the highways (see map of the Asian Highway network in [Priority Investment Needs (2006)]):

- (Beijing, Seoul): potential route AH1 Beijing to Shenyang and existing route AH1 Shenyang to Seoul,
- (Beijing, Hong Kong): route AH1 Beijing to Hong Kong via Zhengzhou, Xinyang and Xianglan,
- (Beijing, Delhi): route A1 Beijing to Zhengzhou, potential routes AH34 Zhengzhou to Xi'an, AH5 Xi'an to Langzhou, AH42 Langzhou to Lhasa and finally existing routes AH42 Lhasa to Zhangmu and AH2 Zhangmu to Delhi,
- (Beijing, Tehran): comprising segments of the following highways (ordered from Beijing to Tehran): AH1, AH34, AH5, AH4, AH65, AH62, AH76 and AH1,
- (Beijing, Moscow): AH3 Beijing to Ulan Ude, AH5 Ulan Ude to Moscow via Irkutsk, Novosibirsk, Omsk, Petropavlovsk, Chelyabinsk and Samara,
- (Seoul, Tokyo): maritime route Pusan to Fukuoka and overland route AH1 Fukuoka to Tokyo,
- (Hong Kong, Ha Noi): potential route AH1 Guangzhou to Nanning, AH1 Nanning to Hanoi,
- (Ha Noi, Ho Chi Minh): AH1,
- (Ha Noi, Kolkata): AH14 Hanoi to Kunming to Mandalay and AH1 Mandalay to Dispur to Kolkata,
- (Ho Chi Minh, Jakarta): AH1 Ho Chi Minh to Kabin Bun, AH19 Kabin Bun to Bangkok, AH2 Bangkok to Hat Yai, AH18 Hat Yai to Singapore and maritime route Singapore to Jakarta,
- (Kolkata, Mumbai): AH46 and AH47,
- (Kolkata, Delhi): AH1,

- (Delhi, Karachi): AH1 Delhi to Lahore, AH2 Lahore to Rohn and AH4 Rohn to Karachi,
- (Karachi, Tehran): AH7 Karachi to Kandahar and AH1 Kandahar to Dilaram to Herat to Zhabzevar to Tehran,
- (Tehran, Moscow): AH8 Tehran to Baku to Astahan to Volgograd to Moscow,
- (Tehran, Istanbul): AH1 Tehran to Yerevan to Ankara and AH5 Ankara to Istanbul.

The Asian Highway graph has the same number of edges as the Silk Road graph; however, they are not isomorphic (Fig. 17.14).

Finding 118. *Slime mould* P. polycephalum *approximates over 76% of the Silk Road routes and the Asian Highway routes.*

The Physarum graph $\mathbf{P}(\frac{6}{38})$ (Fig. 17.11b) approximates 13 of 17 edges of the Silk Road graph(Fig. 17.14a) and also 13 of 17 edges of the Asian Highway graph (Fig. 17.14b). The Physarum graph $\mathbf{P}(\frac{14}{38})$ (Fig. 17.11c) approximates 11 of 17 edges of the Silk Road graph and also 11 of 17 edges of the Asian Highway graph.

Finding 119. *Transport links Beijing–Tehran and Beijing– Hong Kong of the Silk Road and the Asian Highway network are never approximated by* P. polycephalum.

The following routes of the Silk Road are never approximated by any Physarum graph: Delhi–Tehran, Beijing–Hong Kong, Tokyo–Hong Kong and Hong Kong–Jakarta. The routes Beijing–Tehran and Beijing–Kolkata are approximated by $\mathbf{P}(\frac{6}{38})$ but not by $\mathbf{P}(\frac{14}{38})$.

The following routes of the Asian highways are not approximated by any Physarum graph: Beijing–Tehran, Beijing–Hong Kong, Beijing–Delhi and Jakarta–Kolkata. The routes Beijing–Moscow and Ho Chi Minh–Kolkata are approximated by $\mathbf{P}(\frac{6}{38})$ but not by $\mathbf{P}(\frac{14}{38})$.

17.4 Summary

To imitate a hypothetical scenario of the world's colonisation and emergence of principal transport networks, we represented the continents with geometrically shaped plates of non-nutrient agar and the major metropolitan areas with sources of nutrients and inoculated plasmodium of *P. polycephalum* in Beijing. We analysed scenarios of the plasmodium propagation and colonisation of the metropolitan areas. We derived weighted Physarum graphs from the protoplasmic networks recorded in the laboratory experiments.

We found that the Physarum graphs include basic proximity graphs — the minimum spanning tree, the relative neighbourhood graph and the Gabriel graph constructed on the metropolitan areas — as their subgraphs. The longest chain of transport links represented in the minimum spanning tree and in the Physarum graph with edge weights exceeding 0.36 is Canberra–Jakarta–Ho Chi Minh–Ha Noi–Kolkata–Delhi–Karachi– Tehran–Istanbul–London–New York–Mexico City. We found that slime mould *P. polycephalum* approximates over 76% of the Silk Road routes and the Asian Highway network. The transport links Beijing–Tehran and Beijing–Hong Kong of the Silk Road and the Asian Highway network are never approximated by *P. polycephalum*.

We believe that our experimental results will inspire further thoughts, paradigms and approaches for re-evaluation of historical findings on the emergence of ancient roads and will help to design future transcontinental pathways. The results could also be applied in analysis of pandemics' dynamics [Beck et al. (2008); Haour-Knipe and Rector (1996)] and in shaping unorthodox approaches to prediction of international military conflicts and battlefield operations [Ilachinski (2004)].

Chapter 18

Biorationality of motorways

Andrew Adamatzky, Selim Akl, Ramon Alonso-Sanz, Wesley Van Dessel, Zuwairie Ibrahim, Andrew Ilachinski, Jeff Jones, Anne V. D. M. Kayem, Genaro J. Martínez, Pedro P. B. de Oliveira, Mikhail Prokopenko, Theresa Schubert, Peter Sloot, Emanuele Strano, Xin-She Yang

What measures, apart from straightforward comparison of edges between motorway and plasmodium networks, are reliable indicators of matching? Which regions have the most 'Physarum friendly' motorway networks, i.e. show highest degree of matching between motorways and protoplasmic networks along several measures? We analyse the results of our experimental laboratory approximation of motorway networks with slime mould *Physarum polycephalum* discussed in the previous chapters. Motorway networks of 14 geographical areas are considered: Australia, Africa, Belgium, Brazil, Canada, China, Germany, Iberia, Italy, Malaysia, Mexico, The Netherlands, mainland UK and USA. The measures employed are the number of independent cycles, cohesion, shortest-path length, diameter, the Harary index and the Randić index. We obtained a series of intriguing results, and found that the slime mould approximates best of all the motorway graphs of Belgium, Canada and China, and that for all entities studied the best match between Physarum and motorway graphs is detected by the Randić index (molecular branching index).

Numbers of the areas selected for each geographical region are as follows: Africa $n = 35$, Australia $n = 25$, Belgium $n = 21$, Brazil $n = 21$, Canada $n = 16$, China $n = 31$, Germany $n = 21$, Iberia $n = 23$, Italy $n = 11$, Malaysia $n = 20$, Mexico $n = 19$, Netherlands $n = 21$, UK $n = 10$, USA $n = 20$.

As with every living creature, the plasmodium of *P. polycephalum* does not

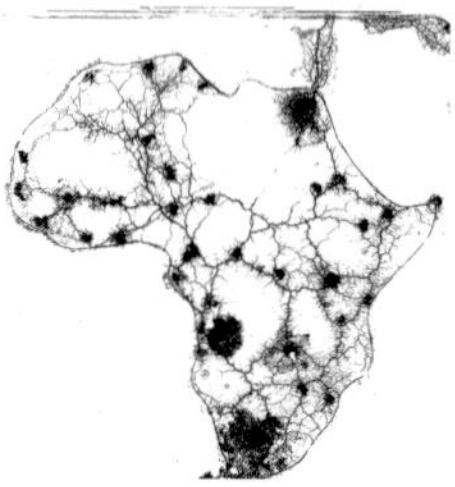

(a) Africa

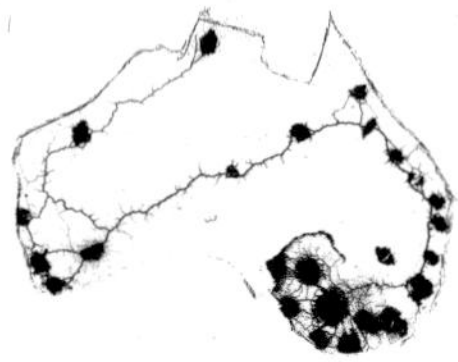

(b) Australia

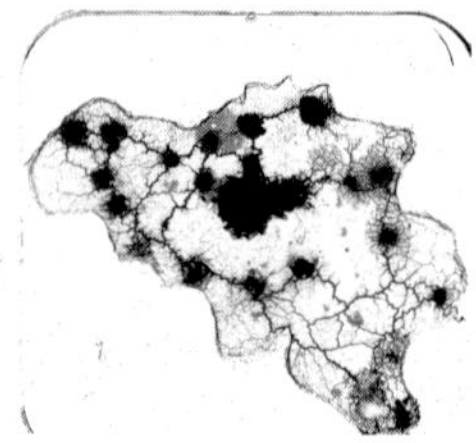

(c) Belgium

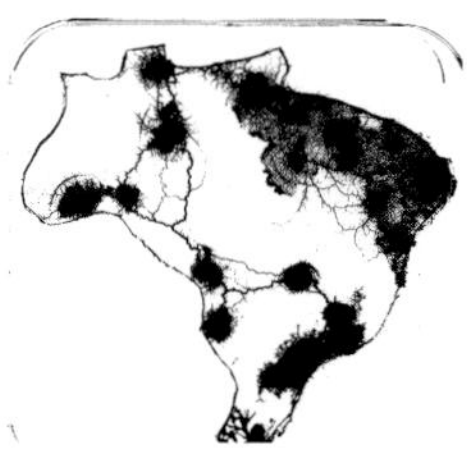

(d) Brazil

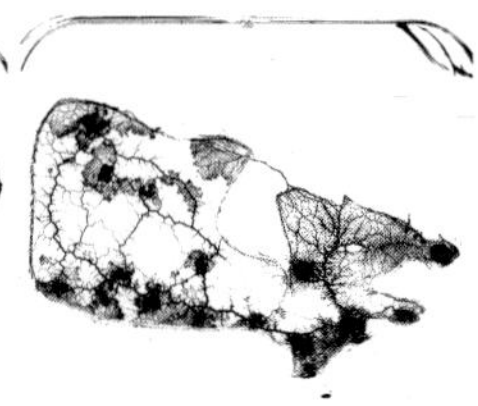

(e) Canada

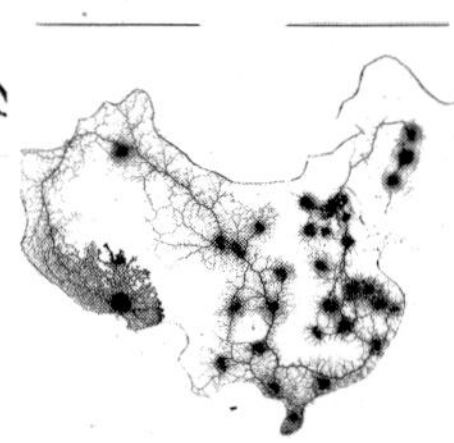

(f) China

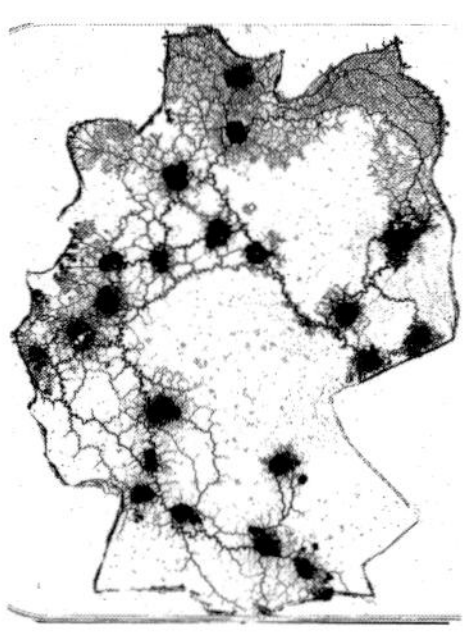

(g) Germany

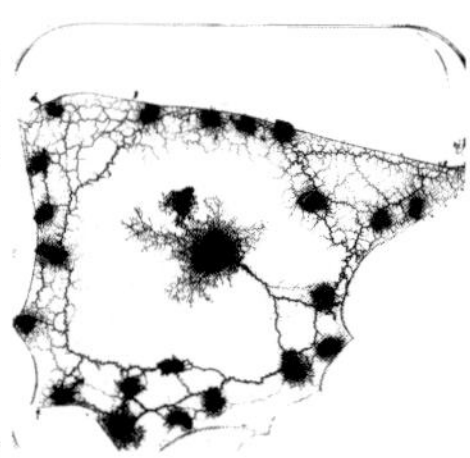

(h) Iberia

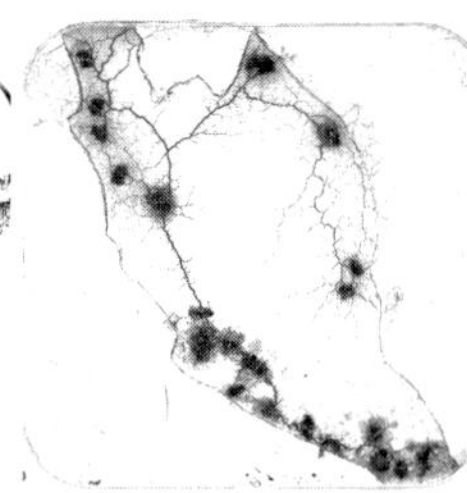

(i) Malaysia

(j) Mexico

(k) The Netherlands

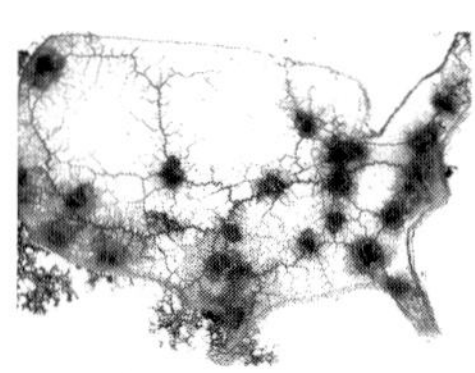

(l) USA

Fig. 18.1 Exemplar configurations of protoplasmic networks developed by slime mould *P. polycephalum* on major urban areas **U** obtained in experimental laboratory studies.

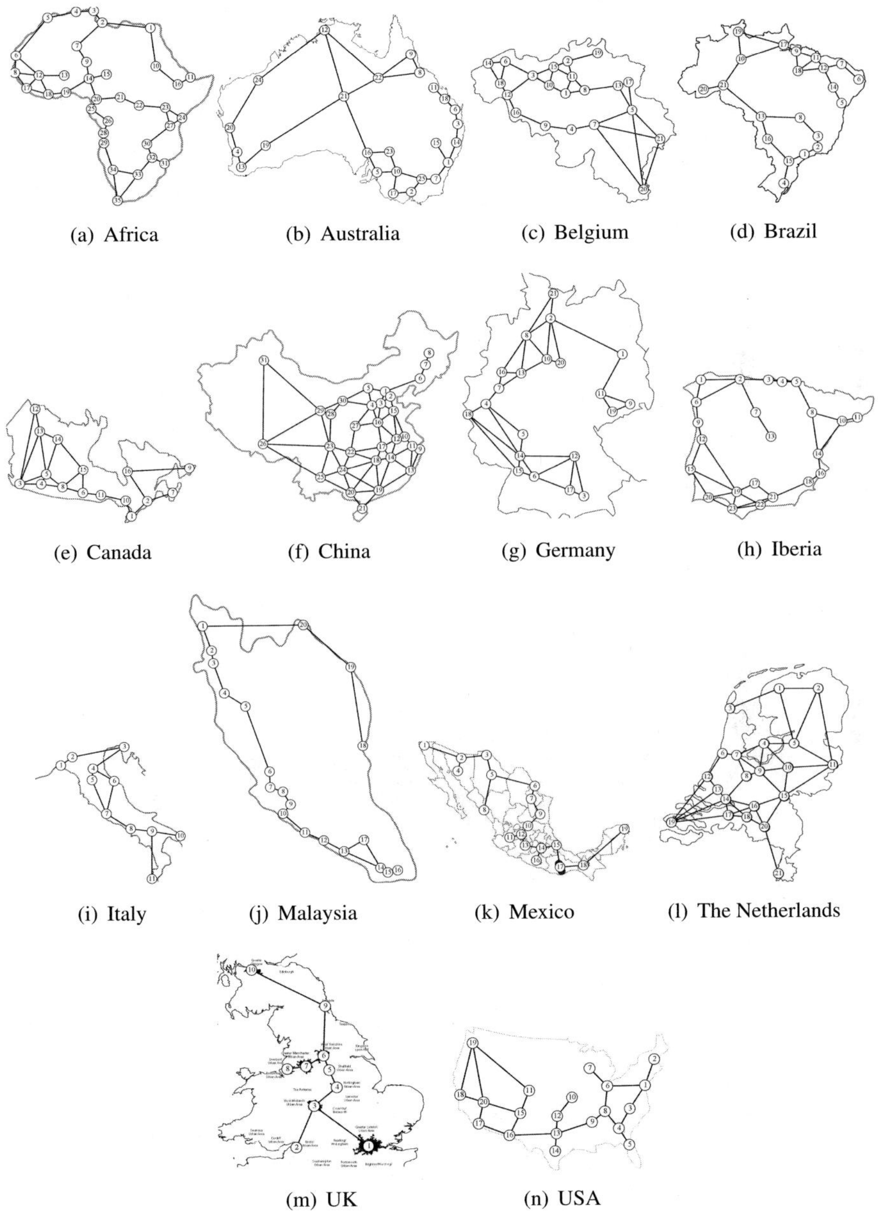

(a) Africa (b) Australia (c) Belgium (d) Brazil

(e) Canada (f) China (g) Germany (h) Iberia

(i) Italy (j) Malaysia (k) Mexico (l) The Netherlands

(m) UK (n) USA

Fig. 18.2 Physarum graphs $\mathbf{P}(\theta')$ for highest values of θ which do not make the graphs disconnected.

always repeat its foraging pattern. To generalise our experimental results, we constructed a Physarum graph with weighted edges.

In the present chapter, we consider only 'stressed' Physarum graphs $\mathbf{P}(\theta')$ (an edge in 'stressed' Physarum graph exists only if it is represented by protoplasmic tubes in at least θ' experiments), which have maximum possible values θ' and yet remain connected: $\theta' = \max\{\theta : \mathbf{P}(\theta) = \text{connected}\}$. These Physarum graphs are shown in Fig. 18.2. Values of θ' for the studied regions are illustrated in Fig. 18.4a.

Motorway graphs extracted from maps of motorway (highway, expressway, autobahn) networks are shown in Fig. 18.3.

Motorway and Physarum graphs were compared directly and using integral measures and indices. Let m be the number of edges in the motorway graph $\mathbf{H}$ and f be the number of edges in the Physarum graph $\mathbf{P}$, let i and j be nodes and let M and F be adjacency matrices. Direct matching between motorway and Physarum graphs is calculated as $\mu = \frac{1}{m}\sum_{ij}\xi(M_{ij}, F_{ij})$, where $\xi(M_{ij}, F_{ij}) = 1$ if $M_{ij} = F_{ij}$ and 0, otherwise. An economy of matching is calculated as $\varepsilon = \frac{\mu}{f}$. Also, we compared the graphs by their average shortest path measured in nodes, average shortest path measured in normalised edge lengths (for each edge $e \in \mathbf{E}$, we normalised its Euclidean length $l(e)$ as $l(e) \leftarrow \frac{l(e)}{max\{l(e') : e' \in \mathbf{E}\}}$), average degrees (sum of degrees of nodes divided by the number of nodes), average edge length (of normalised edges), diameters (longest shortest path) in nodes and normalised edge lengths and maximum number of vertex independent cycles (two cycles are independent of each other if they do not share nodes or edges).

To measure 'compactness' of graphs, we calculated average cohesion: let $\overline{d}$ be an average degree of a graph $\mathbf{G}$, v_{ij} the number of common neighbours of nodes i and j and d_i the degree of node i; then cohesion κ_{ij} between nodes i and j is calculated as $\kappa_{ij} = \frac{v_{ij}}{d_i + d_j}$. Three topological indices were calculated: the Harary index [Plavsic et al. (1993)], Π-index [Ducruet and Rodrigue (2012)] and Randić index [Randić (1975)].

The Harary index is calculated as follows: $H = \frac{1}{2}\sum_{ij}\chi(D_{ij})$, where D is a graph distance matrix, where D_{ij} is a length of a shortest path (in normalised edge lengths) between i and j and $\chi(D_{ij}) = D_{ij}^{-1}$ if $i \neq j$ and 0, otherwise. The Π-index shows a relationship between the total length of the graph $L(\mathbf{G})$ and the distance along its diameter $D(d)$ [Ducruet and Rodrigue (2012)]: $\Pi = \frac{L(G)}{D(d)}$. The Randić index [Randić (1975)] is calculated as $R = \sum_{ij} C_{ij} * (\frac{1}{\sqrt{(d_i * d_j)}})$, where C_{ij} is an adjacency matrix, $C = M$ or $C = F$.

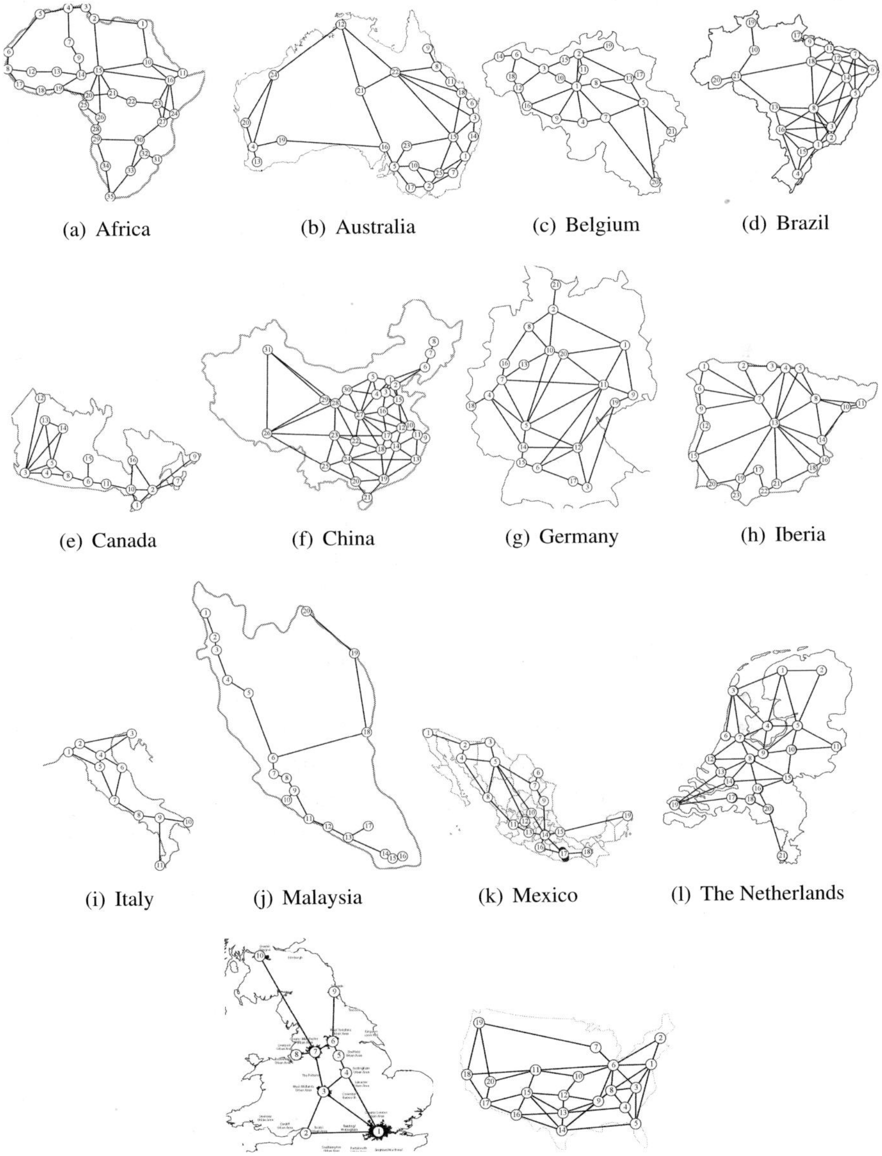

(a) Africa (b) Australia (c) Belgium (d) Brazil

(e) Canada (f) China (g) Germany (h) Iberia

(i) Italy (j) Malaysia (k) Mexico (l) The Netherlands

(m) UK (n) USA

Fig. 18.3 Motorway graphs **H**.

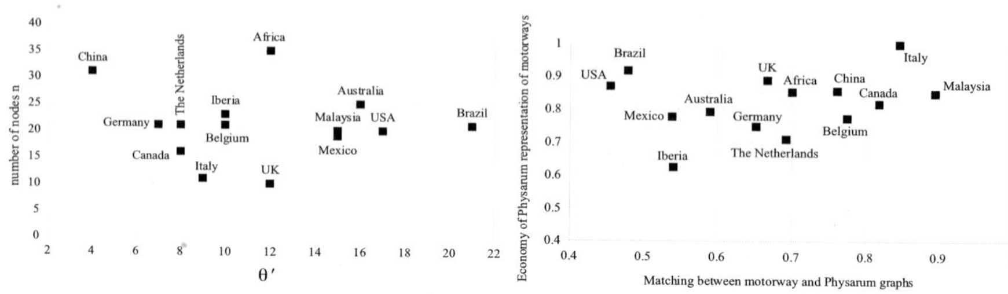

(a) θ' vs number of nodes (b) Matching vs economy

Fig. 18.4 (a) θ' vs number of nodes. (d) Matching μ vs economy ε.

18.1 Matching and economy

The top three regions in the best matches μ between motorway and Physarum graphs are Malaysia, Italy and Canada and the top three most economically ε matches are Italy, Brazil and UK (Fig. 18.4b).

Finding 120. *Let $C_1 \lhd C_2$ if $\mu(C_1) < \mu(C_2)$; then regions can be arranged in the following hierarchy of absolute Physarum matching: USA $\lhd$ Brazil $\lhd$ \{Mexico, Iberia\} $\lhd$ Australia $\lhd$ \{Germany, UK\} $\lhd$ \{Africa, The Netherlands\} $\lhd$ \{China, Belgium\} $\lhd$ Canada $\lhd$ Italy $\lhd$ Malaysia.*

We can consider a product of matching to economy $\omega = \mu \cdot \varepsilon$ as a rough parameter for estimating 'slime optimality' of motorway approximation. By values of ω, regions can be arranged in the descending slime optimality as follows (exact values of ω are in brackets):

(1) Italy (0.85),
(2) Malaysia (0.76),
(3) Canada (0.67), China (0.65), Belgium (0.6), Africa (0.6), UK (0.59),
(4) Netherlands (0.49), Germany (0.48), Australia (0.47), Brazil (0.44), Mexico (0.42), USA (0.4),
(5) Iberia (0.34).

18.2 Average degrees

Averages degrees of motorway graphs are usually lower than degrees of the corresponding Physarum graphs (Fig. 18.5a). This is particularly visible with the US and Brazilian motorway graphs, whose average degrees are nearly 4.5 while

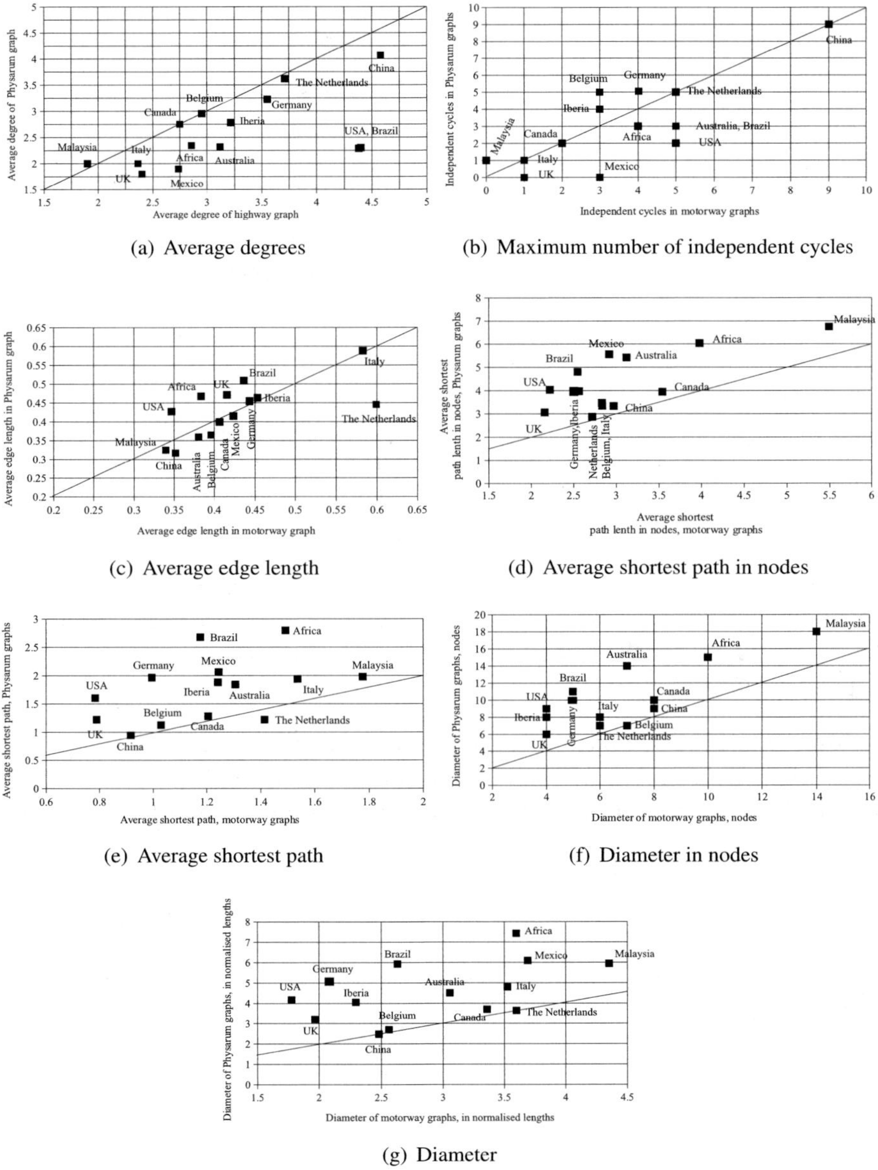

Fig. 18.5 (a) Average degrees. (b) Maximum number of vertex independent cycles. (c) Average edge link. (d) Average length of a shortest path in nodes. (e) Average length of a shortest path in real values. (f) Diameter in nodes. (g) Diameter in real values.

Physarum graphs approximating the motorways have twice less average degree. Belgium, Canada and Malaysia show an almost perfect match between Physarum and motorway graphs in average degrees (Fig. 18.5a).

18.3 Maximum number of independent cycles

Motorway and Physarum graphs of China, The Netherlands, Canada and Italy have the maximum number of independent cycles (Fig. 18.5b) and show a good match. Other regions can be subdivided into two groups:

- number of independent cycles is higher in motorway graphs: Africa, Australia, Brazil, Mexico, UK and USA,
- number of independent cycles is higher in Physarum graphs: Belgium, Berlin, Canada, Germany, Iberia and Malaysia.

The maximum number of independent cycles may characterise two properties of transport networks: fault tolerance (more cycles indicate more chances for transported objects to avoid faulty impassable links, sites of accidents and jams) and locality (distant links increase chances of two cycles sharing a node). Thus, we can propose that the Chinese motorway network is the most fault tolerant and locally connected, while transport networks in Canada, Italy, Malaysia and UK could be sensitive to disasters and overloads.

18.4 Average edge length

The closest match between Physarum and motorway graphs in average edge length is observed for Italy, Iberia, Germany, Mexico, Canada and Malaysia (Fig. 18.5c). Highest mismatch is shown by The Netherlands (edges of motorway graph are longer than edges of Physarum graph) and Brazil, Africa and USA (edges of Physarum graph are longer, on average, than edges of motorway graph).

18.5 Average shortest paths

The average length of a shortest path between nodes shows little match between Physarum and motorway graphs (Fig. 18.5d and e). Usually, Physarum graphs exhibit 1.5–2 times longer average shortest paths, this varies however between regions. Malaysia and Africa are the regions with longest average shortest paths, measured in nodes, in motorway and Physarum graphs (Figs. 18.5d). When shortest paths are measured in normalised edge lengths, Africa and Brazil have longest average shortest paths in Physarum graphs, and Malaysia and Italy in motorway graphs. Countries which show closest match between Physarum and motorway graphs — in average shortest path, measured in nodes (Fig. 18.5d) — are Canada, China and The Netherlands; and — in average shortest path measured in normalised edge lengths (Fig. 18.5e) — are Belgium, Canada and China.

18.6 Diameters

Being the longest shortest path, the diameter shows even less matching between Physarum and motorway graphs (Fig. 18.5f and g) than average shortest path matching. Physarum graphs match motorways in diameter measured in nodes for Belgium, and in diameter measured in normalised edge lengths for Belgium, China and The Netherlands.

18.7 Cohesion

For most regions considered, average cohesion of Physarum graphs is typically higher than that of motorway graphs (Fig. 18.6a). The difference is particularly strong for Brazil and USA, e.g. average cohesion of the Brazilian motorway network is 0.8 while that of the corresponding Physarum graph is 0.4. There is a match between Physarum and motorway graphs of Canada, Italy and The Netherlands. The top four entries with highest cohesion of motorway graphs are Brazil, Canada, China and The Netherlands, and the top three regions with highest cohesion of Physarum graphs are Brazil, China and USA. Cohesion of each edge in the complete graph is $\frac{n-2}{2(n-1)} \sim 0.44$; this limit is nearly approached by **H**(Brazil) (Fig. 18.6a). Motorway and, especially, Physarum graphs of Malaysia and Italy show minimal cohesion, because cohesion is zero in chains and graphs with cycles over three nodes.

Finding 121. *Physarum matches motorway networks of Canada, Italy and The Netherlands in terms of compactness, local density and fault tolerance of transport networks.*

Average cohesion is an indicator of compactness [Egghe and Rousseau (2003)]. A subset of a graph with high cohesion remains connected even when some edges are removed; thus, cohesion may characterise stability [Seidman (1983)] or even fault tolerance of graphs. The cohesion of a node is the minimum number of edges whose deletion makes the node a cut node of the resulting

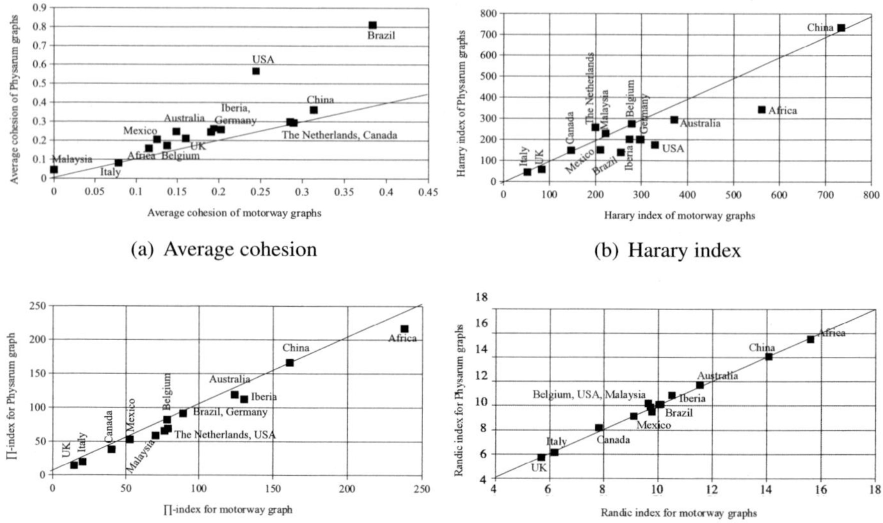

(a) Average cohesion (b) Harary index

(c) Π-index (d) Randić index

Fig. 18.6 (c) Average cohesion. (c) The Harary index. (d) The Π-index. (e) The Randić index.

graph [Ringelsen and Lipman (1983)]; thus, the cohesion is used to characterise a local density of subgraphs [Tibély (2012)] and it is related to centrality [Borgatti and Everett (2006)] and statistical properties of connectivity of graphs [Tainiter (1975)].

18.8 The Harary index

The Harary index [Plavsic et al. (1993)] is well known for its predictive properties in chemistry [Estrada and Rodríguez (1997); Estrada et al. (1997)], and the index is based on "the chemists' intuitive expectation that distant sites in a structure should influence each other less than the near site" [Lucic et al. 2002]. This is probably not the case with slime mould and man-made motorway networks. Only four out of 14 regions satisfy the relation $|1 - \frac{\eta(P)}{\eta(H)}| \leq 0.1$: China, Canada, Belgium and Malaysia, where η is the Harary index. Physarum poorly approximates motorways, in terms of the Harary index, in Brazil and USA (Fig. 18.6b).

Country	$R(\mathbf{H})$	$R(\mathbf{P})$	$1-\frac{R(\mathbf{H})}{R(\mathbf{P})}$	$\frac{n}{2}$
Africa	16.98	16.88	–0.006	17.5
Australia	11.9	12.17	0.022	12.5
Belgium	9.53	10.27	0.071	10.5
Brazil	10.04	10.14	0.01	10.5
Canada	7.25	7.76	0.066	8
China	15.08	15.1	0.002	15.5
Germany	9.94	10.15	0.02	10.5
Iberia	10.62	11.11	0.045	11.5
Italy	5.22	5.22	0	5.5
Malaysia	9.63	9.88	0.025	10
Mexico	8.87	8.95	0.009	9.5
Netherlands	10.09	10.17	0.008	10.5
UK	4.61	4.7	0.021	5
USA	9.71	9.41	–0.031	10

Fig. 18.7 Values of the Randić index.

18.9 The Π-index

Five regions show over 0.1 mismatch, $|1-\frac{\Pi(P)}{\Pi(H)}|>0.1$, between the Π-indices of Physarum and motorway graphs: Germany, Iberia, The Netherlands, Malaysia and USA (Fig. 18.6c). This result is quite interesting. Recall that the Π-index is the ratio of the total length of all normalised edges of a graph to the distance along the graph's diameter. Physarum graphs neither match motorways in diameters (Fig. 18.5f and g), nor do we witness a good match in average edge length or shortest path (Figs. 18.5c–e). However, when these factors are considered in proportion, the match between the graphs occurs.

18.10 The Randić index

Finding 122. *Physarum perfectly approximates motorway networks in terms of the Randić index.*

The Randić index R shows an impeccable match between Physarum and motorway graphs (Figs. 18.6d and 18.7). The largest value $1-\frac{R(\mathbf{H})}{R(\mathbf{P})}=0.07$ is for the Belgian motorway graph and the corresponding Physarum graph. The star graph has a minimum Randić index $\sqrt{n-1}$ [Bollobás and P. Erdös (1998)]. For an arbitrary graph **G**, the boundaries are $\sqrt{n-1} \leq R(\mathbf{G}) \leq \frac{n}{2}$. As we can see in Fig. 18.7, indices for all motorway and Physarum graphs are very close to the upper boundary. The highest values for motorway graphs are in Italy and USA, and for Physarum graphs are in Belgium, Malaysia and Canada.

The Randić index R (originally called by Milan Randić the molecular branching index) [Randić (1975)] characterises relationships between structure, property and activity of molecular components [Estrada (2001)]. The index relates to diameter [Dvořák et al. (2011)] and is actually the upper boundary of diameter [Yang and Lu (2011)]. It also relates to chromatic numbers of graphs and eigenvalues of adjacency matrices [Li et al. (2008)]. There are proven linear relations between the Randić index and molecular polarisability, cavity surface areas calculated for water solubility of alcohols and hydrocarbons, biological potencies of anaesthetics [Kier et al. (1975)], water solubility and boiling point [Hall et al. (1975)] and even bioconcentration factor of hazardous chemicals [Sabljić and Protić (1982)]. Estrada [Estrada (2002)] suggested the following structural interpretation: the Randić index is proportional to an area of molecular accessibility, i.e. the area 'exposed' to outside environment. Or, we can say that the index is inversely proportional to areas of overlapping between spheres of specified radius enclosing the nodes. The more overlapping, the less the Randić index. In terms of transport networks, we can interpret external accessibility as transport inaccessibility, proportional to areas of the country not served by existing motorway links.

Finding 123. *Physarum well approximates motorway graphs in terms of transport accessibility.*

Along the above discourse we can speculate that UK, Italy and Canada (first three regions with the smallest Randić indices) have better transport coverage of their territories than Africa, China and Australia (top three regions with the highest Randić indices).

18.11 Extremal regions

We call a region extremal if it displays minimum or maximum values of at least one measure over its motorway or Physarum graph. The extremal regions are listed in Fig. 18.8.

- Africa shows maximum Π-index and Randić index on both **H** and **P**, and maximum average shortest path and diameter on **P**. This might indicate crit-

Measure μ	$\max \mu(\mathbf{H})$	$\min \mu(\mathbf{H})$	$\max \mu(\mathbf{P})$	$\min \mu(\mathbf{P})$
Average degree	China	Malaysia	China	UK
Number of independent cycles	China	Malaysia	China	UK, Mexico
Average cohesion	Brazil	Malaysia	Brazil	Malaysia
Average edge length	The Netherlands	Malaysia	Italy	China
Shortest path, nodes	Malaysia	UK	Malaysia	The Netherlands
Shortest path	Malaysia	UK, USA	Africa	China
Harary index	China	Italy	China	Italy
Π-index	Africa	UK	Africa	UK
Randić index	Africa	UK	Africa	UK
Diameter, nodes	Malaysia	Iberia, UK, USA	Malaysia	UK
Diameter	Malaysia	USA	Africa	China

Fig. 18.8 Geographical regions with extremal values of measures over motorway graphs **H** and Physarum graphs **P**.

ical dependences between geographically close urban areas, large territorial spread of transport networks and relatively higher density of urban areas along coasts (Fig. 18.3a).

- Brazil shows maximum average cohesion on **H** and **P**. This is because Brazil has the highest (amongst regions studied) number of locally connected subgraphs with large number of dependent cycles (Fig. 18.3d).
- China shows maximum average degree, number of independent cycles and Harary index on **H** and **P**; minimum average edge length, average shortest path and diameter on **P**. These indicate high accessibility of major urban areas in China, and fault tolerance of Chinese motorways at a large scale (Fig. 18.3f). The expressway network in China, known as the national trunk highway system, has been developed very recently, and it is a high-standard transport system planned by the central government. The system is designed to be optimal and many factors were properly taken into account including terrains and landscapes.
- Iberia has minimum diameter in nodes on **H**. The man-made transport network structure resembles a wheel with an 'axle' at Madrid, most major urban areas around the coast forming a 'rim' linked to Madrid by 'spokes' (Fig. 18.3h).
- Italy shows minimum Harary index on **H** and **P** and maximum average edge length on **P**. This is due to the tree-like structure of the transport networks and constraining of the urban areas in the prolonged shape of the country (Fig. 18.3i).
- Malaysia shows maximum shortest path in nodes and diameter in nodes on **H** and **P**; minimum average cohesion on **H** and **P**; maximum average shortest path and diameter on **H**; minimum average degree, number of independent cycles and average edge length on **H**. This is because the Malaysian transport network does not have cycles and consists of a chain connecting urban areas along the western coast (Fig. 18.3j).
- Mexico shows minimum number of independent cycles on **P** because the slime mould approximation is a tree, almost a chain with few branches (Fig. 18.2k).
- The Netherlands shows maximum average edge length on **H** and minimum average shortest path in nodes on **P**, due to the relatively compact location of urban areas with high density of local transport links (Figs. 18.3l and 18.2l).
- UK shows minimum Π-index and Randić index and diameter in nodes on **H** and **P**; minimum average shortest path in nodes, shortest path on **H** and minimum average degree and number of independent cycles on **P** (Figs. 18.3m and 18.2m);

- USA shows minimum average shortest path, diameter in nodes and diameter on **H**. This is because the motorway system in the USA was built with optimality yet efficiency in mind.

18.12 Biorationality of measures

In Fig. 18.9, we provide a binary evaluation $M(C,\mu)$ of matching between Physarum **P** and motorway **H** graphs calculated for each country C and the measure μ as follows: $M(C,\mu) = 1$ if $|1 - \frac{\mu(\mathbf{H}(C))}{\mu(\mathbf{P}(C))}| \leq 0.1$ and 0, otherwise. A biorationality β of a measure μ is the number of regions C for which $\mu(\mathbf{H}(C)) = \mu(\mathbf{P}(C))$ (Fig. 18.9, bottom row).

Finding 124. *Hierarchy of biorationality of measures: Randić index* $>_\beta \Pi >_\beta$ *average edge length, Harary index* $>_\beta$ *average degree, number of independent cycles* $>_\beta$ *shortest path, diameter* $>_\beta$ *average cohesion* $>_\beta$ *diameter in nodes* $>_\beta$ *shortest path in nodes.*

Matching between motorway graphs and Physarum graphs is most strongly expressed in the Randić index; further measures amongst the top ones are Π-index, Harary index and edge length. Thus, we can enhance Finding 122 as follows.

Finding 125. *The Randić index is the most biocompatible measure of transport networks.*

The Randić index is used to characterise relations between structure, property and activity of chemical molecules [Estrada (2001)]; thus, we can speculate that in terms of structure–property–activity Physarum almost perfectly approximates motorway networks in all regions!

18.13 Biorationality of motorways

We calculate a biorationality of a motorway graph as follows: $\rho(C) = \sum_\mu |\xi(|1 - \frac{\mu(\mathbf{H}(C))}{\mu(\mathbf{P}(C))}| \leq \varepsilon)$, $\xi(p) = 1$ if predicate p is true, and 0 otherwise. We choose $\varepsilon = 0.1$. Values of biorationality of motorways are shown in the last column in Fig. 18.9.

Finding 126. *Hierarchy of biorationality of regions is as follows:*
$\{$*Belgium, Canada* $\} >_\rho$ *China* $>_\rho$ $\{$ *Italy, Malaysia* $\} >_\rho$ *The Netherlands* $>_\rho$ $\{$ *Brazil, Germany, Mexico, UK* $\} >_\rho$ $\{$ *Africa, USA* $\}$.

	Average degree	Cycles	Cohesion	Edge length	SP in nodes	SP	Harary index	Π-index	Randić index	Diameter, nodes	Diameter	Biorationality of motorways
Africa									1			1
Australia								1	1			2
Belgium	1					1	1	1	1	1	1	7
Brazil			1					1	1			3
Canada	1	1		1		1	1	1	1			7
China		1				1	1	1	1		1	6
Germany				1				1	1			3
Iberia				1					1			2
Italy		1		1			1	1	1			5
Malaysia	1		1	1			1		1			5
Mexico				1				1	1			3
The Netherlands	1	1							1		1	4
UK							1	1	1			3
USA									1			1
Biorationality of measures	4	4	2	6	0	3	6	9	14	1	3	

Fig. 18.9 Matching *M* between Physarum and motorway graphs for each country and each measure.

Comparing hierarchies of absolute matching, see Finding 120, with the above hierarchy of biorationality we find that Belgium, Canada, China and Italy are at the intersection of the first three levels of the hierarchies. We omit Italy because its shape is intrinsically restrictive and invokes rather trivial architectures of protoplasmic networks.

Finding 127. *Motorway networks in Belgium, Canada and China are most affine to protoplasmic networks of slime mould* P. polycephalum.

18.14 Summary

Based on the results of our previous laboratory experiments with slime mould imitating the development of transport networks in 14 regions, we undertook a comparative analysis of the motorway and protoplasmic networks. We found that in terms of absolute matching between slime mould networks and motorway networks the regions studied can be arranged in the following order of decreasing matching: Malaysia, Italy, Canada, Belgium, China, Africa, The Netherlands, Germany, UK, Australia, Iberia, Mexico, Brazil and USA. We compared the Physarum and the motorway graphs using such measures as average and longest shortest paths, average degrees, number of independent cycles, the Harary index, the Π-index and the Randić index. We found that in terms of these measures motorway networks in Belgium, Canada and China are most affine to protoplasmic networks of slime mould *P. polycephalum*. With regard to measures and topological indices, we demonstrated that the Randić index could be considered as the most biocompatible measure of transport networks, because it matches incredibly well the slime mould and man-made transport networks, yet efficiently discriminates between transport networks of different regions.

Conclusion

We developed a simple and user-friendly technique for evaluating man-made transport systems using slime mould *P. polycephalum*. The experimental laboratory methods are cost efficient and require little if any specialised equipment. Experiments on imitation of transport networks with the slime mould can be conducted by almost anyone with primary school knowledge of physics, chemistry and biology. A few hand-picked facts of general interest are below.

The slime mould *P. polycephalum* approximates best of all motorways in Belgium, Canada and China. The regions studied form the following order of descending **biorationality**: Belgium, Canada, China, Italy, Malaysia, The Netherlands, Brazil, Germany, Mexico, UK, Africa and USA.

All segments of **trans-African highways** not represented by the slime mould have components of non-paved roads. The roads connecting Kampala and Nairobi with Dodoma, and Lubumbashi and Lusaka with Harare, are approximated by the slime mould's protoplasmic tubes in almost all experiments.

The east coast transport chain from the Melbourne urban area in the south to the Mackay area in the north, and the highways linking Alice Springs and Mount Isa and Cloncurry, are represented by the slime mould's protoplasmic tubes in almost all experiments on approximation of **Australian** highways.

If the two parts of **Belgium** were separated with Brussels in Flanders, the Walloon region of the Belgian transport network would be represented by a single chain from Tournai in the north-west to the Liège area in the north-east and down to southernmost Arlon. Motorway links connecting Brussels with Antwerp, Tournai, Mons, Charleroi and Namur, and links connecting Leuven with Liège and

Antwerp with Genk and Turnhout, are redundant from the slime mould's point of view.

Physarum approximates almost all major transport links in **Brazil** apart from those connecting Cuiabá with Belo Horizonte and Campo Grande with Belo Horizonte. Slime mould *P. polycephalum* overdoes the number of transport links in all urban areas of Brazil, especially the northern region.

The slime mould approximates all but the Vancouver to Calgary road links of the **Canadian** highway networks. A core component of the slime mould and the Canadian transport network is a chain along the southern border from the Halifax-Moncton area to Edmonton, and a fork attached to Edmonton; the southern branch of the fork is the Edmonton–Calgary–Vancouver area and the northern branch is Edmonton–Yellowknife–Wrigley.

Slime mould represents 90% of man-made motorways in **China** by its protoplasmic tubes at least once. The transport links less likely to be represented are routes from Beijing to Shenyang, Chengdu to Lanzhou and Urumqi to Lhasa.

Slime mould *P. polycephalum* imitates the 1947 year separation of **Germany** into East Germany and West Germany. The autobahn links represented by the slime mould in almost all trials are Hamburg to Bremen, Hanover to Bielefeld, Cologne to Dortmund, Mannheim to Karlsruhe to Münster and Augsburg to Munich. Autobahn routes between Frankfurt and Braunschweig, Frankfurt and Hanover, Frankfurt and Leipzig and Münster and Leipzig are never represented by protoplasmic tubes of *P. polycephalum*.

The slime mould retraced historical road development in **Italy**. The plasmodium demonstrated that at the initial stage Roma is linked with Corfinium, Capua and Reate, then the roads are built to Populonia, Clusium and Ariminum. The colonisation continues northward until Roma is connected with Genua, Placentia, Verona, Aquileia and Pola. The slime mould imitates southward development of roads from Capua to Brundisiu and Rhegnium.

Slime mould approximates almost all edges of the **Malaysian** expressway graph. Transport routes from Rawang to Port Dickson and from Seremban to Bandar Melaka are never represented by protoplasmic tubes. The federal road from Kuantan to Kota Bahru, and expressways from Alor Start to Ipoh, Bandar Melaka to Skudai and Rawang to Port Dickson, are approximated by *P. polycephalum* in over 80% of experiments.

The slime mould reconstructs principal routes between Central and South America and **Mexico**. The principal routes imitated by *P. polycephalum* include the routes from Chiapas to Ciudad Juárez and from Chiapas to Nuevo Laredo. Transport links to the Tijuana and Mazatlán urban regions are weakly represented by slime mould.

The protoplasmic network forms a subnetwork of the man-made motorway network in **The Netherlands**, i.e. every transport link represented by Physarum can also be found as a segment of the motorway network. The motorway network is not a subnetwork of the protoplasmic network. Transport links Amsterdam to Den Helder, Zwolle to Apeldoorn and Rotterdam to Dordrecht are never represented by the slime mould. By physically imitating flooding of some parts of The Netherlands, we predicted that a real flooding will lead to substantial increase in traffic at the boundary between flooded and non-flooded areas, paralysis and abandonment of the transport network and migration of population from The Netherlands to Germany, France and Belgium.

Plasmodium of *P. polycephalum* segregates transport networks in **Spain** and **Portugal**, and it does not treat Madrid and Valladolid as a part of the connected transport network in the **Iberian** peninsula. From a man-made motorway perspective, Physarum underdevelops transport links in the north-east of Iberia and overdevelops transport links in the south-west of Iberia.

The slime mould approximates the motorway network in the **United Kingdom**, except for the motorway connecting Manchester and Glasgow.

The slime mould well approximates the **USA** interstate network, especially the route from the San Jose area to Las Vegas; the chain of interstates connecting Denver to Albuquerque, Phoenix and Los Angeles; from Kansas City to Oklahoma City, Dallas and Houston; from Milwaukee to Chicago, Nashville and Memphis; and from Boston to New York, Charlotte, Atlanta and Jacksonville. The interstates connecting Denver to San Jose, Houston to Albuquerque and Jacksonsville, New York to Nashville and Boston to Chicago are never imitated by the slime mould.

The slime mould approximates almost 80% of the **Silk Road** routes and the Asian Highway routes. Transport links from Beijing to Tehran and Hong Kong are never imitated by the slime mould.

Bibliography

Adamatzky A. Physarum machine: implementation of a Kolmogorov–Uspensky machine on a biological substrate. Parallel Processing Letters 17 (2007) 455–467.

Adamatzky A. From reaction–diffusion to Physarum computing. Invited talk at Los Alamos Lab workshop. Unconventional Computing: Quo Vadis?, Santa Fe, NM, March 2007.

Adamatzky A. Physarum machines: encapsulating reaction–diffusion to compute spanning tree. Naturwissenschaften 94 (2007) 975–980.

Adamatzky A. Developing proximity graphs by Physarum polycephalum: does the plasmodium follow Toussaint hierarchy? Parallel Processing Letters 19 (2008) 105–127.

Adamatzky A. If BZ medium did spanning trees these would be the same trees as Physarum built. Physics Letters A 373 (2009) 952–956.

Adamatzky A. Hot ice computer. Physics Letters A 347 (2009) 264–271.

Adamatzky A. Routing Physarum with repellents. The European Physical Journal E 31 (2010) 403–410.

Adamatzky A. Physarum Machines: Making Computers from Slime Mould (World Scientific, Singapore, 2010).

Adamatzky A., De Lacy Costello B. and Asai T. Reaction–Diffusion Computers (Elsevier, Amsterdam, 2005).

Adamatzky A., De Lacy Costello B. and Shirakawa T. Universal computation with limited resources: Belousov–Zhabotinsky and Physarum computers. International Journal of Bifurcation and Chaos 18 (2009) 2373–2389.

Adamatzky A. and Alonso-Sanz R. Rebuilding Iberian motorways with slime mould. Biosystems 105 (2011) 89–100.

Adamatzky A. and Akl S. G. Trans-Canada slimeways: slime mould imitates the Canadian transport network. International Journal of Natural Computing Research (2012), accepted.

Adamatzky A., De Baets B. and Van Dessel W. Slime mould imitation of Belgian transport networks: redundancy, bio-essential motorways, and dissolution. arXiv:1112.4507v1 [nlin.AO].

Adamatzky A. and de Oliveira P. P. B. Brazilian highways from slime mold's point of view. Kybernetes 40 (2011) 1373–1394.

Adamatzky A., Ibrahim Z., Abidin A. F. Z. and Muhammad B. Slime mould approximates

Malaysian expressways: is there a bio-logic behind the transport network? Submitted (2012).

Adamatzky A. and Ilachinski A. Slimy interstates: slime mould imitates the USA interstates system. Complex Systems (2012), accepted.

Adamatzky A. and Jones J. Road planning with slime mould: if Physarum built motorways it would route M6/M74 through Newcastle. International Journal of Bifurcation and Chaos 20 (2010) 3065–3084.

Adamatkzy A. and Kayem A. V. D. M. Biological evaluation of trans-African highways. European Journal of Physics ST (2012), in press.

Adamatzky A., Lees M. and Sloot P. Bio-development of motorway networks in the Netherlands: slime mould approach. Advances in Complex Systems (2012), in print.

Adamatzky A., Martinez G. J., Chapa-Vergara S. V., Asomoza-Palacio R. and Stephens C. R. Approximating Mexican highways with slime mould. Natural Computing 10 (2011) 1195–1214.

Adamatzky A. and Prokopenko M. Slime mould evaluation of Australian motorways. International Journal of Parallel, Emergent and Distributed Systems (2011). `DOI: 10.1080/17445760.2011.616204`

Adamatzky A. and Schubert T. Schlauschleimer in Reichsautobahnen: slime mould imitates motorway network in Germany. Submitted (2012).

Adamatzky A., Yang X.-S. and Zhao Y.-X. Slime mould imitates transport networks in China. Submitted (2011).

Alexander C. Note of the Synthesis of Form (Harvard University Press, Cambridge, MA, 1964).

Alpkokin P. Historical and critical review of spatial and transport planning in the Netherlands. Land Use Policy 29 (2012) 536–547.

Anderson O. R. A fine structure study of Physarum polycephalum during transformation from sclerotium to plasmodium: a six-stage description. Journal of Eukaryotic Microbiology 39 (2007) 213–215.

Andersson K. G. (Ed.) Airborne Radioactive Contamination in Inhabited Areas (Elsevier, Amsterdam, 2009).

`http://www.autobahn-online.de/netz1940.gif`

Achenbach F. and Weisenseel M. H. Ionic currents traverse the slime mould Physarum. Cell Biology International Reports 5 (1981) 375–379.

Barrat A., Barthelemy M. and Vespignani A. Dynamical Processes in Complex Networks (Cambridge University Press, Cambridge, 2008).

Barthelemy M. and Flammini A. Modeling urban street patterns. Physical Review Letters 100 (2008) 138702.

Batty M. and Longley P. A. Fractal Cities: A Geometry of Form and Function (Academic Press, San Diego, CA and London, 1994).

Beck E. J., Mays N. and Whiteside A. W. (Eds.) The HIV Pandemic: Local and Global Implications (Oxford University Press, Oxford, 2008).

Beijing Municipal Bureau of Statistics (2011). `http://www.bjstats.gov.cn/esite/`

Belloc H. The Road (Fisher Unwin, London, 1924). Cited by [Bogart (2007)].

Beuthe M., Himanen V., Reggiani A. and Zamparini L. (Eds.) Transport Developments and Innovations in an Evolving World (Springer, New York, 2004).

Blackwell M., Waa J. V. and Reynolds M. Survival of myxomycetes sclerotia after exposure to high temperature. Mycologia 76 (1984) 752–754.

Boel R. and Mihaylova L. A compositional stochastic model for real time freeway traffic simulation. Transportation Research Part B: Methodological 40 (2006) 319–334.

Bogart D. Neighbors, networks, and the development of transport systems: explaining the diffusion of turnpike trusts in eighteenth century England. Journal of Urban Economics 61 (2007) 238–262.

Bollobás B. and Erdös P. Graphs of extremal weights. Ars Combinatoria 50 (1998) 225–233.

Bonabeau E. From classical models of morphogenesis to agent-based models of pattern formation. Artificial Life 3 (1997) 191–211.

Borgatti S. P. and Everett M. G. A graph-theoretic perspective on centrality. Social Networks 28 (2006) 466–484.

Bose P. and Morin P. Competitive online routing in geometric graphs. Theoretical Computer Science 324 (204) 273–288.

Bose P., Cardinal J., Collette S., Demaine E. D., Polop B., Taslakian P. and Zehk N. Relaxed Gabriel graphs. In: 21st Canadian Conference on Computational Geometry (Vancouver, BC, 17–19 August 2009).

Brilon W. Traffic engineering and the new German highway capacity manual. Transportation Research Part A: Policy and Practice 28 (1994) 469–481.

Bull L. J., Kirkby M. J., Shannon J. and Hooke J. M. The impact of rainstorms on floods in ephemeral channels in southeast Spain. Catena 38 (2000) 191–209.

CalPhotos, a project of BSCIT, University of California, Berkeley. `http://calphotos.berkeley.edu/`

Camarasa Belmonte A. M. and Segura Beltrun F. Flood events in Mediterranean ephemeral streams (ramblas) in Valencia region, Spain. Catena 45 (2001) 229–249.

Canada's National Highway System. Condition Report 2008 (Council of Ministers Responsible for Transportation and Highway Safety, 2009).

Chet I. and Henis Y. Sclerotial morphogenesis in fungi. Annual Review of Phytopathology 13 (1975) 169–192.

Chet I., Naveh A. and Henis Y. Chemotaxis of *Physarum polycephalum* towards carbohydrates, amino acids and nucleotides. Journal of General Microbiology 102 (1977) 145–148.

Crowther G. and Wheeler T. Malaysia, Singapore and Brunei. Lonely Planet Country Guide (Lonely Planet Publications, 2001).

Davies H. From Trackways to Motorways. 5000 Years of Highway History (Tempus, Stroud, 2006).

De Lathauwer W. Scenarios for the transport infrastructure of Europe: planning and financing. Tunnelling and Underground Space Technology 10 (1995) 45–51.

Desequilibrios demográficos. Ministerio de Fomento (2012). `http://www.ign.es/espmap/desdem_bach.htm`

Discover life. Physarum polycephalum (2009). `http://www.discoverlife.org/mp/20q?search=Physarum+polycephalum`

Dorigo M. and Stutzle T. Ant Colony Optimization (MIT Press, Cambridge, MA, 2004).

Ducruet C. and Rodrigue J.-P. Graph Theory: Measures and Indices (2012). `http://people.hofstra.edu/geotrans/eng/ch1en/`

meth1en/ch1m3en.html

Dvořák Z., Lidický B. and Škrekovski R. Randić index and the diameter of a graph. European Journal of Combinatorics 32 (2011) 434–442.

Egghe L. and Rousseau R. BRS-compactness in networks: theoretical considerations related to cohesion in citation graphs, collaboration networks and the internet. Mathematical and Computer Modelling 37 (2003) 879–899.

Emsland Nuclear Power Plant Identifier (2012). http://globalenergyobservatory.org/geoid/3071

Estrada E. and Rodríguez L. Matrix algebraic manipulations of molecular graphs. 2. Harary- and MTI-like molecular descriptors. Communications in Mathematical Chemistry 35 (1997) 157–167.

Estrada E., Rodríguez L. and Gutiérrez A. Matrix algebraic manipulations of molecular graphs. 1. Graph theoretical invariants based on distances and adjacency matrices. Communications in Mathematical Chemistry 35 (1997) 145–156.

Estrada E. Generalization of topological indices. Chemical Physics Letters 336 (2001) 248–252.

Estrada E. The structural interpretation of the Randić index. Internet Electronic Journal of Molecular Design 1 (2002) 360–366.

Froment A. and Wildmann B. Landscape ecology and rural restructuring in Belgium. Landscape and Urban Planning 14 (1987) 415–426.

Gabriel K. R. and Sokal R. R. A new statistical approach to geographic variation analysis. Systematic Zoology 18 (1969) 259–278.

Galam S. Sociophysics: A Physicist's Modeling of Psycho-political Phenomena (Cambridge University Press, Cambridge, 2012).

Garnowski M. and Manner H. On factors related to car accidents on German autobahn connectors. Accident Analysis and Prevention 43 (2011) 1864–1871.

Goemans T. and Visser T. The delta project: the Netherlands experience with a mega project for flood protection. Technology in Society 9 (1987) 97–111.

Hall L. H., Kier L. B. and Murray W. J. Molecular connectivity. II. Relationship to water solubility and boiling point. Journal of Pharmaceutical Sciences 64 (1975) 1974–1977.

Haour-Knipe M. and Rector R. (Eds.) Crossing Borders: Migration, Ethnicity and AIDS (Social Aspects of AIDS) (Taylor & Francis, London and New York, 1996).

Hassan F. Africa, north: Egypt, pre-pharaonic. In: D. M. Pearsall (Ed.) Encyclopedia of Archaeology (Elsevier, Amsterdam, 2008), pp. 45–49.

History of Portugal (2012). http://en.wikipedia.org/wiki/History_of_Portugal History of Spain (2012). http://en.wikipedia.org/wiki/History_of_Spain

Huisman P., Cramer W., van Ee G., Hooghart J. C., Salz H. and Zuidema F. C. Water in the Netherlands, Special issue no. 3. Delft: Netherlands Hydrological Society (NHV) (1998). Cited by [Wesselink (2007)].

Ilachinski A. Artificial War: Multiagent-Based Simulation of Combat (World Scientific, Singapore, 2004).

Ing B. The Myxomycetes of Britain and Ireland: An Identification Handbook (Richmond Publishing Company, Slough, 1999).

Jaromczyk J. W. and Toussaint G. T. Relative neighborhood graphs and their relatives.

Proceedings of the IEEE 80 (1992) 1502–1517.

Jaromczyk J. W. and Kowaluk M. A note on relative neighbourhood graphs. In: Proceedings of the 3rd Annual Symposium on Computational Geometry, 1987, pp. 233–241.

Jarrett T. C., Ashton D. J., Fricker M. and Johnson N. F. Interplay between function and structure in complex networks. Physical Review E 74 (2006) 026116.

Jones J. The emergence and dynamical evolution of complex transport networks from simple low-level behaviours. International Journal of Unconventional Computing 6 (2010) 125–144.

Jones J. Characteristics of pattern formation and evolution in approximations of Physarum transport networks. Artificial Life 16 (2010) 127–153.

Karnes T. L. Asphalt and Politics: A History of the American Highway Systems (McFarland, Jefferson, NC, 2009).

Kier L. B., Hall L. H., Murray W. J. and Randić M. Molecular connectivity. I. Relationship to nonspecific local anesthesia. Journal of Pharmaceutical Sciences 64 (1975) 1971–1974.

Kirkpatrick D. G. and Radke J. D. A framework for computational morphology. In: G. Toussaint (Ed.) Computational Geometry (North-Holland Publishing Company, Amsterdam, 1985), pp. 217–248

Larsen J., Urry J. and Axhausen K. Mobilities, Networks, Geographies (Ashgate Publishing, London, 2006)

Lewis T. Divided Highways: Building the Interstate Highways, Transforming American Life (Diane Publishing, New York, 1997).

Li S.-M. and Shum Y. M. Impacts of the National Trunk Highway System on accessibility in China. Journal of Transport Geography 9 (2001) 39–48.

Li X.-Y. Application of computational geometry in wireless networks. In: Cheng X., Huang X. and Du D.-Z. (Eds.) Ad Hoc Wireless Networking (Kluwer Academic Publishers, Dordrecht, 2004), pp. 197–264.

Li X., Shi Y. and Wang L. An updated survey on the Randić index. Mathematical Chemistry Monograph No. 6 (2008) 9–47

Liu H. and Lu M. On the Randić index. Journal of Mathematical Chemistry 38 (2005) 345–354

Liu X. The Silk Road in World History (Oxford University Press, Oxford, 2010).

Lučić B., Miličević A., Nikolić S. and Trinajstić N. Harary index — twelve years later. Croatica Chemica Acta 75 (2002) 847–868.

Lydon G. On Trans-Saharan Trails: Islamic Law, Trade Networks, and Cross-Cultural Exchange in Nineteenth-Century Western Africa (Cambridge University Press, Cambridge, 2009).

Makeham J. and Buell P. D. China: The World's Oldest Living Civilization Revealed (Thames & Hudson, London, 2008).

McNichol D. The Roads that Built America: The Incredible Story of the U.S. Interstate System (Sterling Publishing, New York, 2005).

Morgan A. P. The myxomecetes of the Miami Valley, Ohio. Journal of the Cincinnati Society of Natural History October 1892–January 1893. `http://www.gutenberg.org/files/29534/29534-h/29534-h.htm`

`maps.google.com` (2012)

Matsumoto K., Ueda T. and Kobatake Y. Propagation of phase wave in relation to tactic responses by the plasmodium of *Physarum polycephalum*. Journal of Theoretical Biology 122 (1986) 339–345.

Matsumoto K., Ueda T. and Kobatake Y. Reversal of thermotaxis with oscillatory stimulation in the plasmodium of Physarum polycephalum. Journal of Theoretical Biology 130 (1980) 175–182.

Matula D. W. and Sokal R. R. Properties of Gabriel graphs relevant to geographical variation research and the clustering of points in the same plane. Geographical Analysis 12 (1984) 205–222.

Mouson A. V. Urban morphology as an emerging interdisciplinary field Urban Morphology 1 (1997) 3–10.

Muhammad R. B. A distributed graph algorithm for geometric routing in ad hoc wireless networks. Journal of Networks 2 (2007) 49–57.

Nacu A. Roman Empire 125 (2012). `http://en.wikipedia.org/wiki/File:Iberian_Peninsula_in_125.svg`

Nakagaki T., Yamada H. and Ueda T. Modulation of cellular rhythm and photoavoidance by oscillatory irradiation in the Physarum plasmodium. Biophysical Chemistry 82 (1999) 23–28.

Nakagaki T., Yamada H. and Ueda T. Interaction between cell shape and contraction pattern in the *Physarum plasmodium*. Biophysical Chemistry 84 (2000) 195–204.

Nakagaki T. Smart behavior of true slime mold in a labyrinth. Research in Microbiology 152 (2001) 767–770.

Nakagaki T., Yamada H. and Toth A. Path finding by tube morphogenesis in an amoeboid organism. Biophysical Chemistry 92 (2001) 47–52.

Nakagaki T., Iima M., Ueda T., Nishiura Y., Saigusa T., Tero A., Kobayashi R. and Showalter K. Minimum-risk path finding by an adaptive amoeba network. Physical Review Letters 99 (2007) 068104.

National Bureau of Statistics of China (2012). `http://www.stats.gov.cn/tjgb/rkpcgb/qgrkpcgb/t20110429_402722510.htm`

`http://www.nationsonline.org/oneworld/capitals_africa.htm`

Nesetril J., Milkova E. and Nesetrilova H. Otakar Boruvka on minimum spanning tree problem. Discrete Mathematics 233 (2001) 3–36.

`http://world-nuclear.org/NuclearDatabase/Default.aspx?id=27232` `http://en.wikipedia.org/wiki/List_of_nuclear_reactors`

Okuyama Y. and Chang S. E. Modeling Spatial and Economic Impacts of Disasters (Advances in Spatial Science) (Springer, New York, 2004).

Olariu S. and Weigle M.C. Vehicular Networks: From Theory to Practice. (Chapman and Hall/CRC, London and New York, 2009).

Parry R. Map of Trans-African Highways based on data 2000 to 2003. (17 July 2007). `http://upload.wikimedia.org/wikipedia/commons/0/03/Map_of_Trans-African_Highways.PNG`

Poelmans L. and Van Rompaey A. Detecting and modelling spatial patterns of urban sprawl in highly fragmented areas: a case study in the Flanders–Brussels region. Landscape and Urban Planning 93 (2009) 10–19.

Pointer G. Focus on People and Migration 2005. Chap. 3: The UKs major urban areas. (UK Statistics Authority, Newport, 2005). `www.statistics.gov.uk`

Plavsic D., Nikolic S., Trinajstic N. and Mihalic Z. On the Harary index for the characterization of chemical graphs. Journal of Mathematical Chemistry 12 (1993) 235–250.

Statistics Belgium. Belgian Federal Government (2010). `http://statbel.fgov.be/nl/binaries/bevolkindichtheid_gemeente_Belgie_01012010_web_tcm325-120352.jpg`

Strano E., Adamatzky A. and Jones J. Vie Physarale: evaluation of Roman roads with slime mould. International Journal of Nanotechnology and Molecular Computation (2011), in press.

Priority Investment Needs for the Development of the Asian Highway Network. Economic and Social Commission for Asia and the Pacific. (United Nations, New York, 2006), p. 2.

Randić M. Characterization of molecular branching. Journal of the American Chemical Society 97 (1975) 6609–6615.

Regions of Brazil. `http://en.wikipedia.org/wiki/Regions_of_Brazil`. Latest access: April 2011.

Road Maps of Australian Highways. Australian Travel and Tourism Network (2011). `http://www.atn.com.au/maps/`

Reyes D. R., Ghanem M. G. and George M. Glow discharge in micro fluidic chips for visible analog computing. Lab on a Chip 1 (2002) 113–116.

Ringelsen R. D. and Lipman M. J. Cohesion and stability in graphs. Discrete Mathematics 46 (1983) 191–198.

Rothengatter W. Motorways and motorway finances in Germany and Austria. In: Ragazzi G. and Rothengatter W. (Eds.) Procurement and Financing of Motorways in Europe (JAI Press, Greenwich, 2005), pp. 75–91.

Runions A., Fuhrer M., Lane B., Federl P., Rolland-Lagan A. G. and Prusinkiewicz P. Modeling and visualization of leaf venation patterns. ACM SIGGRAPH 2005 (2005) 702–711.

Sabljić A. and Protić M. Molecular connectivity: a novel method for prediction of bioconcentration factor of hazardous chemicals. Chemico–Biological Interactions 42 (1982) 301–310.

Salingaros N. Principles of Urban Structure (Techne Press, Amsterdam, 2005).

Santi P. Topology Control in Wireless Ad Hoc and Sensor Networks (Wiley, Chichester, 2005)

Scarre C. (General Editor). Past Worlds. The Times Atlas of Archaeology (Times Books, 1996), pp. 190–191.

Schumann A. and Adamatzky A. Physarum spatial logic. In: Proceedings of the 1st International Symposium on Symbolic and Numeric Algorithms for Scientific Computing, Timisoara, Romania, 26–29 September 2009.

Seidman S. B. Internal cohesion of LS sets in graphs. Social Networks 5 (1983) 97–107.

Sevenant M. and Antrop M. Transdisciplinary landscape planning: does the public have aspirations? Experiences from a case study in Ghent (Flanders, Belgium). Land Use Policy 27 (2010) 373–386.

Shirakawa T., Adamatzky A., Gunji Y.-P. and Miyake Y. On simultaneous construction of Voronoi diagram and Delaunay triangulation by Physarum polycephalum. International Journal of Bifurcation and Chaos 19 (2009) 3109–3117.

Solnon C. Ant Colony Optimization and Constraint Programming. (Wiley-ISTE, 2010)

Song W.-Z., Wang Y. and Li X.-Y. Localized algorithms for energy efficient topology in wireless ad hoc networks. In: Proceedings of MobiHoc 2004, Roppongi, Japan, 24–26 May 2004.

Snead J. E., Erickson C. L and Darling A. J. Making Human Space. The Archeology of Trails, Paths, and Roads. Landascape of Movement. Trails, Paths, and Roads in Anthropological Perspective (University of Pennsylvania Press, Philadelphia, 2009).

State of Transport Sector Development in Africa. 1st Session of the Conference of African Ministers of Transport, Addis Ababa, 21–25 April 2008.

Statistische Ämter des Bundes und der Länder (2012). `http://ims.destatis.de/indikatoren/Default.aspx`

Stephenson S. and Stempen H. Myxomeceters. A Handbook of Slime Molds (Timber Press, Portland, OR, 1994).

Supowit K. J. The relative neighbourhood graph, with application to minimum spanning tree. Journal of the ACM 30 (1988) 428–448.

Tainiter M. Statistical theory of connectivity I: basic definitions and properties. Discrete Mathematics 113 (1975) 391–398.

Takada J. Nuclear Hazards in the World: Field Studies on Affected Populations and Environments (Springer, New York, 2010).

Taplin J. H. E., Qiu M. and Han R. Cost–Benefit Analysis and Evolutionary Computing: Optimal Scheduling of Interactive Road Projects (Edward Elgar Publishing, Cheltenham, 2005).

Taylor C. Roads and Tracks of Britain (Dent, London, 1979). Cited by [Bogart (2007)].

The African Development Bank. Review of the Implementation Status of the Trans-African Highways and the Missing Links. SWECO International AB, Sweden, Nordic Consulting Group AB, Sweden. Stockholm (2003). `http://www.afdb.org/fileadmin/uploads/afdb/Documents/Project-and-Operations/00473235-EN-TAH-FINAL-VOL1.PDF`

Terayama K., Ueda T., Kurihara K. and Kobatake Y. Effect of sugars on salt reception in true slime mold Physarum polycephalum. Physicochemical interpretation of interaction between salt and sugar receptions. Journal of Membrane Biology 1977 (34) 369–381.

Tero A., Kobayashi R. and Nakagaki T. Physarum solver: a biologically inspired method of road-network navigation. Physica A 363 (2006) 115–119.

Tero A., Takagi S., Saigusa T., Ito K., Bebber D. P., Fricker M. D., Yumiki K., Kobayashi R. and Nakagaki T. Rules for biologically inspired adaptive network design. Science 327 (2010) 439–442.

Thubron C. Shadow of the Silk Road (Harper Perennial, London, 2008).

Tibély S. Criterions for locally dense subgraphs. Physica A 391 (2012) 1831–1847.

Toroczkai Z. and Guclu H. Proximity networks and epidemics. Physica A 378 (2007) 68. `arXiv:physics/0701255v1`

Toussaint G. T. The relative neighborhood graph of a finite planar set. Pattern Recognition 12 (1980) 261–268.

Toussaint G. T. Some unsolved problems on proximity graphs. In: 1st Workshop on Proximity Graphs, Las Cruces, NM, December 1989. `http://cgm.cs.mcgill.ca/~godfried/publications/openprox.pdf`

Transport in Malaysia, Including: Federal Highway, Malaysia, Smart Tag, Southern Integrated Gateway, Integrated Transport Information System, National Speed Limits (Malaysia), Valet Boy, Bandar Utama Bus Hub, Malaysian Maritime Academy, Puspakom, The Saint Bookstore (Hephaestus Books, 2011).

Tsuda S., Aono M. and Gunji Y.-P. Robust and emergent Physarum logical-computing. Biosystems 73 (2004) 45–55.

Ueda T., Muratsugu M., Kurihara K. and Kobatake Y. Chemotaxis in *Physarum polycephalum*: effects of chemicals on isometric tension of the plasmodial strand in relation to chemotactic movement. Experimental Cell Research 100 (1976) 337–344.

Vandenbulcke G. and Thomas I. Mapping accessibility in Belgium: a tool for land-use and transport planning? Journal of Transport Geography 17 (2009) 39–53.

Guia Quatro Rodas (in Portuguese). `http://mapas.viajeaqui.abril.com.br/guiarodoviario/guia_Rodoviario_viajeaqui.aspx`. Latest access: December 2010.

Villa C. Le Strade Consolari. Storia, itinerari, vicende secolari degli indistruttibili monumenti della potenze di Roma. (Newton and Compton Editori, Roma, 1995).

Wan P.-J. and Yi C.-W. On the longest edge of Gabriel graphs in wireless ad hoc networks. IEEE Transactions on Parallel and Distributed Systems 18 (2007) 111–125.

Wang C. J., Ding J. X. and Yang W. Policy and spatial effect of expressway planning network in China. Acta Geographica Sinica 66(8) (2011) 1076–1088.

Watanabe D. A study on analyzing the road network pattern using proximity graphs. Journal of the City Planning Institute of Japan 40 (2005) 133–138.

Watanabe D. Evaluating the configuration and the travel efficiency on proximity graphs as transportation networks. Forma 23 (2008) 81–87.

`http://wegen-routes.be/hist/histn.html`

Wesselink A. J. Flood safety in the Netherlands: the Dutch response to Hurricane Katrina. Technology in Society 29 (2007) 239–247.

Williams J. Australia State of the Environment Report 2001 (Theme Report). Published by CSIRO on behalf of the Department of the Environment and Heritage (2001). `http://www.environment.gov.au/soe/2001/publications/theme-reports/biodiversity/biodiversity04-1a.html#figure5`

Yamada H., Nakagaki T., Baker R. E. and Maini P. K. Dispersion relation in oscillatory reaction–diffusion systems with self-consistent flow in true slime mold. Journal of Mathematical Biology 54 (2007) 745–760.

Yang Y. and Lu L. The Randić index and the diameter of graphs Discrete Mathematics 311 (2011) 1333–1343.

Zeller T. Driving Germany: The Landscape of the German Autobahn, 1930–1970. (Berghahn Books, Oxford and New York, 2007).

Index